Multivariate Statistical Methods in Physical Anthropology

Multivariate Statistical Methods in Physical Anthropology

A Review of Recent Advances and Current Developments

Edited by

G. N. VAN VARK

Department of Anatomy and Embryology, University of Groningen, The Netherlands

and

W. W. HOWELLS

Peabody Museum of Archeology and Ethnology, Harvard University, U.S.A.

D. REIDEL PUBLISHING COMPANY

A MEMBER OF THE KLUWER ACADEMIC PUBLISHERS GROUP

DORDRECHT / BOSTON / LANCASTER

Library of Congress Cataloging in Publication Data

CIP

Main entry under title:

Multivariate statistical methods in physical anthropology.

Includes index.
1. Physical anthropology–Statistical methods–Congresses.
I. Vark, G. N. van (Gerrit Nanning), 1931- II. Howells, William White, 1908-
GN56.M8 1984 573'.072 84-2007
ISBN 90-277-1734-6

Published by D. Reidel Publishing Company,
P.O. Box 17, 3300 AA Dordrecht, Holland.

Sold and distributed in the U.S.A. and Canada
by Kluwer Academic Publishers,
190 Old Derby Street, Hingham, MA 02043, U.S.A.

In all other countries, sold and distributed
by Kluwer Academic Publishers Group,
P.O. Box 322, 3300 AH Dordrecht, Holland.

Printed in The Netherlands

CONTENTS

PREFACE

Physical anthropologists, like other research workers, are recognizing that the standard multivariate statistical techniques of recent decades are in need of refinement and greater precision. Increasingly it is felt that more sophisticated methods are called for, specifically designed for the materials and problems at issue.

To this end the editors were asked by organizers of the First Intercongress of the International Union of Anthropological and Ethnological Sciences to develop a symposium on this general subject. With the title of this book, the symposium was held in Amsterdam on April 23-25, 1981. Invited were mathematical statisticians who were known to have an acquaintance with and interest in anthropological problems, together with anthropologists and human geneticists who consider multivariate methodology essential for their research.

This volume constitutes an updated and revised selection from among the papers presented, together with a few supplementary papers by authors who were not present but whose work fills out the intended coverage and makes the volume more complete with respect to the state of affairs in the field.

The papers are devoted both to new methodology and to its practical application. Mathematical statisticians may wish to know more about the biological nature and the kinds of materials and samples on which mathematical thinking can be exercised. Anthropologists as practitioners may not be fully aware of the possibilities and limitations in particular mathematical models and methods. Our purpose has been to bring the two groups together, for personal discussions across disciplinary lines as well as within disciplines.

The papers provide instances of practical application of new methodology, showing how these may lead to new interpretations and hypotheses of significance. They should also indicate the directions in which methods have been moving. It is hoped that the book will serve as a practical guide to physical anthropologists and kindred researchers, and as a stimulus to further understanding and refinement.

The editors wish to thank all those who have contributed to making this publication possible, in particular Mr. Don Kuizenga and Drs. S.K. Hazewindus who took a considerable part of all of the varied tasks editors turned out to have, Mrs. G.T. Hoogenberg for her assistance with the extensive correspondence, and Mrs. Th. Deddens and K. van Linschoten for their assistance with making the drawings.

G.N. van Vark W.W. Howells

LIST OF CONTRIBUTORS

A.W. Ambergen, Stichting Mathematisch Centrum, Universiteit van Amsterdam, Kruislaan 413, 1098 SJ Amsterdam, The Netherlands.

Jean-Pierre Bocquet-Appel, Centre National de la Recherche Scientifique, Laboratoire d'Informatique pour les Sciences de l'Homme. 54, Boulevard Raspail, 75270 Paris Cédex 06, France.

Alan Bilsborough, Department of Physical Anthropology, University of Cambridge, Downing Street, Cambridge CB2 3DZ, U.K.

N.A. Campbell, Division of Mathematics and Statistics, C.S.I.R.O., Private Bag, P.O. Wembley, W.A. 6014, Australia.

Robert S. Corruccini, Department of Anthropology, Southern Illinois University, Carbondale, Illinois 62901, U.S.A.

Elisabeth Defrise-Gussenhoven, Centrum voor Biomatematika, Vrije Universiteit Brussel, Pleinlaan 2, 1050 Brussels, Belgium.

P.G.N. Digby, Rothamsted Experimental Station, Harpenden, Herts, AL5 2JQ, U.K.

M. Finnegan, Osteology Laboratory, Kansas State University, Manhattan, Kansas 66506, U.S.A.

W.H.V. de Goede, Afdeling Mathematische Statistiek, Rijksuniversiteit Groningen, Postbus 800, 9700 AV Groningen, The Netherlands.

J.C. Gower, Rothamsted Experimental Station, Harpenden, Herts, AL5 2JQ, U.K.

Jean Hiernaux, Équipe d'Écologie Humaine, Université de Paris, Equipe de Recherche du C.N.R.S., Tour 16, 3e étage, 2, Place Jussieu, 75230 Paris Cedex 05, France.

W.W. Howells, Peabody Museum of Archaeology and Ethnology, Harvard University, 11 Divinity Avenue, Cambridge, Massachusetts 02138, U.S.A.

Rosine Orban-Segebarth, Laboratoire d'Anthropologie et de Génétique Humaine, C.P. 192, 50 Avenue F. Roosevelt, B-1050 Brussels, Belgium.

Charles E. Oxnard, Office of the Dean, University of Southern California, University Park - ADM. 302, Los Angeles, California 90007, U.S.A.

C. Radhakrishna Rao, University of Pittsburgh, Center for Multivariate Analysis, 913 Schenley Hall, Pittsburgh, PA 15260, USA

Dwight D. Read, Department of Anthropology, University of California, 405, Hilgard Avenue, Los Angeles, California 90024, U.S.A.
John G. Rhoads, Department of Anthropology, Yale University, Box 2114, Yale Station, New Haven, Connecticut 06520, U.S.A.
R.M. Rubison, Department of Statistics, Kansas State University, Manhattan, Kansas 66506, U.S.A.
W. Schaafsma, Afdeling Mathematische Statistiek, Rijksuniversiteit Groningen, Postbus 800, 9700 AV Groningen, The Netherlands.
Ilse Schwidetzky, Anthropologisches Institut der Johannes Gutenberg-Universität, Saarstrasse 21, 6500 Mainz, Germany.
Torstein Sjøvold, Osteologiska Forskningslaboratoriet, Ulriksdals Kungsgård, S-17171 Solna, Sweden.
C. Susanne, Laboratorium voor Anthropogenetica, Vrije Universiteit, Pleinlaan 2, 1050 Brussels, Belgium.
Hilde Uytterschaut, Laboratorium voor Anatomie en Embryologie, Rijksuniversiteit Groningen, Oostersingel 69, 9713 EZ Groningen, The Netherlands.
G.N. van Vark, Laboratorium voor Anatomie en Embryologie, Rijksuniversiteit Groningen, Oostersingel 69, 9713 EZ Groningen, The Netherlands.
F.W. Wilmink, Laboratorium voor Anatomie en Embryologie, Rijksuniversiteit Groningen, Oostersingel 69, 9713 EZ Groningen, The Netherlands.
Susan R. Wilson, Department of Statistics, The Australian National University, Mathematical Science Building, P.O.Box 4, Canberra ACT 2600, Australia.

INTRODUCTION

W.W. Howells

Peabody Museum of Archeology and Ethnology,
Harvard University, Cambridge Massachusetts, USA

In the mid-eighteenth century Dr. Johnson (quoted by D'Arcy Thompson, 1942) wrote: "The mathematicians are well acquainted with the difference between pure science, which has only to do with ideas, and the application of laws to the use of life, in which they are constrained to submit to the imperfections of matter and the influence of accident". Two hundred years have not outdated this observation although, in specifics, both mathematical statistics and the anthropological study of populations have progressed long distances in parallel. Each has done what it was able to do, halting at times because of limits in ideas or techniques, and again moving into fresh pastures when such barriers were removed. Anthropology has, as is well known, depended more and more on mathematical statistics and on statisticians, while statisticians have, as is also well known, found human material, expecially crania, particularly suited to their purposes. Such reciprocal advance is necessary but difficult: the mathematics are now rather demanding for most untrained anthropologists and repellent to some. And statisticians do not always see biological questions with the eyes of an anthropologist.

It has been the purpose of our symposium to reflect present day connections and possible future ones. Various of those present have in recent years published important critical papers on anthropological uses of statistics, papers which regrettably cannot be reprinted here. However, their present contributions serve as extensions along the same lines, and this is what we have sought.

G. N. van Vark and W. W. Howells (eds.), Multivariate Statistical Methods in Physical Anthropology, 1–11.

A little history

Karl Pearson, presiding over the Biometric Laboratory at University College, London, interested himself and his associates in craniology well beyond the needs of statistical exercise, publishing on such matters as the fit of Darnley's skull to contemporary portraits of him. He brought about the collection of many series of ethnically differing skulls, a boon to his successors who today are more squeamish about assembling other people's crania. Among his further gifts to anthropology was the Coefficient of Racial Likeness, the first attempt at a multivariate measure of distance. This of course called for a great deal of hand computation, something which hindered its wide adoption. More immediately influential were Pearson's coefficient of correlation and the statistics of sampling and variance. By the latter, samples could be tested in univariate distributions. Some anthropologists were led into rather rigid notions of absolute limits of "statistical significance". In any case it remained impossible to make satisfying comparisons of analyses among populations simply by going from trait to trait one at a time. As to correlation, the germ of multivariate treatment was present here but escaped attention.

By the 1930's the measuring of skulls seemed to have come to a dead end. There were derogatory mutterings about "caliper anthropologists". By 1952 S.L. Washburn, in a perceptive historical critique, could prophesy that, in general, measuring would form a far smaller part of anthropological activity in the future. That events did not bear this out was due (in addition to the fact that measurement is the necessary and objective quantification of variation) to the certainty that statistics would evolve to the point where the content of meaningful measurements could be much more fully revealed. Otherwise, as a comment on the often purposeless measuring of the time, the remark was fair enough when it was made. The anthropologists were in a rut. R.A. Fisher, in a well known article on "The Coefficient of Racial Likeness and the future of craniometry" (1936b) said that such work had still not clarified its aims, and that the "majority of anthropologists, as of biologists, feel so unfamiliar with statistical reasoning as to accept, in some cases, alleged statistical conclusions with something akin to credulous awe, or in others to reject them with indignation, as introducing unnecessary confusion into otherwise plain issues".

The new statistics were in fact already in being; anthropologists had simply been lagging behind. Psychologists had for some time been developing factor analysis; and Fisher had introduced analysis of variance in 1923 and discriminant functions by 1935 (Barnard, 1935; Fisher, 1936a). Anthropologists were making modest use of Pearson's Coefficient of Racial Likeness. Otherwise, in the tedium of hand computation of means and standard deviations, and in the lack of a possible multivariate point of view, anthropologists had maintained a tradition of visual, or "gestalt", appraisal of the multivariate character of crania, that is, the tradition of seeing individuals as types, or perhaps as blends of basic racial elements, assumed to compose the populations under study. The one kind of work in which prevailing statistical uses were productive was that in which t-tests could check sample differences due to some special factor under examination, such as age or climate. But comparative studies, constrained to univariate methods, virtually ended, or perhaps died of inanition, about a human generation ago.

Multivariate statistics provided the escape. To begin with, since they employed covariance, not variance only, they allowed the mathematical analogue of the gestalt approach: the reintegration of morphology, or of whatever biological aspects were under study. (Let us note here as elsewhere that we deal with <u>aspects</u> -- that is, with observational data, not directly with biology, and the sharper the mathematical tools become the more important this fact becomes, so that speaking of a "gestalt" becomes less appropriate.) Therefore, when multivariate methods reached anthropology only shortly before the availability of high speed computers, the adoption of the new methods was the more rapid because there was so much ground to cover. Now, about thirty years after early uses, a look at almost any issue of the American Journal of Physical Anthropology will show the extent of the adoption. In 1952 there appeared a general work, Rao's "Advanced Statistical Methods in Biometric Research", at roughly the time that electronic computers were exchanging vacuum tubes for semiconductors, achieving accuracy and high speed. Following this, packaged programs made computers much more accessible to anthropologists, and are today quite explicit for workers, offering alternative forms of solutions for a given procedure and incorporating various secondary tests and options. The hazards of too easy use - tempting workers to run their data through programs which may be inappropriate - have frequently been pointed out. On the other hand, unlike the old days, anthropologists and statisticians often work in pairs, in which case there is no need to fall into traps. At any rate, this second generation of statistical usages has been generally digested by anthropologists, notably by younger workers, better at mathematics and untrammeled by prejudices and older outlooks.

Where, judging by this symposium, do we stand today? That is to say, apart from tools put to direct anthropometric uses, how have the statisticians influenced the anthropologists (this being the main direction of influence)? In the earlier day the anthropologists finally learned, from genetics and from understanding the normal curve of error, not to carve up populations of skulls or of tribesmen into components which were in fact simply sections of the normal variation. In the later generation, multivariate statistics have provided a much better sort of model for variation within and between populations, in the kinds of overlapping groupings which can be detected in a multivariate space. These are satisfying, but we must always inquire how true such fits are. We wish art to imitate nature, but it is up to the anthropologists in particular to see that nature does not imitate art. Thus the mathematicians must control the uncertainties and traps, while the anthropologists must have in hand the complexities of genetic variation - the genetic constitution of the gene pool - and those of the environment, to detect where a satisfyingly sharp mathematical result may nevertheless be departing from biological reality.

Distance, classification, interpretation

Overarching other questions are the twin problems of group differences or distances and of grouping and classification. Here the biological side provides the pitfalls. When dealing with species the problems are small - we can confidently try to arrive at the maximum segregation of individuals or groups. The same holds true with zoological subspecies, which are satisfactory OTU's (see below) having some natural geographic as well as morphological separation. So also might be "races" of man, if they existed, as they did conceptually in the old days of anthropology. But, although there are highly visible contrasts between the most removed peoples (e.g. Africans and Eskimos) the essential pattern of more continuous variation in modern man has been very difficult for anthropologists to articulate and agree upon. The only accepted perception is one of breeding populations - ideally approaching the demes of zoology - genetic entities of varying isolation, whose gene pools and morphology overlap almost continuously. This makes difficult any kind of taxonomic segregation, or search for limits, and suggests the acceptance of an inherent looseness, for existing populations, between the statistical determination of affiliation on one hand and actual biological affiliation on the other. Statistically, the problem is the identification of groups, from the specific data, before attempting to arrange them.

Given the purposes of the Symposium it is natural that distance and classification are the subjects of most of the papers, specifically those of Campbell, Rao, and Ambergen and Schaafsma. All serve the stated purpose of advancing and refining both theory and methodology from recent levels (with attention of course to anthropology). Rao's in particular considers the approach to non-metric (genetic) materials, both as to distance and as to classification.

For distances, Pearson's Coefficient of Racial Likeness (Pearson, 1926; Tildesley, 1921) was, as we have seen, the first step. Pearson standardized the variables but ignored correlation among them (the objection also to Penrose's size and shape distances). Mahalanobis, Pearson's student, pointed this out to him without effect, and himself eventually formulated the basic solution in D^2 (Mahalanobis, Majumdar and Rao, 1947), from which stem the variations available today. In fact, as already suggested and as noted by Corruccini, the only, and the critical, difference between D and simpler distances is the use of correlation. However, D^2 or D lends itself to mathematical refinement and controls - see below.

Distances are important as leading to clustering procedures and grouping methods (Wilmink and Uytterschaut), and are thus the materials of taxonomy for mathematically-minded anthropologists. For them, it is necessary that the populations concerned be specified as narrowly as possible as to natural homogeneity, as distinguished from being convenient conglomerations of individuals, accepted perhaps because of small numbers, the usual case with fossil hominids. And it is important that taxonomic levels be considered: clustering procedures are like the Sorcerer's Apprentice, and may go on to the bitter end, joining all taxonomical units no matter how absurdly. The first anthropological caution is to recognize the distinction between distances among species, (and all higher taxonomic categories), which are genetically isolated, and populations, which are not thus isolated and which may show all degrees of genetic and/or morphological overlap, in time as in space. In anthropological studies, populations are the most usual items, and it is with these that anthropologists have been more comfortable in using the above treatments (see Corruccini, 1975). The footing appears to change when fossil hominids are included, and still so more with genus and species differences entering, as when other primates are subjects of analysis.

This distinction is a real problem, which has been examined across three classes of vertebrates (frogs, lizards and mammals) by Cherry et al. (1982). These workers used eight linear measures only, taken on 184 different taxa (species or higher). From these measures, after standardizing size on the mean of each individual's linear measures, they computed four kinds of distances: 1) H, the simple mean difference over all the measurements, 2) Delta, or proportional distance, in which each measurement difference is divided by the sum of the two means for the measurement concerned in the two species, 3) M, similar but corrected by the standard deviation for each measurement (it is seen as similar to the Coefficient of Racial Likeness), and 4) D, the Mahalanobis distance which is, of course, further corrected for covariance of the measurements. In the result, the first two, simplest distances appeared as the most satisfactory. They were the least variable when very small samples were used; the least biassed by small sample size (the Mahalanobis distance was the most seriously biassed); they were free from violations of the rule of "triangle inequality" (with three species no distance can be less than the sum of the other two - the Mahalanobis distances violated the rule in 8% of cases tested); and correlation with standard judgements of taxonomic distance was highest.

This reads like a serious indictment of D, the Mahalanobis distance. In addition, D requires equivalence of covariance matrices, and thus demands large samples for this reason alone. The fact remains that D does encompass more information than the other statistics used here, and is capable of being subjected to corrections and controls, which are currently being developed. It seems best to meet the problems of differences and classification among human populations (i.e. within a single species), which is the anthropological problem. But this is an example of where mathematical refinements should before long - i.e. in the present phase - resolve matters in favor of correct rather than simple procedures. The study just cited does suggest the hazards of multivariate analysis when small samples are used (something also emphasized by Van Vark in this volume), together with groups widely separated taxonomically. We need to know what confidence to place in results on closely related populations (i.e. human) when many dimensions are available of which only a few may be functionally significant. This is the case for which Mahalanobis distances and other informative multivariate statistics are appropriate. Just now we would like to know how reliable the scaling is when the morphologically more discrete samples of earlier hominids are involved.

To return to the important point, it is to be remembered that clustering is not the same as phylogeny, however well the two seemingly or actually correspond. Clustering is simply and arrangement of data, and by the same token, no multivariate analysis will automatically recognize species boundaries.

The other side of the coin is classification or allocation of individuals, actually the purpose for which discriminant functions were first formulated. Recognizing groups and assigning group membership is important in medical diagnosis as it is in anthropology. In the papers herein, Campbell uses the predictive method to obtain posterior and typicality probabilities. Ambergen and Schaafsma express the uncertainties involved in estimating posterior and typicality probabilities by providing asymptotic variances. Both papers contain an application to the problem of assigning the important Pleistocene Border Cave skull of South Africa, a problem considered earlier by Rightmire (1979).

A frequently occurring complication is that of missing data. De Goede shows that elegant classical optimality theory is applicable if one may assume that the covariance matrix is known. Without this assumption everything seems to become very complicated. Certain other problems demand, and are getting, further mathematical attention, in addition to the main one, of classifying individuals of unknown group affiliation. One such problem is the effect of including or not including such an individual or population in the basic computation, something which has been important in discussions among anthropologists studying primate species especially. Composition and size of the design series, and choice or limitation of variables have been treated by van Vark (1976). And again there enters the question of the actual biological distinction among design series and between them and the test cases: if the differences are great, as between species, apparent classification may be specious. Corruccini (1975) has expressed grave reservations about such uses of D. It is an important subject, fortunately one dealt with in several of the symposium papers. Campbell discusses how to reduce the sensitivity of D to distortions from outlying cases, and Oxnard (1983) illustrates the detection of such cases due either to plain error or to markedly varied data from different primate genera. Van Vark meets head-on some problems of judging distances among widely different, and usually very incomplete, human fossils. Defrise-Gussenhoven treats the case of an individual distance from a single reference population, touched on by others and used in the past (see e.g. Wilson, 1981; Thorne and Wilson, 1977). These contributions and refinements are an obvious goal of considerable value.

Interpretation of results is an old quest but one still needing much work. Placing populations or individuals in a hyper-space, as above, is one thing; reading the axis which position them is another, and is a basic anthropological purpose. Corruccini herein poses the problems in simple form but with pointers to various difficulties. All along the way we have such questions as how many axes are biologically meaningful, how the selection of variables or of populations may have dictated results -- a set of breadths can tell nothing about lengths; also, major axes will reflect such more numerous related variables while minor, later ones are likely to represent single, isolated measures -- whether a rotation is legitimate or justified, and so on. The possible effects of including or leaving out a morphologically or taxonomic-ally peripheral group have been mentioned; again, various corrective procedures are becoming available.

One treatment involves tracing back to original univariate measures (Corruccini), without which nothing is likely to be legitimate; this can easily find significance in differences which would not have been noticed from the univariate statistics alone (Oxnard). Another recourse is repetition of analysis with changed input or by different forms of analysis; Oxnard (1983 and many earlier papers) and colleagues have compared analyses of parallel morphological systems - e.g. shoulder and hip - represented also by different sets of measurements. Comparisons of different studies in the past have often been disappointing as to coincidence of results; but this was due largely to the lack of determinacy in factor analysis, the procedure used in most of such earlier studies but now generally avoided. A further example of breaking down the measures into appropriate sets, and doing parallel but separate analyses on them, is the paper of Bilsborough and Gower in which the several configurations are linked by Procrustean methods. This is also something Oxnard and colleagues have practised in the past, on different regions of the primate skeleton.

Special aids to interpretation are very welcome. The usual display of distances is by plotting axes or components two at a time on graphs. Andrews' method (1972) of plotting data in higher dimensions is most useful in segregating populations or individuals graphically, but tracing the source of the differences back to the important variables has heretofore been difficult. Oxnard, in a forthcoming book (personal communication), will attack the problem, a welcome contribution. A fresh idea, of a somewhat similar kind is the lively faces, musical notes, etc. proposed in this volume by Wilson.

Secondary problems

Certain subsidiary matters pertain particularly to interpretation. One is the problem of comparing different kinds of analyses or distances, on the same populations, essentially as between continuous and discontinuous variables. A typical case is that in which anthropometry on the living is compared with blood gene distributions or dermatoglyphics (and both of them with language and geography.) Most often the comparison is between metric versus non-metric traits of the skull. Some workers using the two last kinds of data have arrived at totally contrary opinions. Corruccini (1974) has examined this case from all angles. He finds that small scale correlation exists among discrete or non-metric skull traits; he concludes that superiority of non-metric traits is questionable and quotes with approval from Jantz that "much more work is needed before non-metric traits can afford a basis for definitive statements about population relationships". This particular problem is at base the difference in scaling the data: measurements are continuous variables which are readily transformed, and from which the essential covariance matrices can be derived; while non-metric traits, especially single-locus allelic characters like blood genes, are not systematically correlated, and algorithms for handling them are more difficult and less obviously satisfying. Non-metric traits, in which variations are expressed as percentages, require large samples for reliability of the percentages. Gower (1972) has considered all this at length, and expresses himself as unaware of any satisfactory distance equivalent, for non-metric traits, of D^2 for measurements. Beyond all this, however, close correspondence is not really to be expected, because the mathematics for the two kinds of data are different, so that likenesses or differences found are partly mathematical artefacts. And this is a reminder that, though we are thinking of populations and biology, we are dealing with specific and limited kinds of data.

A further special problem is size versus shape, the problem being primarily that of size (see Corruccini, 1975). As Healy and Tanner say (1981): "The distinction between size and shape is one of those intuitively obvious notions which become less and less clear the longer we look at them". That is the situation: we have no actual criterion of size when we talk about it, whether volume or weight or length. Stature, for example, is as much shape as size, when we come to deal with Dinkas and Eskimos. And in studies of allometry, linear distances are typically used in both sides of the equation. So we are not readily reminded of the vagueness at the root of the matter, nor of whether it is justified to truncate size from shape outright.

Penrose (1954), with his size and shape distances, cut the knot by standardizing and summing the deviations of each individual (or sample) from the average. If skull A is exactly 3% larger than skull B in all of its measures, then it has exactly the same shape, and we have a number for size, while shape consists of the deviation among the measurements as they depart from the above perfect agreement in size differences. As distances, Penrose's size and shape have been extensively used; but they ignore correlation among measures, allometry, the choice of measures used, and above all the likelihood that there will exist more than one shape factor. In fact, in general we seem to try to control size simply by somehow getting rid of it. But should we? That is to ignore what size means biologically.

Healy and Tanner (op.cit.) recommend using log transforms of the original data, so that size differences are better scaled, i.e. additively rather than multiplicatively (well suited, I should think, for their study of growth). Corruccini (op.cit.) more empirically guided than Penrose, double-centers the data. I have followed this last kind of procedure for human crania, neglecting any force of allometric relations in skulls of present-day populations. This consisted of 1) standardizing all measurements to a grand mean (across all series) of zero with unit standard deviation, followed by 2) taking the mean of such scores, for each individual, as his measure of size, and subtracting this figure again from each of the measurements, supposedly to sweep size out of the residual figures in this way. Such a single size figure, however, does not operate as well as expected over all populations. It has positive correlations with original measures of cranial lengths and, above all, facial projections, but lower correlations with other measures, especially cranial breadths. Here are sample figures:

Nasion radius	0.82	Cranial length	0.73
Basion-nasion length	0.77	Cranial breadth	0.28
Orbit breadth	0.53	Orbit height	0.24
Bizygomatic breadth	0.65	Biauricular breadth	0.52

The component of population differences in <u>shape</u> would appear to be responsible for the low association with "size", which otherwise does reflect known population differences in size (e.g. Polynesians being very large, Andamanese small).

It has been thought that the first component in a principal components analysis, with all positive loadings, would provide a natural vector of absolute size, but in the result such a component appears, irregularly, only as an expression of size rather than a full measure of its extraction. Corruccini (1983) shows that in fact the coefficients of a first principal component are algebraically related to measures of body size allometrically, not linearly, so that the hope is a vain one. Gelvin (1983) reports using, on data from ossa coxae and from teeth, multiple regression, canonical variates and principal components. A general size factor was detected as lying <u>between</u> the first two canonical variates. This suggests how size is probably distributed in many analyses, i.e. over more than one axis. Unfortunately this is essentially a biological problem, to which statisticians can only bring suggestions, and cautions as to which procedures may be unsuitable.

It is apparent, in the results of the Symposium and in other ways, that mathematical solutions are indeed forthcoming at important points, points which we recognize from anthropological problems that have already been attacked repeatedly and unsuccessfully. I have attempted to outline matters which concern anthropologists, and particular questions which need the best possible statistical answers, such as the placing of isolated human fossils. I close with these sobering but encouraging words from Gower (1972): "New techniques devised to help solve new problems or clarify old ones generate further problems that may take many years to assimilate before some implications are understood. New methods cannot be properly evaluated until they have been tried out on real sets of data."

References

Barnard, M.M.: 1935, The secular variations of skull characters in four series of Egyptian skulls. Annals of Eugenics, 6, pp. 352-372.

Campbell, N.A.: 1978, Multivariate analysis in biological anthropology: some further considerations. Journal of Human Evolution, 7, pp. 197-203.

Cherry, Lorraine M., Susan M. Case, Joseph G. Kunkel, Jeff S. Wyles and Allan C. Wilson: 1982, Body shape metrics and organismal evolution. Evolution, 38/5, pp. 914-933.

Corruccini, Robert S.: 1974, An examination of the meaning of cranial discrete traits for human skeletal biological studies. American Journal of Physical Anthropology, 40/3, pp. 425-445.

Corruccini, Robert S.: 1975, Multivariate analysis in biological anthropology: some considerations. Journal of Human Evolution, 4, pp. 1-19.

Corruccini, Robert S.: 1983, Principal components for allometric analysis. American Journal of Physical Anthropology, 60/4, pp. 451-453.

Fisher, R.A.: 1936b, The use of multiple measurements in taxonomic problems. Annals of Eugenics, 7, pp. 179-188.

Fisher, R.A.: 1936a, "The Coefficient of Racial Likeness" and the future of craniometry. Journal of the Royal Anthropological Institute, 66, pp. 57-63.

Gelvin, B.V.: 1983, Effects of removing size variation in multivariate morphometrics. Abstract of paper at American Association of Physical Anthropologists. American Journal of Physical Anthropology, 60/2, pp. 195-196.

Gower, J.C.: 1972, Measures of taxonomic distance and their analysis. In The Assessment of Population Affinities in Man, ed. by J.S. Weiner and J. Huizinga, ch. 1, pp. 1-24. Oxford: Clarendon Press.

Healy, M.J.R. and J.M. Tanner: 1981, Size and shape in relation to growth and form. Symposia (?) of the Zoological Society of London, no. 46, pp. 19-35.

Mahalanobis, P.C., D.N. Majumdar and C.R. Rao: 1949, Anthropometric Survey of the United Provinces, 1941. Sankhya, 9, pp. 89-324.

Oxnard, Charles E.: 1983, Multivariate statistics in physical anthropology: testing and interpretation. Zeitschrift für Morphologie und Anthropologie, 73/3, pp. 237-278.

Pearson, Karl: 1926, On the Coefficient of Racial Likeness. Biometrika, 18, pp.105-117.

Penrose, L.S.: 1954, Distance, shape and size. Annals of Eugenics, 18, pp. 337-343.

Rightmire, G.P.: 1979, Implications of Border Cave skeletal remains for Later Pleistocene human evolution. Current Anthropology, 20/1, pp. 23-35.

Van Vark, G.N.: 1976, A critical evaluation of the application of multivariate statistical methods to the study of human populations from their skeletal remains. Homo, 27/2, pp. 94-114.

Washburn, S.L.: 1952, The strategy of physical anthropology. In Anthropology Today, ed. by A.L. Kroeber et al., pp. 714-728. Chicago: University of Chicago Press.

Wilson, S.R.: 1981, On comparing fossil specimens with population samples. Journal of Human Evolution, 10, pp. 207-214.

INTERPRETATION OF METRICAL VARIABLES IN MULTIVARIATE ANALYSIS

Robert S. Corruccini

Department of Anthropology, Southern Illinois University
Carbondale, IL 62901 USA

Oftentimes statisticians and anthropologists communicate ineffectively, the former speaking mathematical language while the latter speak biological language. Therefore it is possible for the two to misinterpret each other's goals and methods, occasionally resulting, among some biological users of multivariate analysis, in attraction to theory and statistical formulation rather than to morphological understanding or functional interpretation.

The multivariate methods symposium and volume constitute an opportunity for me to raise one fairly specific point for discussion among biologically and mathematically trained users of multivariate statistics. That issue is the interpretation of character variation in canonical variate or discriminant function analysis of morphometric data. Frequently encountered is the attribution of direct significance to the quantities constituting the linear discriminant function or variate. In other words, characters carrying a large positive discriminant coefficient are interpreted as strongly indicating the variate's significance and interpretation in the direction of the positive pole, negatively weighted characters the opposite, and negligibly loaded characters are not functionally implicated in the morphological pattern revealed or described by the variate. As emphasized by Howells (1972, 1973), this can be misleading, and it is better to employ correlation coefficients (across-group, not pooled within-group) between characters and canonical variate scores to interpret the meaning of group separations. It seems few have noted these warnings. One problem is that this literal imputation of meaning to canonical coefficients is actively encouraged by the user manuals for the major canned computer packages.

One common tendency is to interpret a major canonical variate as an indicator of shape contrasts because the coefficients for

G. N. van Vark and W. W. Howells (eds.), Multivariate Statistical Methods in Physical Anthropology, 13–19.

the variate show both positive and negative values. I have found repeatedly that such variates, based on raw size measures, are purely driven by body size despite the coefficient pattern. This is shown both by uniform, high, positive correlations of individual variables (including body size) with the variate, and by commonsense inspection of the pattern of discrimination. On the other hand, a discriminant function composed solely of positive values does not necessarily signal a size variate; I obtain such variates regularly from normalized data with row and column means of zero, when the data and results reflect shape (size having been partialled out by using residuals from regression against size).

The weightings given to individual measurements on a discriminant function result from complex interplay between the pattern of within-group correlation and between-group separation, and these cross-cutting factors are probably responsible for the imprecise correspondence between statistically-produced coefficient patterns and biologically inferred morphological patterns. Lubischew (1962) illustrates that in two dimensions discriminant coefficients relate to the difference in intraspecific versus interspecific variability, and to the contrast in signs (or direction) between intraspecific correlation and interspecific correlation. Such discriminant coefficients can be fairly literally interpreted, but the situation changes with three or more dimensions.

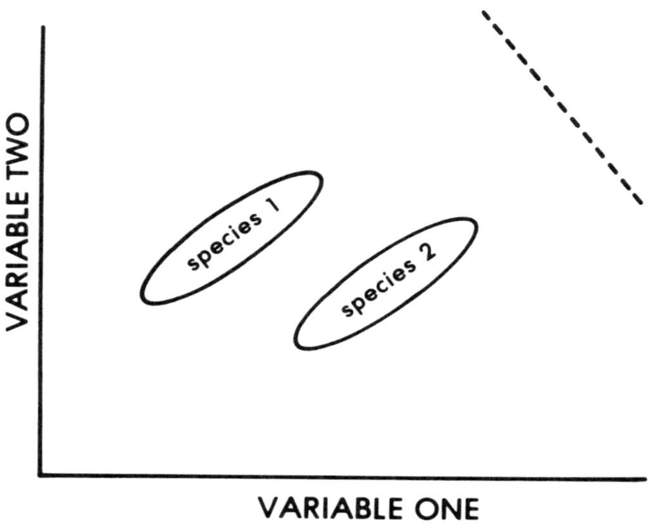

Figure 1. Schematic diagram of covariation ellipses of two imaginary variables in two species

Consider the theoretical situation in Figure 1. Discriminant coefficients of different signs will separate the species (projected into the dashed vector), even though they overlap in both univariate ranges, because variable one maintains a relatively larger relation to variable two in species two.

Imagine a third variable, which is redundant with variable one and slightly less correlated with variable two than is variable one. A discriminant analysis based on all three variables, as shown by simulations, will assign contrasting-sign standardized coefficients to variables one and two and a near-zero weight to variable three. Despite this, an analysis based only on variables two and three will produce very nearly as effective discrimination as that based on variables one and two and again assign substantial opposite-sign coefficients. If a new case added to the samples now causes variable three to be slightly <u>more</u> correlated than variable one with variable two, a new 3-variable analysis will reverse or exchange the coefficients originally assigned to variables one and three. Now, based on these discriminant weights one would state that in the first case the interspecies differences result from variable one being relatively larger and variable two relatively smaller in species two compared to species one, while variable three does not figure in the differences (negligible loading). With a slight change in intraspecific correlation structure, the interpretation would be that it is variable <u>three</u> that counteracts two to produce discrimination while <u>one</u> is trivial. However, the correct biological interpretation is that variables one and three both relatively increase (from species one to two) compared to variable two.

Howells (1972, 1973) recommended method for morphological interpretation of canonical functions is to rely on correlation (across cases or means) between individual measurement values and the cases' orthogonally-projected axis position or canonical score. Experience and common sense show that this method does in fact facilitate functional anatomical interpretation. The correspondence between these correlation coefficients, and analogous standardized canonical variate or discriminant function coefficients, is compared over 24 analyses for which adequate data sets were available to me in Table 1.

The correspondences are very erratic. Furthermore, I can detect little pattern to the deviations from good correspondence. Analyses based on raw measurements, indices, and logarithms seem not to differ, all are unpredictable. There seems to be some tendency for the higher correspondences to result from studies with small numbers of entered variables. Only about 25% (coefficient of determination) of the variation is shared on the average between these two methods of character interpretation.

One example is broken down in Table 2, giving the results of our (Corruccini and Ciochon, 1976) multivariate analysis of primate shoulder measurements. This is one instance of moderate correspondence between correlation and canonical discriminant

Table 1. Across-variable correlations among discriminant coefficients, and correlation coefficients between original variables and the position of cases on discriminant axes

Study	Raw measurements		Dimensionless variables (shape)	
	Axis I	Axis II	Axis I	Axis II
Shoulder (Corruccini and Ciochon, 1976)	-	-	0.50	-
Hip (Burr et.al., 1977)	0.16	0.19	-	-
Maxillary teeth (Henderson, 1976)	0.39	0.13	-	-
Mandibular teeth (Henderson, 1976)	0.03	0.35	-	-
Primate hip (McHenry and Corruccini, 1976)	0.73	0.40	0.36	0.07
Wrist joint (Corruccini, 1978)	0.65	0.42	-	-
Primate elbow (McHenry and Corruccini, 1975)	0.33	0.55	0.30	0.53
Pelvic sexing (Richman et.al., 1979)	0.06	0.20	-	-
Primate dietary (Corruccini unpublished)	0.89	0.35	-	-
M1 primates (Kay, 1975)	0.64	0.38	-	-
Hominoid dentition (C.T. Madden data)	0.25	0.59	-	-
M2 cusp size (Corruccini, 1977a)	-	-	0.67	0.56
Primate DP4 (Corruccini, 1977b)	-	-	0.03	0.22
Neandertal cranium (unpublished)	0.09	0.41	0.48	0.07
Human skull (Howells, 1973) males	0.20	0.23	-	-
Human skull (Howells, 1973) females	0.20	0.14	-	-
Primate cranium (Fleagle, 1975)	-	-	0.28	0.92
Sacrum (Flander and Corruccini, 1980)	0.14	-	0.14	-
Fossil shoulder (Ciochon and Corruccini, 1976)	-	-	0.71	0.65
Nubian skull (Carlson and van Gerven, 1977)	0.03	0.14	-	-
Fossil pelvis (McHenry and Corruccini, 1978)	-	-	0.03	0.07
Primate pelvis (McHenry and Corruccini, 1978)	-	-	0.09	0.07
Hand (Susman and Creel, 1979) phalanx I	0.91	0.94	-	-
Hand (Susman and Creel, 1979) phalanx III	0.84	0.94	-	-

coefficients (r = 0.50). Whenever both correlation and discriminant coefficient values are of high absolute value, they have same signs, indicating codirectional variation. However, there are some noteworthy exceptions that are very important to functional interpretation. The articular surface extent measurement was the most discriminating in a univariate sense, being the only single

Table 2. Contrast between trait correlations and canonical variate coefficients for the major axis in a study of the primate shoulder joint

Traits	Combined sample correlation coefficient	Canonical discriminant coefficient
Spinoglenoid notch depth	0.98	0.54
Intertubercular width	-0.96	0.32
Articular surface extent	-0.96	0.00
Acromion length	0.93	1.32
Glenoid cavity height	-0.87	0.67
Clavicle length	0.80	0.89
Coracoid length	0.79	0.33
Humerus head m-l	-0.78	-0.19
Scapula length	-0.77	-4.15
Infraspinous fossa length	-0.76	-0.87
Humerus length	0.76	3.43
Axillary border length	-0.65	0.31
Humerus head a-p	0.65	3.14
Scapular spine thickness	-0.54	1.39
Acromion width	0.43	0.60
Sternal facet height	0.35	0.39
Supraspinous fossa length	0.15	0.40
Greater tubercle height	-0.01	2.33

trait that completely separated the groups that were discriminated by the first canonical variate. This fact would not be detected merely from examining the canonical coefficients - articular surface has a zero on the discriminant vector! Yet from my biological point of view I would argue that this would be one of the most important individual variables, certainly for understanding the variate's functional significance. Evidently the trait's shared variance with other variables results in its negligible weighting. A different sort of departure from expectations takes place at the other end of the list of correlations. Here the variable least predictive (r = -0.01) of the results of the discriminant axis, which is the position of the greater tubercle, carries a heavy coefficient on the canonical vector. An interpretation that this variable is functionally important is incorrect (I would say) <u>even though</u> it may be statistically important in improving the within-groups to between-groups variance ratio. Inspection shows it had low to nonexistent correlations with the other variables within groups, just as it did between groups. In other words, it is a somewhat randomly distributed, idiosyncratic, and hard-to-interpret variable. Perhaps this very independence helps to sharpen group discrimination, compared against the other variables (like a ratio divisor that is functionally independent). However, I would hesitate to attach morphological significance to that statistical utility.

In conclusion, I believe Professor Howells was right in observing that it is at best hazardous to attempt biological interpretation of statistical discriminant coefficients, and at worst unjustified. It may be possible to use loadings in combination with other information, as done successfully by Oxnard (1973). A fine example of using univariate difference patterns in combination with multivariate coefficients for interpretation is given by Carlson and van Gerven (1977).

Acknowledgments

I profited from discussion with Dennis van Gerven; Karen Schmitt drew the figure. This research was supported in part by NSF (grant DEB 7922186) and a NATO Fellowship.

References

Burr, D.B., Van Gerven, D.P., and Gustav, B.L. 1977, Sexual dimorphism and mechanics of the human hip: a multivariate assessment. Am. J. Phys. Anthrop. 47:273-278.

Carlson, D.S., and Van Gerven, D.P. 1977, Masticatory function and post-Pleistocene evolution in Nubia. Am. J. Phys. Anthrop. 46:495-506.

Ciochon, R.L., and Corruccini, R.S. 1978, Shoulder joint of Sterkfontein Autralopithecus. S. Afr. J. Sci. 72:80-82.

Corruccini, R.S. 1977a, Crown component variation in hominoid lower third molars. Zeit. Morph. Anthrop. 68:14-25.

Corruccini, R.S. 1977b, Cartesian coordinate analysis of the hominoid second lower deciduous molar. J. Dent. Res. 56:699.

Corruccini, R.S. 1978, Comparative osteometrics of the hominoid wrist joint, with special reference to knuckle-walking. J. Hum. Evol. 7:307-321.

Corruccini, R.S., and Ciochon, R.L. 1976, Morphometric affinities of the human shoulder. Am. J. Phys. Anthrop. 45:19-38.

Flander, L.B., and Corruccini, R.S. 1980, Shape differences in the sacral alae. Am. J. Phys. Anthrop. 52:399-403.

Fleagle, J.G. 1975, A small gibbon-like hominoid from the Miocene of Uganda. Folia Primat. 24:1-15.

Henderson, A.M. 1976, Dental field theory: an application to primate dental evolution. Ph.D. dissertation, Colorado University, Boulder.

Howells, W.W. 1972, Analysis of patterns of variation in crania of recent man. In: The Functional and Evolutionary Biology of Primates. R. Tuttle, ed. Aldine-Atherton, Chicago, pp. 123-151.

Howells, W.W. 1973, Cranial Variation in Man. Papers Peabody Museum, 67:1-269.

Kay, R.F. 1975, The functional adaptations of primate molar teeth. Am. J. Phys. Anthrop. 43:195-216.

Lubischew, A.A. 1962, On the use of discriminant functions in taxonomy. Biometrics 18:455-462.

McHenry, H.M., and Corruccini, R.S. 1975, Distal humerus in hominoid evolution. Folia Primat. 23:227-244.

McHenry, H.M., and Corruccini, R.S. 1976, Fossil hominid femora and the evolution of walking. Nature 259:657-658.

McHenry, H.M., and Corruccini, R.S. 1978, Analysis of the hominoid os coxae by cartesian coordinates. Am. J. Hum. Evol. 48:215-226.

Oxnard, C.E. 1973, Form and Pattern in Human Evolution. University of Chicago Press.

Richman, E.A., Michel, M.E., Schulter-Ellis, F.P., and Corruccini, R.S. 1979, Determination of sex by discriminant function analysis of postcranial skeletal measurements. J. Foren. Sci. 24:159-167.

Susman, R.L., and Creel, N. 1979, Functional and morphological affinities of the subadult hand (O.H. 7) from Olduvai Gorge. Am. J. Phys. Anthrop. 51:311-331.

SOME RECENT ADVANCES IN MULTIVARIATE ANALYSIS APPLIED TO ANTHROPOMETRY

J.C. Gower and P.G.N. Digby

Rothamsted Experimental Station, Harpenden, Herts, AL5 2JQ, U.K.

1. INTRODUCTION

With fossil hominids, skeletal evidence is often fragmentary. Certain regions of the skull may be missing for whole populations. Missing populations are fatal to the use of some methods but others, discussed below, can cope with this difficulty. With fossil material it is common for some populations to be represented by a single example - or at most very few. Whether it is admissible to calculate generalised (or any other) distance based on such small samples, and using a covariance matrix calculated from modern data, is noted but not otherwise discussed here. But it may be possible to evaluate generalised distances for some parts of the skull - the lower jaw or the articular region, for example. The question immediately arises as to how generalised distances between n populations, based on measurements from one region of the skull, relate to generalised distances between the same n populations based on measurements from another region of the skull. If the two sets of generalised distances seem to agree well, how should they be combined to give a joint analysis?
Such questions immediately generalise to more than two skull regions, and in the following, m denotes the number of different regions. Of course problems of this kind need not relate only to fossil material. For example, it may be of general interest to investigate the stability of generalised distances based on different sets of measurements or different definitions of distance; the techniques discussed below are also appropriate in such circumstances. Of course all estimated distances are subject to sampling fluctuations and one would like to be able to compare sampling errors of distances evaluated from different regions of the skull.

G. N. van Vark and W. W. Howells (eds.), Multivariate Statistical Methods in Physical Anthropology, 21–36.

Unfortunately the sampling theory associated with the methods discussed below is largely unknown; the possibility that differences might be attributed to sampling variations rather than to something systematic should be remembered when interpreting results.

2. OUTLINE OF METHODOLOGY

Thus we shall be concerned with m symmetric matrices $D_1, D_2, \ldots, D_m$ with elements d_{ijk} giving the estimated (generalised) distances between populations i and j, for the kth region of the skull, where $i,j=1,2,\ldots,n$ and $k=1,2,\ldots,m$. Associated with each distance matrix D_k is a matrix X_k whose n rows give the coordinates of the n populations, and generate the distances between them. With generalised distances X_k will usually be a matrix of canonical variate means. Often X_k will generate distances that approximate D_k rather than fit it exactly. In what follows there is no reason why methods other than canonical variate analysis should not be used to derive X_k from D_k - various forms of multidimensional scaling (metric or non-metric) may be appropriate, for example. The dimensions of X_k will be $n \times r_k$, so that the number of its columns, r_k, may vary from region to region - indeed with canonical variates $r_k = p_k$, the number of variables in the kth region, except where fewer canonical variables are used in an acceptable approximation.

The methods about to be discussed are not new but have been developed over recent years, mainly for use in psychometry. Anthropometric applications have been few but the methods are ripe for exploitation. Sections 2.1 and 2.2 describe two main choices of method. There is space in section 3 to illustrate only one method from each of these sections, but we hope that enough information is given for readers to try for themselves some of the other methods that are mentioned.

2.1. Procrustes Methods

To compare two distance matrices D_k and D_ℓ which refer to regions k and ℓ of the skull, Procrustes methods operate on the matrices X_k and X_ℓ. This is not the place for a detailed mathematical exposition of these methods; a comprehensive review is given by Sibson (1978). Geometrically we can regard the ith row of X_k as giving the coordinates in r_k dimensions of a point P_i that represents the ith population. Similarly the ith row of X gives a point Q_i. Each matrix gives n such points, where the distance $\Delta(P_iP_j)$ approximates d_{ijk} and $\Delta(Q_iQ_j)$ approximates $d_{ij\ell}$. So the problem of comparing D_k with D_ℓ is equivalent to comparing the configurations of points arising from X_k and X_ℓ. Although the coordinate axes of these two configurations may have some interest, they are not normally comparable. What is wanted is some measure $m_{k\ell}$, for the disagreement between D_k and D_ℓ, or X_k and X_ℓ, for regions k and ℓ such that m_{k1} does not depend on the particular axes used for the two configurations, but is a function only of the distances. The rigid body motions of translation, rotation and reflection are the only ones which maintain distance; Procrustes methods use these transformations to fit one configuration to the other in such a way that

$m_{k\ell}^2 = \sum_{i=1}^{n} \Delta^2(P_i Q_i)$ is minimised. This criterion is itself a function

of distance. In the following, whenever $m_{k\ell}$ is mentioned, the minimum value of the criterion is implied. A little care has to be taken when $r_k \neq r_\ell$; this case is discussed below.

The first part of the solution to this minimisation problem is to choose the translation that makes the centroids of the two configurations coincide. For simplicity it is convenient to choose the joint centroid to be the origin, and in the following we assume that this has been done, so the column sums of $X_k (k=1,\ldots,m)$ are all zero. Rotation and reflection are expressed in terms of an orthogonal matrix H (say). Thus for the second part of the solution we wish to find H such that $X_\ell H$ fits X_k optimally, in the sense of minimising the chosen criterion.

For this to have meaning X_k and X_ℓ must have the same number of dimensions. When $r_k \neq r_\ell$ extra zero columns must be added to the smaller matrix to make it match the bigger. The rotation H then occurs in the higher dimensional space. This problem has a straightforward algebraic solution. Of course the inverse rotation H' fits X_k to X_ℓ giving the same value of $m_{k\ell}^2$. Often the different distance matrices are not strictly commensurable and some scaling factor ρ has to be allowed for. This is simple to do, but now

$m_{k\ell}^2 \neq m_{\ell k}^2$, which prohibits the construction of the symmetric

matrix M (see below). One simple way to overcome this, is first to scale every X_i to have unit sum-of-squares. If this is done additional scaling ρ may be allowed for, but now $m_{k\ell}^2 = m_{\ell k}^2$.

After a Procrustes analysis, the configurations X_k and $X_\ell H$ (or $\rho X_\ell H$) may be plotted. If more than two dimensions are involved it is useful to refer everything to the principal axes of one set of points, or even to a combination of the two sets, such as their pairwise centroids as is discussed below for Generalised Procrustes Analysis. The fit of the two configurations can be inspected and the relative distances of P_i from Q_i (after rotation) assessed to indicate which populations, if any, are acting in different ways in the two regions k and ℓ. When the number of regions m is large, the above analysis becomes difficult to handle because all pairs of regions have to be compared.

One simplification is to focus attention on the values $m_{k\ell}^2$ which can be built into an m x m symmetric matrix M. It can be shown that these values form a metric, though not necessarily a Euclidean metric (see Sibson, 1978). The matrix M may be analysed by metric or non-metric multidimensional scaling methods. Gower

(1971) gives an example of such an analysis in an anthropometric context. In the multidimensional scaling, m points are plotted, one for each region. Regions giving rise to similar matrices D_k and D_ℓ will correpond to neighbouring points P_k and P_ℓ. When D_k and D_ℓ differ strongly then P_k and P_ℓ will be distant.

The methods so far discussed do not offer any kind of average configuration of points that combines the information in all the matrices D_k (k=1,2,...,m). Generalised Procrustes analysis (Gower, 1975) gives one way of doing so. In this method all the matrices X_k are simultaneously translated, rotated and, optionally scaled to fit an, unknown, average configuration X. Full mathematical details can be found in Gower (1975) with improvements by Ten Berge (1977); here we confine ourselves to a brief non-technical description of the method.

Translation is again best when all the centroids are made to coincide (at the origin). There is no longer a closed-form algebraic solution for the orthogonal rotation matrices H_k, which themselves have to be found by iterative methods, each step of which is a simple two-matrix orthogonal Procrustes problem as outlined above. The estimation of scaling is more simple than previously, because each X_k is scaled now to fit the unique X so that problems associated with what is fitted to what, vanish. At the end of the fitting process we have

$$X = \frac{1}{m}\sum_{k=1}^{m} \rho_k X_k H_k$$

for the coordinates of the average configuration, from which combined distance values can be evaluated. Finally to obtain a unique representation and to aid interpretation all coordinates are expressed relative to the principal axes of the average configuration X. When m=2 this approach gives a neat way of handling the basic two-matrix Procrustes problem. The average configuration can be plotted, together with the rotated and scaled configurations for all the regions. Each population is represented by m points, one for each region. Tight clusters of points indicate populations behaving similarly for all regions, but a point far from the centroid indicates a region that is behaving aberrantly. The whole analysis can be summed up in an analysis of variance but with no associated significance tests. An illustration of this kind of analysis is given in section 3.

The methods so far discussed cannot accommodate any missing population, which would contribute a missing row in one or more of the matrices X_k. However weighted generalised Procrustes analysis described by Everitt & Gower (1981) is available for this situation. The essential idea is that a weight w_{ik} is associated with the ith row of X_k. When the ith population is absent for the kth region, then $w_{ik} = 0$, otherwise $w_{ik} = 1$ for equal weighting, or w_{ik} might be set equal to the number of samples of region k of the ith population. The average configuration is now calculated as

$$X = \left(\sum_{k=1}^{m} W_k \right)^{-1} \sum_{k=1}^{m} \rho_k W_k X_k H_k$$

where W_k = diag $(w_{1k}, w_{2k}, \ldots, w_{nk})$. Similar adjustments are made when evaluating the sums-of-squares criterion. This method has not yet been used in anthropometry but Everitt & Gower (1981) give a biomedical example.

2.2. Individual Scaling and Allied Methods

The methods described in this sub-section focus attention directly on the distance matrices $D_k (k=1,2,\ldots,m)$ rather than the coordinates X_k that generate the distances. The various methods give different ancilliary information, or sometimes similar information obtained by fitting the same model but optimising a different criterion of goodness of fit, but they all provide an average configuration X in p dimensions, the number of its columns. We shall write x_i for the ith row of X.

The oldest of these methods is individual differences scaling of Carroll & Chang (1970), so called because it assesses the differences between m (human) individuals in the way they perceive the relationships between all pairs of n stimuli. In the current context the "individuals" are the regions of the skull and the "stimuli" are the populations. The model fitted is:

$$\delta_{ijk}^2 = (x_i - x_j) W_k (x_i - x_j)' \quad i,j=1,2,\ldots,n; \; k=1,2,\ldots,m$$

where the fitted values δ_{ijk} are to approximate the observed values d_{ijk} and W_k is a diagonal matrix of positive weights. The expression may be written

$$\delta_{ijk}^2 = \sum_{r=1}^{p} w_{kr} (x_{ir} - x_{jr})^2$$

This is to be interpreted as an ordinary Euclidean distance in p dimensions between the ith and jth populations weighted on the rth dimension differently for each of the kth regions. Thus the average configuration X may be plotted, and associated distances computed to give an overall assessment. The points represent the n populations. An additional plot may be made of the weights, with one point for each region. Regions that weight the dimensions equally will be on the 45° line (in a two dimensional representation) whilst points far from this line will weight one dimension more than the other. Such diagrams can give insight into the data.

In fact Carroll & Chang (1970) do not fit the above model directly, but fit to matrices B_k derived from D_k $(k=1,2,\ldots,m)$ a model which has the form

$$B_k \simeq XW_kX'.$$

The matrix B_k is derived by first defining n x n matrices E_k with elements $-d^2_{ijk}/2$. Then $B_k = (I-N)\ E_k\ (I-N)$ where N has dimensions n x n with all elements set to 1/n. B_k is said to be a doubly-centred matrix and its elements are $d_{ik}\ d_{jk}\ \cos\theta_{ijk}$ where d_{ik} is the distance of the ith population from the centroid of the kth region, and θ_{ijk} is the angle subtended at the centroid by the ith and jth populations. The model can be fitted by using a special case of a general three-way canonical decomposition algorithm (CANDECOMP) that fits models of the form $y_{ijk} = \sum_{r=1}^{p} a_{ir}b_{jr}c_{kr}$ by least squares. More recent work on fitting this same model is discussed by De Leeuw and Pruzansky (1978).

An important property of individual differences scaling is that unlike the equivalent matrix in Procrustes methods, here the matrix X is unique. Although rotation of axes would not alter the distances between the rows of X (i.e. the populations) it would generate a completely new set of sub-optimal weights. It follows that not only is the solution for a given value of p unique, but also that solutions for different values of p are not simply related. An example of individual difference scaling is given in section 3. A generalisation, termed idiosyncratic scaling, has also been suggested. The matrix W_k can now be any positive-definite symmetrics matrix. Thus each skull-region has its own "idiosyncratic" metric space as governed by W_k. X may be plotted as before, but the comparison of the matrices W_k no longer admits to any simple plot, even for p = 2 dimensions. Comparison of the matrices W_k itself needs individual-scaling-like methods and this may explain why the method has not been much used. However there does seem to be a case for considering matrices W_k intermediate in structure between the diagonal matrices of individual differences scaling and the general positive definite symmetric matrices of idiosyncratic scaling.

Takane, Young and De Leeuw (1977) fit the individual differences scaling model by directly minimising the so-called STRESS (squared stress) criterion $\Sigma(d^2_{ijk} - \delta^2_{ijk})^2$, allowing for missing values, replicated information and a non-metric scaling mode of operation.

Heiser & De Leeuw (1979) minimise the STRESS criterion $\Sigma w_{ijk}(d_{ijk} - \delta_{ij})^2$. Here the weights w_{ijk} are given and hence differ from the unknown weights w_{kr} used above. The quantities δ_{ij} are the distances between the ith and jth rows (populations) of the average matrix X. This form admits simplification because it turns out that it is necessary only to fit δ_{ij} to the weighted average d_{ij} of the observed distances.

2.3. Hierarchic Methods

Most of what has been written above is directly applicable to comparing trees derived by hierarchic classification methods. The matrices D_1, $D_2,\dots,D_m$ are now the matrices of ultrametric distances derived from the m different dendrograms (see e.g. Hartigan (1967)). It has been shown that the distances are Euclidean in (n-1) dimensions (Holman (1972)) so that real matrices of coordinates $X_1,X_2,\dots,X_m$ can be found that generate the ultrametric distance matrices. Gower & Banfield (1975) give a concise method for specifying and calculating these coordinates. The only problem is that the average matrix X will not, in general, generate ultrametric distances and hence a tree. If one wishes to fit an average tree to the m given trees, then an additional step is needed to approximate X (or at least the distances it generates) by fitting a tree using some preferred method. Usually this will be the same method that generated the m given trees.

3. EXAMPLES

In this section we use data for 10 populations of skulls and 6 regions of the skull. Table 1 gives the number of samples of each population for each region and the number of features measured on each region. Preliminary canonical variate analyses were done for the ten populations on each of the six skull regions separately, giving 6 sets of canonical variate means for the 10 populations. It was considered adequate to keep only the first 3 dimensions of the canonical variate analysis solutions. Thus the matrices $X_i(i=1,2,\dots,6)$ each have 10 rows and 3 columns: in the notation of section 2, n=10, m=6, and $r_k=3$ $(k=1,2,\dots,m)$. To the same degree of approximation six distance matrices $D_k(k=1,2,\dots,m)$ were calculated directly from the canonical variate means.

3.1. Example of Generalised Procrustes Analysis

The six matrices of canonical variate means were analysed using Generalised Procrustes analysis as described in section 2.1. The centroid X, expressed relative to its principal axes, is given in Table 2 and its first two dimensions are shown in Figure 1; Tables 3 and 4 are summary analyses of variance. Table 3 shows for each population the allocation of the total sum of squares into the concensus, or centroid, part and the residual part, i.e. deviations from the centroid. None of the residual sums of squares is large which indicates that all the regions accord fairly well with the centroid for all the populations. However, the large concensus sums of squares for East Rudolf Hominids No. 406 and Olduvai Hominids No. 5 indicate that these populations are about twice as far from the centroid than are the others.

Table 4 gives a similar analysis of variance, but for the skull regions. It can be seen that 5 of the regions have similar

Table 1. Population sample sizes for 6 skull regions and number of features in each region

Code	Population	Articular	Basi-cranium	Face	Upper Jaw	Balance	Cranial Vault
A	Modern Homo Sapiens	50	50	50	50	50	50
B	Late Pleistocene Homo Sapiens	4	5	5	5	5	6
C	Neanderthal	7	3	4	1	2	4
D	Pekin Homo Erectus	3	2	1	2	1	4
E	Australopithecus Africanus	4	3	1	3	1	1
F	East Rudolf Hominids No. 406	1	1	1	1	1	1
G	East Rudolf Hominids No. 1813	1	1	1	1	1	1
H	East Rudolf Hominids No. 3733	1	1	1	1	1	1
I	Olduvai Hominids No. 5	1	1	1	1	1	1
J	Olduvai Hominids No. 24	1	1	1	1	1	1
	Number of features	7	5	20	11	6	14

Table 2. Average configuration from Generalised Procrustes Analysis

Population	Dimensions 1	2	3
Modern Homo Sapiens	.208	.001	-.067
Late Pleistocene Homo Sapiens	.219	-.083	-.084
Neanderthal	.223	-.125	-.016
Pekin Homo Erectus	.201	-.110	.055
Australopithicus Africanus	-.128	.173	-.058
East Rudolf Hominids No. 406	-.544	-.090	-.083
East Rudolf Hominids No. 1813	.014	.141	.011
East Rudolf Hominids No. 3733	.099	.052	.094
Olduvai Hominids No. 5	-.327	-.090	.117
Olduvai Hominids No. 24	.035	.130	.029

Table 3. Analysis of Variance for 10 populations

Population	Concensus	Residual	Total
Modern Homo Sapiens	0.287	0.078	0.365
Late Pleistocene Homo Sapiens	0.371	0.072	0.443
Neanderthal	0.394	0.149	0.543
Pekin Homo Erectus	0.332	0.147	0.480
Australopithecus Africanus	0.299	0.110	0.409
East Rudolf Hominids No. 406	1.860	0.096	1.956
East Rudolf Hominids No. 1813	0.122	0.115	0.237
East Rudolf Hominids No. 3733	0.129	0.151	0.280
Olduvai Hominids No. 5	0.774	0.188	0.962
Olduvai Hominids No. 24	0.113	0.212	0.326

Table 4. Analysis of Variance for 6 skull regions

Region	Concensus	Residual	Total	Scaling
Articular	0.863	0.194	1.057	0.808
Basicranium	0.970	0.154	1.124	0.904
Face	0.884	0.180	1.064	1.266
Upper Jaw	0.886	0.181	1.067	1.150
Balance	0.224	0.420	0.644	1.099
Cranial Vault	0.855	0.190	1.045	1.288

values; however attention is focussed on the Balance region which does not accord well with the others, having low concensus and high residual sums of squares. The final column of Table 4 gives, for each region, the scaling ρ_k mentioned in section 2.1; these do not differ greatly, most probably because the sets of canonical variate means are already standardised.

Figure 1 shows the first 2 dimensions of the Generalised Procrustes solution. The dotted line, joining the centres of the 'stars', shows the centroid configuration; the centres of the stars being the centroids of the locations of the different regions for the 10 populations. It can be seen that 8 of the 10 centroid locations lie roughly on a straight line, whereas the points for the two aberrant populations are remote from the others. For each population the unbroken lines join the locations for different skull regions to their centroid; thus they show the residuals mentioned in section 2. For clarity only three such lines are marked for eight of the populations; these are for the region that fits best to the centroid (Basicranium), the region which fits worst (Balance) and one of the intermediate regions (Cranial Vault). Figure 1 confirms the high residual sum of squares for the Balance

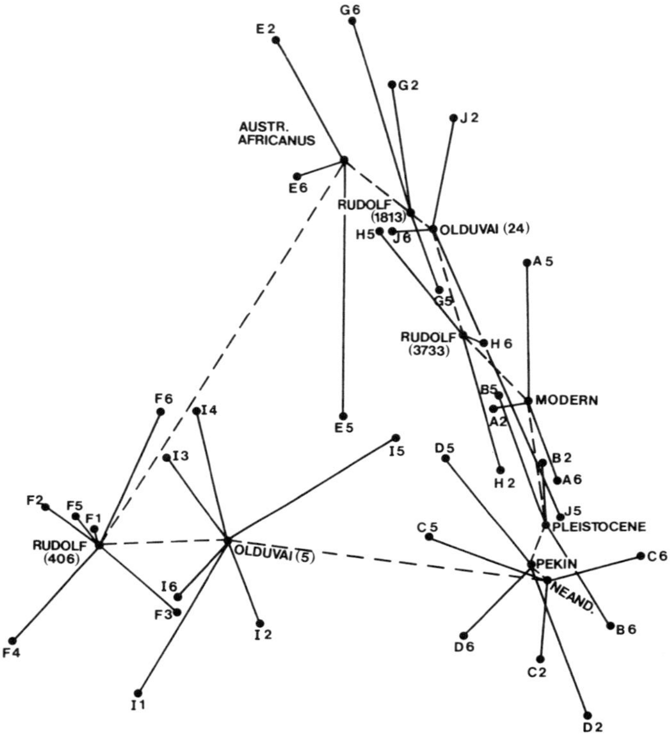

Figure 1. First two dimensions of a generalized Procrustes analysis. The labelled points are for populations in specific regions, according to the codes given in Table 1, e.g. D5 is the point for the Pekin population from the Balance configuration

region (Table 4) as many of the residuals shown for that region are large. The group of four points near the centre of Figure 1 contribute to the low concensus sum of squares mentioned above.

Table 5(a) gives the distances between the centroids, in three dimensions, for the 10 populations, scaled so that the average distance is one. The large distances between the two aberrant populations (East Rudolf Hominids No. 406 and Oldovai Hominids No. 5) and the other eight can be seen; however the distance between those two populations (0.81) is larger than might be expected from Figure 1. This is a consequence of using a two-dimensional display of a three-dimensional analysis.

Table 5. Distances (scaled to average one) between the fitted points for 10 populations from (a) three dimensional generalized procrustes analysis (b) two dimensional individual scaling analysis (c) three dimensional individual scaling analysis; each analysis combining data from six skull regions (Code as in Table 1)

(a)

	A	B	C	D	E	F	G	H	I
B	.24								
C	.38	.22							
D	.46	.39	.21						
E	1.04	1.19	1.28	1.24					
F	2.09	2.10	2.13	2.09	1.36				
G	.69	.88	.94	.87	.44	1.69			
H	.55	.70	.67	.54	.82	1.88	.41		
I	1.58	1.60	1.56	1.47	1.03	.81	1.17	1.24	
J	.65	.84	.88	.81	.52	1.74	.08	.33	1.19

(b)

	A	B	C	D	E	F	G	H	I
B	.20								
C	.44	.24							
D	.24	.14	.32						
E	1.18	1.24	1.41	1.11					
F	2.09	2.02	2.04	1.89	1.36				
G	.85	.96	1.16	.84	.38	1.66			
H	.32	.37	.58	.25	.88	1.80	.59		
I	1.63	1.60	1.67	1.46	.77	.59	1.09	1.32	
J	.99	1.10	1.31	.99	.32	1.66	.15	.74	1.07

(c)

	A	B	C	D	E	F	G	H	I
B	.20								
C	.41	.21							
D	.40	.33	.34						
E	1.02	1.15	1.29	1.20					
F	1.98	1.98	1.99	1.99	1.31				
G	.73	.88	1.04	.88	.44	1.65			
H	.40	.46	.56	.35	.86	1.74	.55		
I	1.54	1.56	1.56	1.40	1.08	1.00	1.17	1.17	
J	.82	.97	1.13	.91	.60	1.77	.21	.60	1.18

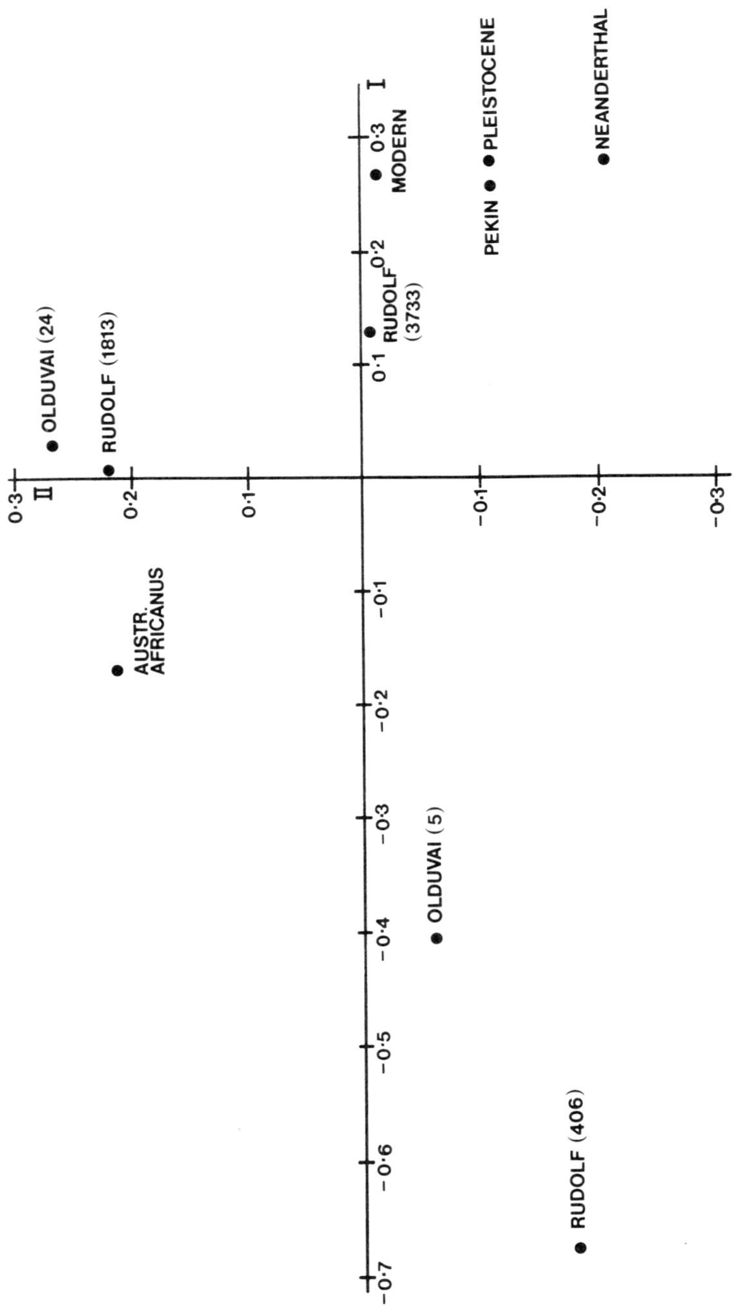

Figure 2. Average configuration of 10 populations from an individual scaling analysis

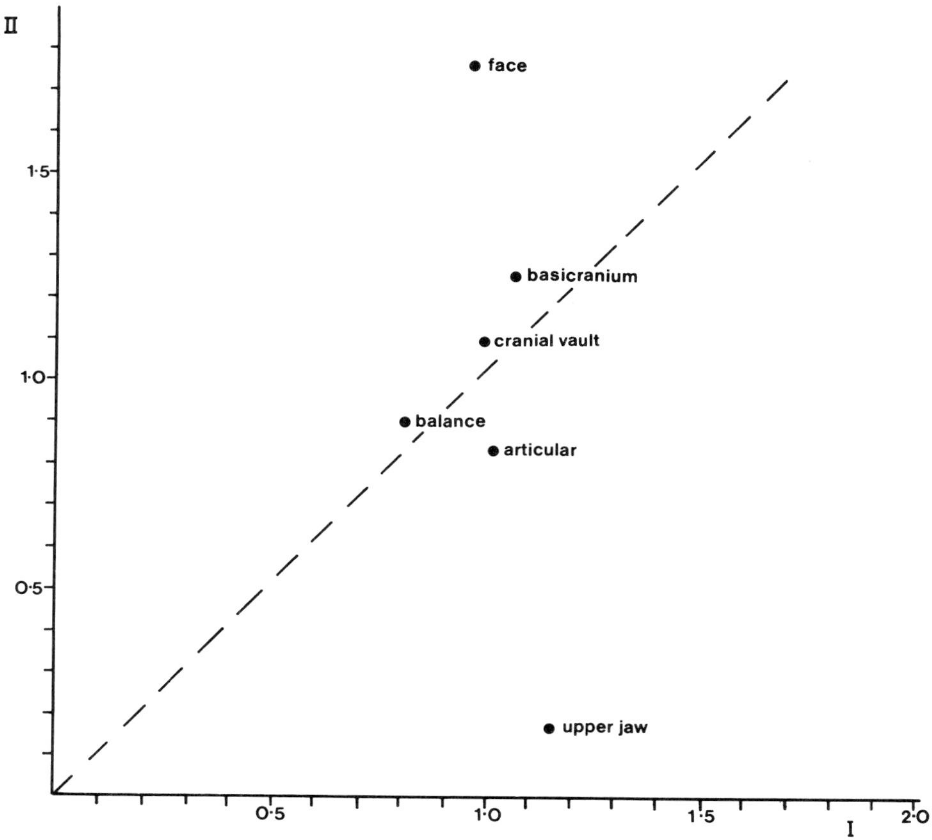

Figure 3. Weights for six skull regions from an individual scaling analysis. Points on the dotted line denote regions that give equal weight to the axes of the average configuration of Figure 2.

3.2. Example of Individual Differences Scaling

The six matrices D_k described at the beginning of section 3 were analysed by the Individual Differences Scaling method outlined in section 2.2, to give the two-dimensional solution shown in Figures 2 and 3. Figure 2 shows the average configuration for the 10 populations and may be compared with the Generalised Procrustes centroid configuration of Figure 1. Although the Figures are similar, the individual scaling solution gives a stronger suggestion of a separate sub-group of Australopithecus Africanus, East Rudolf Hominids No. 1813 and Olduvai Hominids no. 24 as well as the previously found disparate points for East Rudolf Hominids nNo. 406, and Olduvai Hominids No. 5.

Figure 3 shows the weights for the six skull regions. It can be seen that the weights for the first axis are all roughly equal to one; thus the skull regions contribute equally to the first axis of the average configuration (Figure 2). In contrast for the second axis of Figure 2, the Upper Jaw contributes little and the Face contributes most. A plot of the populations for the Face alone would be similar to Figure 2, but with the second axis stretched. This would have the effect of increasing the separation of the sub-group of three populations from the others; thus we might expect that those three populations differ from the others with particular regard to their facial features.

Table 5(b) gives the distances between the points of Figure 2, scaled to average one to permit comparison with Table 5(a). Differences in the results of the two analyses could be investigated by plotting the values of the Tables against each other: points far from the 45° line would indicate discrepancies. Comparison of Tables 5(a) and (b) shows general agreement between the analyses, although the distances for East Rudolf Hominids No. 3733(H) and Olduvai Hominids No. 24 (J) do indicate some discrepancy; this may be because the dimensionality of the analyses is different. Although we do not show the analysis here, the resulting distances, again scaled to average one, from a three dimensional individual scaling analysis are given in Table 5(c): comparison with Table 5(a) shows that the pair of three dimensional analyses agree.

4. COMPUTING AND CONCLUSION

Orthogonal Procrustes Analysis for two matrices is easy to program using widely available numerical algorithms for singular value decomposition or symmetric matrix eigenvalue calculations. Genstat (1977) has a special directive ROTATE. Generalised Procrustes Analysis is available as a Genstat macro in the Macro Library and is also available as one of the options of PINDIS (see, e.g. Borg (1979)); it is not very difficult to program for oneself. Weighted generalised Procrustes Analysis exists as a Genstat macro.

Individual Differences Scaling has a special program INDSCAL that is widely available and Idiosyncratic Scaling has an associated program IDIOSCAL (Carroll & Chang (1972)). The program ALSCAL has been written for minimising STRESS (Takane, Young and De Leeuw (1977)) and the program SMACOFF-I for minimising STRESS (Heiser & De Leeuw (1979)).

These methods are brought to the attention of anthropologists who will probably find many uses for them. Like all statistical methods they should be used with care. In particular we again draw attention to the paucity of distributional results that might form the basis of significance tests. The only distributional result known to us in this field is given by Davies (1978) who evaluated the asymptotic mean and variance of the m^2 statistic in the null case when the regions k and ℓ have the same generalised distances $\delta_{ijk} = \delta_{ij\ell}$ between all pairs (i,j) of the n populations. The difficulty of formulating plausible null hypotheses in this area is immense and the mathematical consequences of anything formulated are likely to be intractable. This causes us little concern.

We thank Dr. A. Bilsborough for letting us use his data.

References

Borg, I.: 1979, Some basic concepts of facet theory. In: Geometric Representations of Relational Data (ed. Lingoes). Ann Arbor: Mathesis Press.

Carroll, J.D. and Chang, J.J.: 1970, Analysis of individual differences in multidimensional scaling via an n-way generalization of "Eckart-Young" decomposition. Psychometrika, 35, pp. 283-319.

Carroll, J.D. and Chang, J.J.: 1972, IDIOSCAL (Individual differences in orientation scaling). Paper presented at the Spring meeting of the Psychometric Society, Princeton, New Jersey, April 1972.

Davies, A.W.: 1978, On the asymptotic distribution of Gower's m^2 goodness-of-fit criterion in a particular case. Ann.Inst. Statist.Math., 30, pp. 71-79.

De Leeuw, J., and Pruzansky, S.: 1978, A new computational method to fit the weighted Euclidean distance model. Psychometrika, 43, pp. 479-490.

Everitt, B.S. and Gower, J.C.: 1981, Plotting the optimum positions of an array of cortical electrical phosphenes. In: Interpreting Multivariate Data (ed. Barnett). Wiley.

Genstat Manual: 1977, GENSTAT, a general statistical program. Oxford: Numerical Algorithms Group.

Gower, J.C.: 1971, Statistical methods of comparing different multivariate analyses of the same data. In: Mathematics in the Archaeological and Historical Sciences (ed. Hodson, Kendall and Tauto). Edinburgh: University Press, pp. 138-149.

Gower, J.C.: 1975, Generalized Procrustes analysis. Psychometrika, 40, pp. 33-51.

Gower, J.C. and Banfield, C.F.: 1975, Goodness-of-fit criteria for hierarchical classification and their empirical distributions. In: Proceedings of the 8th International Biometric Conference (ed. Corsten and Postelnicu). Romania: Editura Academiei, pp. 347-362.

Hartigan, J.A.: 1967, Representation of similarity matrices by trees. J. Am. Stat. Assoc., 62, pp. 1140-1158.

Heiser, W.J. and de Leeuw, J.: 1979, How to use SMACOF-I. A program for multidimensional scaling. Dept. of Datatheory, Faculty of Social Sciences, University of Leiden, Wassenaarseweg 80, Leiden, The Netherlands, pp. 1-63.

Holman, E.W.: 1972, The relation between hierarchical and euclidean models for psychological distances. Psychometrika, 37, pp. 417-423.

Sibson, R.:1978, Studies in the robustness of multidimensional scaling: Procrustes statistics. J.R. Statit.Soc.(B), 40, pp. 234-238.

Takane, Y., Young, F., and de Leeuw, J.: 1977, Non-metric individual differences multidimensional scaling: an alternating least squares method with optimal scaling features. Psychometrika, 42, pp. 7-67.

Ten Berge, J.M.F.: 1977, Orthogonal Procrustes rotation for two or or more matrices. Psychometrika, 42, pp. 267-276.

INCOMPLETE SAMPLES FROM MULTIVARIATE NORMAL DISTRIBUTIONS WITH THE SAME, KNOWN COVARIANCE MATRIX

W.H.V. de Goede

Department of Mathematics, University of Groningen,
Postbox 800, Groningen, The Netherlands

ABSTRACT

In physical anthropology one easily gets entangled in missing data problems, because excavated skulls usually are not complete. In this contribution, under special model assumptions, optimal estimators will be derived for some parameters. The results are useful in Van Vark's approach to classification and discrimination problems in the study of older hominid populations.

0. Introduction and model assumptions

Mathematically oriented statisticians tend to ignore missing data problems because (1) such problems are usually nasty (if the covariance matrix is unknown, then incomplete exponential families appear and best unbiased estimators do not exist) (2) the missing of the data is often caused by neglect, inaccuracy or laziness of the experimentist.
However, it is not the anthropologist's fault that skulls of older hominids are often incomplete and Rao's story of the broken skulls [Rao (1965) 4.g.3] shows that one should do one's utmost to include all available information into the analysis.

We shall develop theory with applications to the following anthropological situation. Let there be k prehistoric populations. A random drawing from population h ($h = 1,2,\ldots,k$) would result in a vector of p (possibly only partly observable) skull measurements having the p-variate normal distribution with vector of means ν_h and covariance matrix Σ_h. A random drawing from the present-day worldpopulation leads to a vector of skull-measurements having the $N_p(\mu,\Sigma)$-distribution. The basic model assumption we shall impose is that μ and Σ are known and also

G. N. van Vark and W. W. Howells (eds.), Multivariate Statistical Methods in Physical Anthropology, 37–48.

$$\Sigma_1 = \Sigma_2 = \ldots = \Sigma_k = \Sigma \tag{0.1}$$

so, Σ_h is known ($h = 1,2,\ldots,k$). Thus only $\nu_1,\nu_2,\ldots,\nu_k$ appear as unknown parameters.

The background of this assumption lies in Van Vark's approach to the problem of older hominids.
Contemplating the phenomenon that opinions based on visual inspection tend to show more coherency than those based on multivariate analytical computations, Van Vark saw as one of the possible reasons that the visualists use their understanding of natural variations within the recent worldpopulation, more or less intuitively, as a common yardstick. He realized that the results of the mathematical approach would be improved considerably if they were based on a reliable estimate for the covariance matrix Σ of the recent world population. To obtain such an estimate, Van Vark used Howells' famous data (Harvard). This corresponds with a stratified sample of a large part of the present-day worldpopulation. An unbiased estimator for the covariance matrix in this part could easily be derived. The corresponding outcome is regarded as the required reliable estimate; it will be used as if it were the true Σ. Sample sizes for older hominids are so small, that the data cannot be used to reject the hypothesis that all covariance matrices are equal to the obtained estimate. Moreover older hominids skulls are usually incomplete. To solve the corresponding missing data problems, Van Vark used the (unbiased estimates of the) mean μ and covariance matrix Σ to compute estimates or rather predictions for the missing older hominid observations (Van Vark, 1984). The applied linear regression technique is obviously inappropriate because incomplete older hominid skulls cannot be considered as random drawings from the recent world population. Campbell criticized Van Vark's approach and Rao emphasized that a solution should be obtained. To obtain such a solution, a few further assumptions have to be added.
Let $X_1,X_2,\ldots,X_n$ be the independent random sample (i.r.s.) from the $N_p(\nu;\Sigma)$ distribution belonging to one of the older hominid populations. The problem is that the X_i's ($i=1,2,\ldots,n$) are only partially observed. Therefore, let Y_i $Y_2,\ldots,Y_n$ with $Y_i = A_iX_i$ denote the vectors of actual observations.
Note that the jth row of the r_i x p matrix A_i consists of p - 1 zeros and one unity, namely at the h-th place if X_{ih} is the j-th observed measurement for individual i. It has to be assumed that there is no dependence between the values of the scores and their possible missing. More precisely: the conditional distribution of $(X_i^1,\ldots,X_n)$ given $(A_1,\ldots,A_n)$ is that of an i.r.s. from $N_p(\nu,\Sigma)$. Moreover, every variable should have been observed at least once (otherwise it is omitted). It is obvious that no theory can be

developed unless the above-mentioned independence assumptions are made. On the other hand, one should note that there are at least two reasons why the assumptions are doubtful: (1) the prehistoric skulls or skull fragments come from a limited number of burial sites, (2) Rao's earlier mentioned story of the broken skulls suggests that large skulls are more often broken; this implies that some dependence exists between the values of the scores and their possible missing.

1. Estimating ν

Let $X_1,\ldots,X_n$ and $Y_i = A_iX_i$ $(i=1,\ldots,n)$ be as defined at the end of the previous section. $Y_1,\ldots,Y_n$ are independent, Y_i having the $N_{r_i}(A_i\nu_i, A_i\Sigma A_i^T)$ distribution. The probability density function (p.d.f.) $p_\nu(y_1,\ldots,y_n)$ of the distribution of $(Y_1,\ldots,Y_n)$ is equal to

$$c_1 \exp\{-\tfrac{1}{2}\sum_{i=1}^{n}(y_i - A_i\nu)^T(A_i\Sigma A_i^T)^{-1}(y_i - A_i\nu)\}$$

which can be written as

$$c_2(y)c_2(\nu)\exp\{\nu^T \sum_{i=1}^{n} A_i^T(A_i\Sigma A_i^T)^{-1}\, y_i\} \; .$$

Hence theory of multiparameter exponential families provides

$$\sum_{i=1}^{n} A_i^T\,(A_i\Sigma A_i^T)^{-1}Y_i$$

as a complete sufficient statistic. Its expectation is equal to

$$\sum_{i=1}^{n} A_i^T\,(A_i\Sigma A_i^T)^{-1}\,A_i\nu$$

so

$$\hat{\nu} = \left[\sum_{i=1}^{n} A_i^T\,(A_i\Sigma A_i^T)^{-1}A_i\right]^{-1}\left[\sum_{i=1}^{n} A_i^T\,(A_i\Sigma A_i^T)^{-1}\,A_iX_i\right] \tag{1.1}$$

is the best unbiased (also efficient and maximum likelihood) estimator for ν. The covariance matrix of the estimator $\hat{\nu}$ is equal to

$$\operatorname{cov}\hat{\nu} = \left[\sum_{i=1}^{n} A_i^T\,(A_i\Sigma A_i^T)^{-1}\,A_i\right]^{-1} \tag{1.2}$$

The rest of this section is devoted to a very special case for which $\hat{\nu}$ is compared with two other estimators, $\tilde{\nu}$ and ν^*, for ν. The quick reader is invited to jump to section 2 because the comparison does not go very deeply. This special case is as follows:
Let $p = 2$ and the i.r.s. $X_1,\ldots,X_n$ from $N_2(\nu,\Sigma)$ be as follows:

$X_1, X_2, \ldots, X_n$ are complete 2-dimensional vectors as before; suppose $n = 2m$, $Y_i = (X_{i1}, X_{i2})^T$ and $Y_{m+i} = X_{m+i1}$ $(i=1,2,\ldots,m)$, so X_{i2} is missing for $i=m+1,\ldots,2m$.

An obvious unbiased estimator for $\nu = (\nu_1, \nu_2)^T$ is

$$\tilde{\nu} = \left((2m)^{-1} \sum_{i=1}^{2m} X_{i1}; \ m^{-1} \sum_{i=1}^{m} X_{i2}\right)^T \tag{1.3}$$

With the notation $\Sigma = \begin{bmatrix} \sigma^2 & \rho\sigma\tau \\ \rho\sigma\tau & \tau^2 \end{bmatrix}$, the covariance matrix of the estimator $\tilde{\nu}$ (1.3) is equal to

$$\operatorname{cov} \tilde{\nu} = m^{-1} \begin{bmatrix} \frac{1}{2}\sigma^2 & \frac{1}{2}\rho\sigma\tau \\ \frac{1}{2}\rho\sigma\tau & \tau^2 \end{bmatrix} \tag{1.4}$$

By specializing the result (1.1) to this case, where $A_i = \begin{bmatrix} 1 & 0 \\ 0 & 1 \end{bmatrix}$ for $i = 1,2,\ldots,m$ and $A_i = [1,0]$ for $i = m+1,\ldots,2m$, one obtaines the estimator $\hat{\nu}$:

$$\hat{\nu} = \left[m\Sigma^{-1} + m \begin{bmatrix} \sigma^{-2} & 0 \\ 0 & 0 \end{bmatrix}\right]^{-1} \left[\Sigma^{-1} \sum_{i=1}^{m} X_i + \sigma^{-2} \sum_{i=m+1}^{2m} (X_{i1}, 0)^T\right]$$

which can be expressed as

$$\left((2m)^{-1} \sum_{i=1}^{2m} X_{i1} \ ; \ m^{-1} \sum_{i=1}^{m} X_{i2} + (2m)^{-1}\rho\tau\sigma^{-1}\left(\sum_{i=m+1}^{2m} X_{i1} - \sum_{i=1}^{m} X_{i1}\right)\right)^T .$$

If we set $m^{-1} \sum_{i=1}^{m} X_{i1} = X_{.1}$; $m^{-1} \sum_{i=1}^{m} X_{i2} = X_{.2}$ and $m^{-1} \sum_{i=m+1}^{2m} X_{i1} = \bar{X}_{.1}$, we can write:

$$\hat{\nu} = \left(\tfrac{1}{2}X_{.1} + \tfrac{1}{2}\bar{X}_{.1}; \ \tfrac{1}{2}X_{.2} + \tfrac{1}{2}(X_{.2} + \rho\tau\sigma^{-1}(\bar{X}_{.1} - X_{.1}))\right)^T \tag{1.5}$$

Now we recognize $X_{.2} + \rho\tau\sigma^{-1}(\bar{X}_{.1} - X_{.1})$ as the mean of the predictions $X_{i2} + \rho\tau\sigma^{-1}(X_{i1} - X_{.1})$ for the X_{i2}'s based on the Y_{i1}'s $(i = m+1,\ldots,2m)$. In the usual predictionformula $\nu_2 + \rho\tau\sigma^{-1}(X_{i1} - \nu_1)$ the parameters ν_2 and ν_1 are estimated with $X_{.2}$ and $X_{.1}$.

The covariance matrix (1.2) of $\hat{\nu}$ in this case is equal to

$$\left[m\ \Sigma^{-1}+m\begin{bmatrix} c^{-2} & 0 \\ c & 0 \end{bmatrix}\right]^{-1} = m^{-1}\begin{bmatrix} \frac{1}{2}\sigma^2 & \frac{1}{2}\rho\sigma\tau \\ \frac{1}{2}\rho\sigma\tau & (1-\frac{1}{2}\rho^2)\tau^2 \end{bmatrix} \tag{1.6}$$

Comparing the covariance matrix (1.6) of $\hat{\nu}$ with that of $\tilde{\nu}$ (1.4) we recognize that the difference

$$\operatorname{cov}\tilde{\nu} - \operatorname{ccv}\hat{\nu} = m^{-1}\begin{bmatrix} 0 & 0 \\ 0 & \frac{1}{2}\rho^2\tau^2 \end{bmatrix}$$

is positive definite, a result also implied by the fact that $\hat{\nu}$ is best unbiased. If $|\rho|$ is not too small, the best unbiased estimator $\hat{\nu}$ prcvides substantial improvement over the obvious unbiased estimator $\tilde{\nu}$.

If the assumption that Σ is known is dropped and the real unknown covariance matrix is

$$\begin{bmatrix} \alpha^2 & \kappa\alpha\beta \\ \kappa\alpha\beta & \beta^2 \end{bmatrix}$$

instead of Σ, the estimators $\hat{\nu}$ and $\tilde{\nu}$ both remain unbiased and the difference $\operatorname{cov}\tilde{\nu} - \operatorname{cov}\hat{\nu}$ is:

$$m^{-1}\begin{bmatrix} \frac{1}{2}\alpha^2 & \frac{1}{2}\kappa\alpha\beta \\ \frac{1}{2}\kappa\alpha\beta & \beta^2 \end{bmatrix} - m^{-1}\begin{bmatrix} \frac{1}{2}\alpha^2 & \frac{1}{2}\kappa\alpha\beta \\ \frac{1}{2}\kappa\alpha\beta & \beta^2+\frac{1}{2}\rho^2\tau^2\sigma^{-2}\alpha^2-\rho\tau\sigma^{-1}\kappa\alpha\beta \end{bmatrix} =$$

$$= m^{-1}\begin{bmatrix} 0 & 0 \\ 0 & \rho\tau\sigma^{-1}\alpha(\kappa\beta-\frac{1}{2}\rho\tau\sigma^{-1}\alpha) \end{bmatrix}$$

This difference will still be positive if the true values of the parameters do not differ too much from ρ, σ and τ.

Van Vark's earlier mentioned missing data technique is based on the proposal tc estimate the missing figures by applying linear regression as if the individual belonged to the recent world population.

Specialized to our case, let $\mu = (\mu_1,\mu_2)^T$ denote the expectation of the vector of scores from the present day world population. The covariance matrix Σ is as before, of course.

Van Vark considers 2m random vectors

$$X_1,\ldots,X_m,\quad X^*_{m+1},\ldots,X^*_{2m} \quad \text{where}$$

$$X^*_i = (X_{i1},\mu_2+\rho\tau\sigma^{-1}(X_{i1}-\mu_1))^T \quad i=m+1,\ldots,2m$$

and he estimates ν with

$$\nu^* = (2m)^{-1} \left(\sum_{i=1}^{m} X_i + \sum_{i=m+1}^{2m} X_i^* \right)$$

This estimator ν^* will in general not be unbiased, its bias being

$$E(\nu^*-\nu) = \tfrac{1}{2}(0;\ \mu_2-\nu_2 + \rho\tau\sigma^{-1}(\nu_1-\mu_1))^T$$

The mean squared error of Van Vark's estimator ν^* is equal to

$$E((\nu^*-\nu)(\nu^*-\nu)^T) = \operatorname{cov} \nu^* + E(\nu^*-\nu)E(\nu^*-\nu)^T =$$

$$= m^{-1} \begin{bmatrix} \tfrac{1}{2}\sigma^2 & \tfrac{1}{2}\rho\sigma\tau \\ \tfrac{1}{2}\rho\sigma\tau & \tfrac{1}{4}(1+\rho^2)\tau^2 + \tfrac{m}{4}(\mu_2-\nu_2+\rho\tau\sigma^{-1}(\nu_1-\mu_1))^2 \end{bmatrix}$$

If the regression lines of the second variable on the first one are the same for the recent world population and the population under study, then Van Vark's estimator is unbiased and the difference $\operatorname{cov} \nu^* - \operatorname{cov} \hat{\nu}$ =

$$= m^{-1} \begin{bmatrix} 0 & 0 \\ 0 & -\tfrac{3}{4}(1-\rho^2)\tau^2 \end{bmatrix}$$

is negative definite, showing that Van Vark's estimator does better than the best unbiased estimator $\hat{\nu}$. This is a matter of course if we recall that $\hat{\nu}$ uses estimators for the regression parameters, whereas ν^* uses true values. However, it is very unlikely that the regressionlines of the recent worldpopulation and the population under study coincide, even if covariance matrices are equal.

That Van Vark's estimator can lead to a nasty bias is illustrated in figure 1.

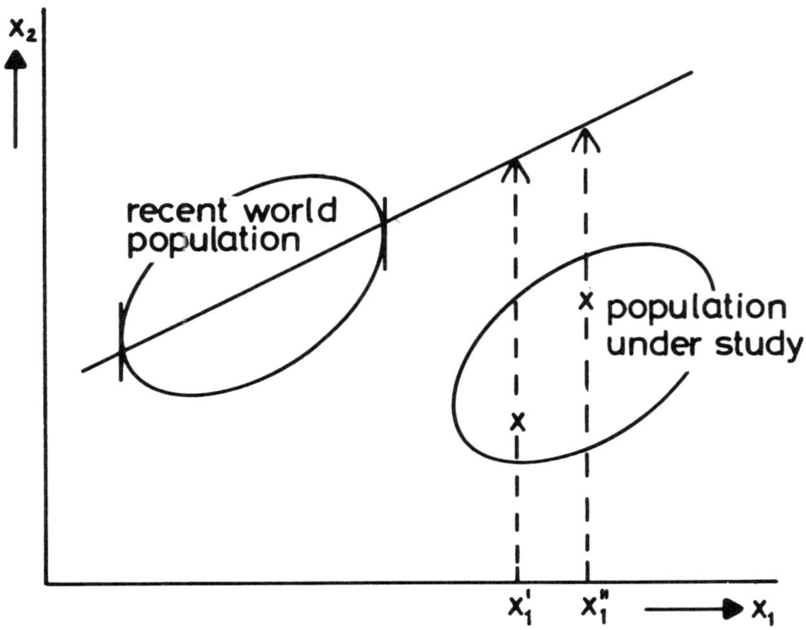

Figure 1. True vectors x with unknown second coordinate are replaced by vectors Δ, thus giving a wrong impression

Summary

The robustness of the estimator $\tilde{\nu}$ is a quite attractive property but if one feels pretty sure about the covariance matrix Σ and the assumption of normality, then one should use $\hat{\nu}$, because it is best unbiased. Brailowsky (1981) considers a two-category classification problem with p-variate normality, $\Sigma_1 = \Sigma_2 = \Sigma$ known and a priori probabilities equal; complications being caused by missing data. Using $\tilde{\nu}$, he observed unusual phenomena in some applications, which according to Barra (1981) disappear when using another estimator. The estimator suggested by Barra appears to be equivalent to our $\hat{\nu}$.
If one expects that the populationsample leads to a $N(\nu,\Sigma)$ distribution, which is very close to the $N(\mu,\Sigma)$ distribution belonging to the recent world population or at least that regressions are almost the same, then one might use ν^*.
In the applications in the following sections, the attention will almost entirely be focussed on $\hat{\nu}$ and $\tilde{\nu}$, because the assumptions concerning ν^* are too much demanding in our opinion.

2. Estimating Mahalanobis' Δ^2

Consider two populations of older hominids, characterized by their vectors of means ν_1 and ν_2. The Mahalanobis' squared distance between these two populations is defined by

$$\Delta^2 = (\nu_1-\nu_2)^T \Sigma^{-1}(\nu_1-\nu_2) \tag{2.1}$$

Plugging in the best unbiased estimators $\hat{\nu}_1$ and $\hat{\nu}_2$, we obtain an estimator for Δ^2 which is a function of the complete sufficient statistic and can be replaced by a best unbiased estimator on the basis of the following standard lemma concerning moments of quadratic forms:

Lemma (2.2)

> If $X \sim N_p(\mu,\Sigma)$ and A is a symmetric p x p matrix, then
>
> $$E(X^TAX) = \mu^TA\mu + \text{trace } A\Sigma$$
>
> and
>
> $$\text{var}(X^TAX) = 4\mu^TA\Sigma A\mu + 2 \text{ trace } (A\Sigma)^2$$

The lemma implies that

$$E((\hat{\nu}_1-\hat{\nu}_2)^T\Sigma^{-1}(\hat{\nu}_1-\hat{\nu}_2)) = \Delta^2 + \text{trace } \Sigma^{-1}\text{cov}(\hat{\nu}_1-\hat{\nu}_2).$$

Hence, keeping in mind that $\text{cov}(\hat{\nu}_1-\hat{\nu}_2) = \text{cov } \hat{\nu}_1 + \text{cov } \hat{\nu}_2$ is determined by 1.2,

$$\hat{\Delta}^2 = (\hat{\nu}_1-\hat{\nu}_2)^T\Sigma^{-1}(\hat{\nu}_1-\hat{\nu}_2) - \text{trace } \Sigma^{-1}(\text{cov } \hat{\nu}_1+\text{cov } \hat{\nu}_2) \tag{2.3}$$

is the best unbiased estimator for Δ^2.
An expression for the variance of this best unbiased estimator can also be given with the help of lemma 2.2:

$$\text{var } \hat{\Delta}^2=4(\nu_1-\nu_2)^T\Sigma^{-1}(\text{cov } \hat{\nu}_1+\text{cov } \hat{\nu}_2)\Sigma^{-1}(\nu_1-\nu_2)+$$
$$+ 2 \text{ trace}(\Sigma^{-1}(\text{cov } \hat{\nu}_1+\text{cov } \hat{\nu}_2))^2. \tag{2.4}$$

Of course with the same lemma 2.2 it is possible to construct an unbiased estimator for Δ^2 based on $\tilde{\nu}$, but this will give no improvement. Eventual robustness advantages of $\tilde{\nu}$ are lost because the assumption concerning the knowledge of Σ is essential in this construction.

With respect to ν^*, Van Vark has certain experiences which seem to show that whereas ν^* may lead to substantial bias in estimating ν, the estimates for Δ^2 based on ν^* are reasonably reliable. We do not study this phenomenon here, because it is too complicated and also because we prefer to use $\hat{\Delta}^2$.

3. Judging the typicality of an individual

The problem to consider in this section is the following: Given an incomplete sample $A_1X_1,\dots,A_nX_n$ of some older hominid population (cf section 0) and a vector of scores X of one individual, the question to be answered is: "Is there evidence to reject the hypothesis that individual X belongs to the population under study?"
Before giving two approaches to this problem we must assume that if $X = (X_1,\dots,X_p)^T$, all variables $1,\dots,p$ are observed at least once in the sample.

A possible solution is the following: Consider X as a sample of size one and formulate a two sample testing problem. So if $X \sim N_p(\lambda;\Sigma)$, the null hypothesis H: $\lambda = \nu$ is to be tested against the alternative hypothesis A: $\lambda \neq \nu$. From section 1 we know that $\hat{\nu} \sim N_p(\nu;\ \mathrm{cov}\ \hat{\nu})$, so $X-\hat{\nu} \sim N_p(\lambda-\hat{\nu};\ \Sigma+\mathrm{cov}\ \hat{\nu})$, where $\Sigma + \mathrm{cov}\ \hat{\nu}$ is known. Under the hypothesis H : $X-\hat{\nu} \sim N_p(0;\Sigma + \mathrm{cov}\ \hat{\nu})$ and so under H: $(X-\hat{\nu})^T(\Sigma+\mathrm{cov}\ \hat{\nu})^{-1}(X-\hat{\nu}) \sim \chi^2_p$.
We will reject the hypothesis H if

$$(X-\hat{\nu})^T(\Sigma+\mathrm{cov}\ \hat{\nu})^{-1}(X-\hat{\nu}) \geq \chi^2_{p;\alpha}$$

($\chi^2_{p;\alpha}$ is found in tables of the χ^2-distribution)

Another approach to the problem is introducing the typicality of individual X with respect to the population under study: $\theta(x,\nu)$. If V is a random variable having the χ^2_p-distribution then

$$\theta(x,\nu) = P_V(V \geq (x-\nu)^T\ \Sigma^{-1}(x-\nu))$$

$\theta(x,\nu)$ will be "large" if individual x belongs to the population under study, otherwise it will be "small". More precisely: if individual x belongs to the population under study, then $\theta(X,\nu)$ is uniform distributed on [0,1]. We shall construct confidence intervals for $\theta(x,\nu)$. If $\hat{\nu}$ is the estimator for ν from section 1, a $(1-\alpha)$ x 100 percents confidence region for ν is found as

$$\{\nu|(\hat{\nu}-\nu)^T(\mathrm{cov}\ \hat{\nu})^{-1}(\hat{\nu}-\nu) \leq \chi^2_{p;\alpha}\}$$

This result might be used to construct a conservative confidence interval for $\theta(x,\nu)$, but this method has a disadvantage: starting with a $(1-\alpha)$ x 100 per cents confidence level, the transformations will increase this level and so, especially if p is not very small, the result may be too large an interval.

An approximate confidence interval can be found as follows: Plugging in the estimator $\hat{\nu}$ for ν in $(x-\nu)^T\ \Sigma^{-1}(x-\nu)$ gives a random variable with expectation:

$$E_\nu((x-\hat{\nu})^T \Sigma^{-1}(x-\hat{\nu})) = (x-\nu)^T \Sigma^{-1}(x-\nu) + \text{trace } \Sigma^{-1} \text{ cov } \hat{\nu}$$

and variance

$$\text{var}_\nu((x-\hat{\nu})^T\Sigma^{-1}(x-\nu))=4(x-\nu)^T\Sigma^{-1}\text{cov } \hat{\nu}\ \Sigma^{-1}(x-\nu)+2\text{trace}(\Sigma^{-1}\text{cov } \hat{\nu})^2$$

(Recall that $\hat{\nu} \sim N_p(\nu;\ \text{cov } \hat{\nu})$, where cov $\hat{\nu}$ is determined by 1.2 and so $x-\hat{\nu} \sim N_p(x-\nu;\ \text{cov } \hat{\nu})$ and lemma 2.2 gives the result). A confidence interval with approximate level $(1-\alpha) \times 100$ per cent for $(x-\nu)^T \Sigma^{-1}(x-\nu)$ now will be

$$E((x-\hat{\nu})^T \Sigma^{-1}(x-\hat{\nu})) \pm u_{\frac{1}{2}\alpha} \{\text{var}(x-\hat{\nu})^T \Sigma^{-1}(x-\hat{\nu})\}^{\frac{1}{2}}$$

where ν is replaced by the estimate for ν and $u_{\frac{1}{2}\alpha}$ is taken from tables of the $N(0,1)$ distribution.

Transformation with tables of the χ^2_p-distribution will lead to an approximate confidence interval for $\theta(x,\nu)$.

4. Estimating posterior probabilities

If the prior probabilities are given by $\rho_1,\dots,\rho_k$ and the population densities by $f_1,\dots,f_k$, then Bayes' Theorem provides the posterior probabilities $\rho_{1|x},\dots,\rho_{k|x}$ of an individual whose vector of scores is given by x. The relevant expression

$$\rho_{h|x} = \rho_h \cdot f_h(x) / \sum_{i=1}^{k} \rho_i \cdot f_i(x)$$

can be reduced considerably if p-variate normality is assumed, together with homogeneity of covariance matrices. By substituting

$$f_h(x) = |2\pi\ \Sigma|^{-\frac{1}{2}} \exp\{-\tfrac{1}{2}\Delta_h^2(x)\}$$

where

$$\Delta_h^2(x) = (x-\nu_h)^T \Sigma^{-1}(x-\nu_h)$$

one obtains

$$\rho_{h|x} = \{1+\sum_{\substack{i=1\\ i\neq h}}^{k} \rho_i\rho_k^{-1} \exp(\tfrac{1}{2}\Delta_h^2(x)-\tfrac{1}{2}\Delta_i^2(x))\}^{-1}$$

Various estimators can be proposed in case $\nu_1,\dots,\nu_k$, Σ are unknown and have to be estimated by training samples. The accuracy of such estimators has been studied by several authors. Asymptotic variances were presented in Schaafsma (1976), Schaafsma-Van Vark (1977) and Ambergen-Schaafsma (1984). This section is devoted to the situation conform section 0: trainingsamples are incomplete and the covariance matrix is known.

The following lemma is useful in trying to find an unbiased estimator for $\rho_{h|x}^{-1}$:

Lemma (4.1)

> If $X \sim N_p(\mu;I_p)$ and A is a symmetric pxp matrix, then under certain conditions with respect to A and the constant c:
>
> $$E_\mu(\exp(c\ X^T AX)) = |I_p-2cA|^{-\frac{1}{2}}\exp((I_p-2cA)^{-1}-I_p)\mu$$
>
> A proof of this lemma can be given by transforming A with orthogonal transformations into a diagonal matrix and subsequent straightforward calculations.

By applying lemma 4.1 and some linear algebra it is shown that the best unbiased estimators for $\exp(\frac{1}{2}\Delta_h^2(x))$ and $\exp(-\frac{1}{2}\Delta_i^2(x))$ exist and are respectively given by

$$E_h = |I_p+\Sigma^{-1}\text{cov}\ \hat{\nu}_h|^{-\frac{1}{2}}\ \exp(\tfrac{1}{2}(x-\hat{\nu}_h)^T(\Sigma+\text{cov}\ \hat{\nu}_h)^{-1}(x-\hat{\nu}_h)) \quad \text{and}$$

$$E_i = |I_p-\Sigma^{-1}\text{cov}\ \hat{\nu}_i|^{-\frac{1}{2}}\ \exp(\tfrac{1}{2}(x-\hat{\nu}_i)^T(\Sigma-\text{cov}\ \hat{\nu}_i)^{-1}(x-\hat{\nu}_i))$$

and so the best unbiased estimator for $\rho_{h|x}^{-1}$ will be

$$1 + \sum_{\substack{i=1\\ i\neq h}}^{k} \rho_i\rho_h^{-1}\ E_hE_i \tag{4.2}$$

Recall that cov $\hat{\nu}$ is determined by 1.2.

Also by applying lemma 4.1 it is possible to find an expression for the exact variance of the estimator 4.2, which can be used to construct approximate confidence intervals for $\rho_{h|x}^{-1}$ and thus for $\rho_{h|x}$.

References

Ambergen, A.W. and Schaafsma, W: 1984, Interval estimates for posterior probabilities, applications to Border Cave. In: Van Vark and Howells.

Anderson, T.W.: 1958, An introduction to multivariate statistical analysis. New York; Wiley.

Barra, J.R.: 1981, A propos d'un résultat de Brailovsky concernant une probabilité d'erreur en analyse discriminante. Ann. Inst. Henri Poincaré, section B. Vol. XVII, no. 1, p. 21-29.

Brailovsky, V.: 1981, On the influence of sample set structure on decision rule quality for the case of a linear discriminant function. IEEE Transactions on pattern analysis and machine intelligence. Vol. PAMI - 3, no. 4, p. 454-459.

Campbell, N.A.: 1984, Some aspects of allocation and discrimination. In: Van Vark and Howells

Ksirsagar, A.M.: 1972, Multivariate analysis. New York; Marcel Dekker.

Rao, C.R.: 1965, Linear statistical inference and its applications. New York; Wiley.

Schaafsma, W.: 1976, The asymptotic distribution of some statistics from discriminant analysis. Report TW-176. Dept. Math. Postbox 800, Groningen, The Netherlands.

Schaafsma, W. and Van Vark, G.N.: 1977, Classification and discrimination problems with applications, part I. Statistica Neerlandica 31, p. 25-45.

Schaafsma, W. and Van Vark, G.N.: 1979, Classification and discrimination problems with applications, part IIa. Statistica Neerlandica 33, p. 91-126.

Van Vark, G.N.: 1984, On the determination of hominid affinities. In: Van Vark and Howells

Van Vark, G.N. and Howells, W.W.: 1984, Multivariate statistical methods in physical anthropology. D. Reidel Publishing Company Dordrecht, Holland.

USE OF DIVERSITY AND DISTANCE MEASURES IN THE ANALYSIS OF QUALITATIVE DATA

C. Radhakrishna Rao*

University of Pittsburgh

SUMMARY

The paper discusses some theoretical and practical considerations in the choice of diversity and distance measures or comparing populations in terms of gene frequencies associated with various characteristics. It develops systematic methods for grouping populations by similarity in genetic diversity, apportioning diversity as between and within populations, grouping of populations by similarity in gene frequencies (cluster analysis) and testing consistency of results by using different diversity and distance measures and subsets of data. The methods are illustrated using the data on nine serological and six biochemical characteristics of the Makiritare Indians. They can be used in the analysis of any qualitative data, such as those that arise in sociological research.

1. INTRODUCTION

In an earlier paper, Rao and Boudreau (1982) developed some systematic methods for an analysis of gene frequencies associated with blood group systems to study diversity within and dissimilarity between human populations. This paper provides a further discussion of these problems and illustrates the methods using the serological and biochemical data on Makiritare Indians, extensively studied by Gershowitz et.al. (1980).

With quantitative measurements, it is generally assumed that

* This work is supported by the Air Force Office of Scientific Research under Contract F 49629-82-K-001.

G. N. van Vark and W. W. Howells (eds.), Multivariate Statistical Methods in Physical Anthropology, 49–67.

variances and covariances within populations are nearly the same, in which case comparisons between populations need only be made in terms of the mean values of the measurements. With qualitative measurements like gene or genotypic frequencies, it is necessary to compare populations both with respect to genetic diversity within and dissimilarity between them. Methods for comparing populations in terms of genetic diversity within were developed in recent papers by Karlin et.al. (1979), Rao (1982a, b, c) and Rao and Boudreau (1982). Methods for grouping populations by similarity of gene structures have been developed over the last forty years, and key references to the literature on the subject can be found in two recent review papers by Jorde (1980) and Lalouel (1980). In this paper, we describe a number of distance measures (satisfying the triangular postulate of a distance function), give their geometrical interpretation and illustrate their use in cluster analysis.

2. ANALYSIS OF DIVERSITY

2.1. Measurement of diversity

The measurement of diversity (variability) in a population with respect to a qualitative characteristic is of some importance in biological work. For instance, we may like to have a quantitative measure of diversity of a characteristic like the eye color of individuals (which has a finite number of possibilities) in a population. Such a quantity would serve as a descriptive measure in terms of which several populations could be compared and the differences interpreted in biological terms. We shall discuss some theoretical and practical considerations for the choice of a diversity measure.

First, let us consider a single attribute with k different categories occurring with relative frequencies $p_1,\dots,p_k$ in a population, and denote by $H(\underline{p}) = H(p_1,\dots,p_k)$ a measure of diversity of the attribute in the population. If all the individuals of a population exhibit only one category of the attribute, then there is no diversity which leads to the condition

$$C_0 : H(\underline{p}) = 0 \text{ iff all the components of } \underline{p} \text{ are zro except one.}$$

If $\underline{p}$ and $\underline{q}$ are the vectors of relative frequencies of the k categories in two populations, then $\lambda p + (1-\lambda)q$ is the vector of relative frequencies in a mixture of the two populations in the ratio λ: $(1-\lambda)$. It is an intuitive requirement that the diversity in a mixture of populations should not be smaller than the average diversity within the individual populations. This leads to the condition

$$C_1 : H(\lambda\underline{p} + (1-\lambda)\underline{q}) \geq \lambda H(\underline{p}) + (1-\lambda)\, H(\underline{q})$$

with equality iff $\underline{p} = \underline{q}$, or the function H is strictly concave in the space of multinomial probability distributions. The conditions C_0 and C_1 do not specify the diversity function, but they seem to be logical requirements. Other criteria have to be brought in for choosing a diversity measure for application in any given problem.

Some examples of diversity functions which received wide applications in biology are entropy functions such as the following.

(1) $H_S(\underline{p}) = -\Sigma p_i \log p_i$ (Shannon).

(2) $H_\alpha(\underline{p}) = (1-\Sigma p_i^\alpha)/(2^{\alpha-1}-1)$, $\alpha > 0$, $\alpha \neq 1$ (α-order entropy of Havrda and Charavát).

(3) $H_P(\underline{p}) = -\Sigma p_i \log p_i - \Sigma(1-p_i)\log(1-p_i)$ (Paired Shannon entropy).

(4) $H_R(\underline{p}) = (1-\alpha)^{-1} \log \Sigma p_i^\alpha$, $0 < \alpha < 1$ (α-degree entropy of Renyi).

(5) $H_\gamma(\underline{p}) = [1-(\Sigma p_i^{1/\gamma})^\gamma]/[1-2^{\gamma-1}]$, $\gamma > 0$ (γ-entropy).

All these entropy functions (1) - (5) satisfy the conditions C_0 and C_1 in the range of the parameters (α or γ) indicated in each case. But, unfortunately, they do not seem to have a simple interpretation in biological terms.
Further, as shown elsewhere (Rao, 1982b and Burbea and Rao, 1982a, 1982b) they do not possess higher order convexity properties necessary for carrying out analyses of diversity (ANODIV) similar to ANOVA (analysis of variance).

2.2. The quadratic entropy

Rao (1982a, b, c) introduced a new measure called quadratic entropy which seems to provide a unified approach to the measurement of diversity and its analysis. This depends on assigning non-negative quantities d_{ij} representing the amount of dissimilarity between the categories i and j of an attribute. Then the average dissimilarity between two random events from a multinomial distribution characterized by $\underline{p} = (p_1,\ldots,p_k)$, the probabilities of the k events, is

$$H_\Delta(\underline{p}) = \Sigma\Sigma d_{ij}p_ip_j = \underline{p}\ \Delta\ \underline{p}' \tag{2.2.1}$$

where Δ is the matrix (d_{ij}). For $H_\Delta(\underline{p})$ to satisfy the conditions C_0 and C_1, the d_{ij} should be chosen such that

$$d_{11} = \ldots = d_{kk} = 0 \tag{2.2.2}$$

and the (k-1) × (k-1) matrix

$$(d_{ik} + d_{jk} - d_{ij} - d_{kk}) \tag{2.2.3}$$

is nonnegative definite. A function $H_\Delta(\underline{p})$ with Δ satisfying the conditions (2.2.2) - (2.2.3) is called the quadratic entropy. Its properties are studied in Rao (1982a, b) and Lau (1983). The quadratic entropy can be interpreted in terms of the chosen d_{ij} in any given problem.

Suppose that we want to measure the diversity of the alleles of a gene in a population. If we assign a score zero if two randomly drawn chromosomes have the same allele and one if they have different alleles, then

$$H_\Delta(\underline{p}) = 1 - \Sigma p_i^2 \text{ (Gini-Simpson index)} \tag{2.2.4}$$

where $p_1,\ldots,p_k$ are the proportions of the different alleles. The expression (2.2.4) is the index of heterozygosity, which is extensively used in genetic work. If we consider a random mating population and score the difference between two individuals (genotypes) as 0,1,2 depending on the number of alleles not common to the genotypes, then

$$H_\Delta(\underline{p}) = 1 - \Sigma p_i^2 - \Sigma p_1^2(1-p_i)^2 \tag{2.2.5}$$

which is the index of diversity introduced by Latter (1973). For other examples, the reader is referred to Rao (1982a).

2.3. Diversity with respect to a number of attributes

If we want to measure the diversity with respect to a number of attributes, there are a number of alternatives. Let there be m attributes with the i-th attribute having k_i alternatives. Then:

(1) We may consider the joint distribution of the m attributes with $k = k_1 x \ldots x k_m$ alternatives in which case we have a multinomial distribution in k classes, and we can use any of the formulae for diversity mentioned in Sections 2.1 and 2.2.

(2) If H_i is the diversity in the i-th attribute as measured by any diversity measure, then we may take a suitable function of $H_1,\ldots,H_m$ as a joint measure of diversity, such as the arithmetic, geometric or harmonic mean

(3) The range of a diversity measure of an attribute may depend on the number of alternatives of the attribute. In such a case, a standardized measure can be obtained by

$$H'(p) = H(\underline{p})/\max_{\underline{p}} H(\underline{p})$$

which lies between 0 and 1. Then the standardized diversities $H_1', \ldots, H_m'$ can be combined as in (2) to obtain a joint measure of diversity.

It is seen that if any of the entropy measures (1) - (5) given in Section 2.1 are used then

$$H'(\underline{p}) = H(\underline{p})/H(\tfrac{1}{k}, \ldots, \tfrac{1}{k}).$$

We have mentioned some of the possible ways in which the joint diversity with respect to a number of attributes can be measured. For practical applications, one has to choose a measure which is appropriate to the problem. If the quadratic entropy described in Section 2.2 is used, then the method (1) can generate a variety of measures depending on how the dissimilarity between two individuals with respect to a number of attributes is defined. (See for example notes 3, 4, 5 on page 28 in Rao, 1982a).

2.4. Apportionment of diversity

The concavity of a diversity function (H) enables us to apportion the diversity in a mixed population as between and within populations. If $\underline{p}$, $\underline{q}$, $\underline{r}$, ... are the relative frequencies in populations 1,2,3,... with a priori probabilities λ, μ, ν, ... then the diversity in the mixture $\lambda\underline{p} + \mu\underline{q} + \nu\underline{r} + \ldots$ can be decomposed as

$$H(\lambda\underline{p}+\mu\underline{q}+\nu\underline{r}+\ldots) = [\lambda H(\underline{p})+\mu H(\underline{q})+\nu H(\underline{r})+\ldots] + \\ + J(\{\underline{p},\underline{q},\underline{r},\ldots\}:\{\lambda,\mu,\nu,\ldots\}) \qquad (2.4.1)$$

$$T = W + B.$$

In (2.4.1), W represents the average diversity within populations and B, which is nonnegative and vanishes only if $\underline{p} = \underline{q} = \underline{r} = \ldots$, denotes the diversity between populations. The expression for between populations diversity

$$J(\{\underline{p},\underline{q},\ldots\}:\{\lambda,\mu,\ldots\}) = H(\lambda\underline{p}+\mu\underline{q}+\ldots)-[\lambda H(\underline{p})+\mu H(\underline{q})+\ldots] \quad (2.4.2)$$

was called the Jensen difference in Rao (1982a) and suggested as a general measure of divergence or dissimilarity between populations. This is an alternative to the measure of divergence defined by Csiszár (1967), which includes several known divergence measures as special cases.

Table 1. Overall diversity within villages and the index of diversity between villages by different diversity measures*

(S = Serological, B = Biochemical)

Diversity measure	Type of data	Within villages A	B	C	E	F	G	H	Average within W	Between B	Ratio G=B/(B+W)
H_S	S	.8640	.9175	.8731	.9058	.8498	.8750	.8849	.8814	.0385	4.18
	B	.3453	.4463	.5313	.4271	.3855	.4437	.4330	.4303	.0291	6.33
H_2	S	.4182	.4456	.4184	.4370	.4115	.4205	.4254	.4252	.0217	4.86
	B	.1456	.1941	.2261	.1813	.1799	.1900	.1869	.1863	.0118	5.94
H_γ	S	.8959	.9412	.9082	.9335	.8816	.9087	.9172	.9123	.0307	3.26
	B	.4167	.5275	.6356	.5158	.4144	.5278	.5130	.5073	.0491	8.83

* The original data are given in Tables 7 and 8 of the Appendix.

Table 2. Index of diversity (G) between villages by diversity measures and individual biochemical or serological characteristics*

Biochemical	H_S	H_2	H_γ	Serological	H_S	H_2	H_γ
Hp	6.52	8.68	4.60	MN	2.81	3.07	2.16
Gc	4.13	4.32	3.17	Ss	5.88	7.86	4.15
Alb	12.86	1.59	27.04	P	1.74	2.34	1.22
Ap	6.49	2.38	9.97	Duffy	1.27	1.50	0.94
PGM	5.70	4.75	4.92	Kidd	2.48	3.16	1.78
6 PDG	21.60	1.90	47.31	Diego	17.36	17.02	14.54
				Lewis	7.91	10.44	5.62
Overall	6.33	5.94	8.83	Ee	0.59	0.80	0.41
				Cc	0.43	0.58	0.30
				Overall	4.18	4.86	3.26

* The original data are given in Tables 7 and 8 of the Appendix.

The ratio G = B/T has been called the index of diversity between populations and used in several studies. Reference may be made to papers by Lewontin (1972), who uses the Shannon entropy (H_S), and Nei (1973) and Chakraborty (1974), who use the Gini-Simpson index (H_2). As mentioned earlier, any strictly concave diversity function could be used for this purpose.

We illustrate the use of diversity measures using serological and biochemical data on Makiritare Indians from seven villages, which has been extensively analysed by Gershowitz et.al. (1970). Table 1 gives the diversity within each village averaged over all the serological or biochemical characteristics, average diversity within villages and the index of diversity between populations according to the three diversity measures H_S, H_2 and H_γ (with $\gamma = \frac{1}{2}$). It is seen that the diversity is of the same order within each village with respect to both the serological and biochemical characteristics. The diversity between populations as measured by G appears to be somewhat larger for the biochemical characteristics, and relatively more so in the case of H_S and H_γ. To examine this discrepancy, the index of diversity was computed for the individual characteristics as shown in Table 2. The large values of G for H_S and H_γ have arisen in cases where one of the alleles of a gene is very rare. In such border line cases, H_S and H_γ are very sensitive and the results may be misleading specially when the estimates of gene frequencies are based on small samples as in the case of the Makiritare study. The Gini-Simpson index seems to be better suited for studying diversity within and between populations over a wide range of gene frequencies (see Rao, 1982a for further remarks).

2.5. Clustering of populations by diversity

In the case of Makiritare Indians, the diversity within villages appears to be of the same order even for individual serological or biochemical characteristics. But in situations where differences in diversity is established in one or more characteristics, it is of some interest to group the populations by similarity or dissimilarity in diversity. For this purpose we need a compound measure of dissimilarity in diversities over the different characteristics between two populations. Let $d_1,\ldots,d_m$ and $d_1',\ldots,d_m'$ be the diversities (using any measure of diversity) in m characteristics for two populations. If we want a simple ordering of the populations by overall diversity we may use the dissimilarity measure $\bar{d}-\bar{d}'$ where $\bar{d} = (d_1+\ldots+d_m)/m$ and $\bar{d}' = (d_1'+\ldots+d_m')/m$. Alternative compound measures are

$$\delta_1 = \sum_1^m \frac{2\sqrt{d_1 d_i'}}{d_i+d_i'} \tag{2.5.1}$$

$$\delta_2 = -\sum_1^m \log \frac{2\sqrt{d_i d_i'}}{d_i+d_i'} \tag{2.5.2}$$

where δ_1 is more suited when the d_i's are small. (Note that δ_2 becomes infinity when any d_i or $d_i' = 0$). With δ_1 or δ_2 calculated for every pair of populations, we have a dissimilarity matrix for making a cluster analysis of the populations. Reference may be made to Rao and Boudreau (1982) where an example of cluster analysis is given based on a matrix of δ_2 values.

3. DISTANCE MEASURES AND CLUSTER ANALYSIS

3.1. Genetic distance measures

The literature on distance (or dissimilarity) measures is reviewed in two papers by Jorde (1980) and Lalouel (1980). Karlin et.al. (1979) and Rao and Boudreau (1982) examined the robustness of a variety of distance functions used in genetic studies. A unified approach to the construction of dissimilarity or similarly measures is given in Rao (1982a).

There are considerable controversies on the choice of dissimilarity measures for studying the genetic structure of a given set of populations. (See Morton, 1973 for a discussion by Balakrishnan, Morton, Nei and others). It is well known that different distance functions emphasize different aspects of genetic differences between populations and, in general, it would be more illuminating to analyse and interpret given genetic data under different possible genetic and environmental models using appropriate distance functions. We describe some distance measures which are found useful in evolutionary studies.

3.2. Distance functions based on Nei's measures of dissimilarity

Let the gene frequencies at the i-th locus in a population (say π_α) be represented by the row vector

$$\underline{p}_i = (p_{i1},\dots,p_{ik_i}),\ i = 1,2,\dots$$

in which case the whole gametic array of gene frequencies at m loci can be represented by the partitioned vector

$$\underline{p} = (\underline{p}_1 : \underline{p}_2 : \dots : \underline{p}_m). \tag{3.2.1}$$

The vector $\underline{p}$ may be represented by a point in $k=k_1+\dots+k_m$ dimensional Euclidean space. (Actually the points will be confined to a subset of the space since all the coordinates are nonnegative and the sum of the coordinates of each subvector is equal to unity. If

$$\underline{q} = (\underline{q}_1 : \underline{q}_2 : \dots : \underline{q}_m) \tag{3.2.2}$$

is the corresponding vector in another population (say π_β), then Nei's minimum distance between π_α and π_β is proportional to the square of the Euclidean distance between $\underline{p}$ and $\underline{q}$

$$\begin{aligned} D^{(1)}_{\alpha\beta} &= (\underline{p} - \underline{q})(\underline{p} - \underline{q})' \\ &= (\underline{p}_1 - \underline{q}_1)(\underline{p}_1 - \underline{q}_1)' + \dots + (\underline{p}_m - \underline{q}_m)(\underline{p}_m - \underline{q}_m)' \\ &= \sum_1^m \underline{p}_i\underline{p}_i' + \sum_1^m \underline{q}_i\underline{q}_i' - 2\sum_1^m \underline{p}_i\underline{q}_i' \\ &= (\sqrt{J_{\alpha\alpha}} - \sqrt{J_{\beta\beta}})^2 + 2\sqrt{J_{\alpha\alpha}J_{\beta\beta}}\,(1 - \cos\theta^{(2)}_{\alpha\beta}) \end{aligned} \tag{3.2.3}$$

where we use the notation

$$J_{\alpha\alpha} = \Sigma\underline{p}_i\underline{p}_i',\quad J_{\beta\beta} = \Sigma\underline{q}_i\underline{q}_i',\quad \cos\theta^{(2)}_{\alpha\beta} = \Sigma\underline{p}_i\underline{q}_i'/\sqrt{J_{\alpha\alpha}J_{\beta\beta}}. \tag{3.2.4}$$

The expression (3.2.3) has two components; the first measures the difference in lengths of the gametic vectors (or difference in homozygosity or diversity within populations) and the second depends on the angle between the vectors (difference in orientation). Thus Nei's minimum distance measures the overall difference both in length and orientation of the gametic vectors.
In practice, it is better to use the equivalent measure

$$\delta_{\alpha\beta}^{(1)} = (D_{\alpha\beta}^{(1)})^{\frac{1}{2}} \tag{3.2.5}$$

for cluster analysis, as $\delta_{\alpha\beta}^{(1)}$ is a regular distance function (i.e., satisfies the triangular postulate).

Nei's standard distance between populations π_α and π_β is

$$D_{\alpha\beta}^{(2)} = -\log \cos \theta_{\alpha\beta}^{(2)} \tag{3.2.6}$$

which is based only on the difference in the orientation of the vectors $\underline{p}$ and $\underline{q}$. The two measures $D_{\alpha\beta}^{(1)}$ and $D_{\alpha\beta}^{(2)}$ are equivalent when the diversities within populations are the same. [When the diversities differ the measure (3.2.6) does not reflect the entire difference between $\underline{p}$ and $\underline{q}$.] Instead of (3.2.6), one could use the equivalent measure

$$\theta_{\alpha\beta}^{(2)} = \cos^{-1}[\exp(-D_{\alpha\beta}^{(2)})] \tag{3.2.7}$$

for cluster analysis as $\theta_{\alpha\beta}^{(2)}$ is a distance function.

Finally, Nei's maximum distance is proportional to

$$D_{\alpha\beta}^{(3)} = -\log \left(\prod_{i=1}^{m} \cos \theta_{\alpha\beta}^{(i)} \right) \tag{3.2.8}$$

where

$$\cos \theta_{\alpha\beta}^{(i)} = p_i q_i' / \sqrt{(p_i p_i')(q_i q_i')}.$$

The measure (3.2.8) is the sum of Nei's standard distances (3.2.6) over the m loci. It is a compound measure of differences in orientation of the gene frequency vectors at the different loci and does not take into account differences in the diversities at the different loci. Instead of (3.2.8), one could use the equivalent measure

$$\theta_{\alpha\beta}^{(3)} = \cos^{-1}\left(\prod_{i=1}^{m} \cos \theta_{\alpha\beta}^{(i)} \right) \tag{3.2.9}$$

for cluster analysis as $\theta_{\alpha\beta}^{(3)}$ is a distance function.

Suppose that the m attributes are independently distributed and we consider the multinomial distribution of the $k = k_1 \times \ldots \times k_m$ combinations of the m attributes. Then Nei's standard distance derived from the combined multinomial is the same as Nei's maximum distance (3.2.8).

3.3. Other measures of dissimilarity

3.3.1. Jensen difference

Consider two multinomial distributions with cell probability vectors $\underline{p}_1$ and $\underline{p}_2$ and H any strictly concave diversity measure. Then Rao (1982a) defined the Jensen difference between the distributions as

$$J(\underline{p}_1,\underline{p}_2) = H\left(\frac{\underline{p}_1+\underline{p}_2}{2}\right) - \tfrac{1}{2}H(\underline{p}_1) - \tfrac{1}{2}H(\underline{p}_2) \tag{3.3.1}$$

which can be used as a dissimilarity measure. If we are comparing two populations π_α and π_β with the gametic vectors $\underline{p}$ and $\underline{q}$ as defined in (3.2.1)-(3.2.2), then we may use the combined expression

$$D_{\alpha\beta} = J(\underline{p}_1,\underline{q}_1) + \ldots + J(\underline{p}_m,\underline{q}_m) \tag{3.3.2}$$

as the dissimilarity between π_α and π_β. Alternatively, one can consider the joint distribution of the m attributes and apply the formula (3.3.1) on the joint distributions. A number of dissimilarity measures can be derived by choosing different diversity measures. It may be convenient to make a monotone transformation of the dissimilarity measure to convert it to a regular distance function. Some examples are given in Rao (1982a).

3.3.2, Matusita's distance and related measures

If $\underline{p}_1 = (p_{11},\ldots,p_{1a})$ are $\underline{p}_2 = (p_{21},\ldots,p_{2a})$ are probability vectors for two multinomial distributions, then Matusita (1957) defined the distance between the distributions as

$$\begin{aligned} M(\underline{p}_1,\underline{p}_2) &= (\sqrt{p_{11}} - \sqrt{p_{21}})^2 + \ldots + (\sqrt{p_{1a}} - \sqrt{p_{2a}})^2 \\ &= 2 - 2(\sqrt{p_{11}p_{21}} + \ldots + \sqrt{p_{1a}p_{2a}}) \\ &= 2(1 - \cos\theta_{\alpha\beta}) \end{aligned} \tag{3.3.3}$$

where $\cos\theta_{\alpha\beta}$ is the angle between the vectors

$$(\sqrt{p_{11}},\ldots,\sqrt{p_{1a}}) \text{ and } (\sqrt{p_{21}},\ldots,\sqrt{p_{2a}}).$$

Instead of (3.3.3), one could use the equivalent distance measure

$$\theta_{\alpha\beta} = \cos^{-1}\,[1 - \tfrac{1}{2}M(\underline{p}_1,\underline{p}_2)] \tag{3.3.4}$$

which is the Hellinger's distance measure proposed by Bhattacharya (1946).

When there is more than one attribute, we can consider the joint distributions and apply the formula (3.3.3) or (3.3.4).

Alternatively one could define the distance between the gametic vectors $p = (\underline{p}_1:\ldots:\underline{p}_m)$, and $q = (\underline{q}_1:\ldots:\underline{q}_m)$ as

$$D_{\alpha\beta}^{(4)} = M(\underline{p}_1,\underline{q}_1) + \ldots + M(\underline{p}_m,\underline{q}_m)$$
$$= 2(1 - \cos\theta_{\alpha\beta}^{(1)}) + \ldots + 2(1 - \cos\theta_{\alpha\beta}^{(m)}) \qquad (3.3.5)$$

which is proportional to the square of the distance function proposed by Cavalli-Sforza and Edward (1967). We denote

$$\delta_{\alpha\beta}^{(4)} = (D_{\alpha\beta}^{(4)})^{\frac{1}{2}} \qquad (3.3.6)$$

which is a distance measure. An equivalent measure is the angle

$$\theta_{\alpha\beta}^{(4)} = \cos^{-1}[m^{-1}(\cos\theta_{\alpha\beta}^{(1)} + \ldots + \cos\theta_{\alpha\beta}^{(m)})] \qquad (3.3.7)$$

which is also a distance measure. It may be noted that if we consider the vectors obtained by replacing each component of $\underline{p}$ and $\underline{q}$ by its square root, then the new vectors are of the same length but may differ in orientation, and the angle between them is as given in (3.3.7). Alternatively one could construct an angle based on the geometric mean of $\cos\theta_{\alpha\beta}^{(i)}$

$$\theta_{\alpha\beta}^{(5)} = \cos^{-1}(\prod_{i=1}^{m} \cos\theta_{\alpha\beta}^{(i)}) \qquad (3.3.8)$$

which is also a distance measure.

3.4. Cluster analysis

We have described a variety of distance measures

$$\delta_{\alpha\beta}^{(1)}, \theta_{\alpha\beta}^{(2)}, \theta_{\alpha\beta}^{(3)}, \delta_{\alpha\beta}^{(4)}, \theta_{\alpha\beta}^{(4)}, \theta_{\alpha\beta}^{(5)} \qquad (3.4.1)$$

which can be used in cluster analysis. Rao and Boudreau (1982) used the first three measures (3.4.1) in the analysis of blood group data on human populations. The results were consistent except in cases where the populations differed markedly in within genetic diversities. In the case of Makiritare Indians, we have already seen that the genetic diversity within villages are of the same order. In such a case, the cluster analyses based on $\delta_{\alpha\beta}^{(1)}, \theta_{\alpha\beta}^{(2)}$ and $\theta_{\alpha\beta}^{(3)}$ will necessarily be consistent. The results based on $\delta_{\alpha\beta}^{(4)}$, $\theta_{\alpha\beta}^{(4)}$ and $\theta_{\alpha\beta}^{(5)}$ will also be consistent. It is therefore of interest to compare the cluster analyses based on one measure, say $\delta_{\alpha\beta}^{(1)}$ chosen from the first three and another measure, say $\theta_{\alpha\beta}^{(5)}$ chosen from the last three. The distances based on these measures between the Indian groups are given in Tables 3 and 4 for the serological and biochemical data respectively.

Table 3. Comparison of genetic distances ($\delta_{\alpha\beta}^{(1)}$) based on serological and biochemical data

Villages	A	B	C	E	F	G	H
A		.644	1.213	.569	.608	.471	.713
B	.319		.655	.413	.610	.486	.434
C	.403	.164		.941	1.015	1.059	.844
E	.412	.157	.133		.406	.329	.429
F	.597	.341	.427	.407		.379	.549
G	.371	.433	.445	.462	.734		.462
H	.280	.323	.355	.366	.633	.116	

Table 4. Comparison of genetic distances ($\theta_{\alpha\beta}^{(5)}$) based on serological and biochemical data

Villages	A	B	C	E	F	G	H
A		31	52	27	25	21	31
B	17		28	18	32	22	19
C	25	15		39	48	45	37
E	22	15	13		23	15	18
F	28	21	31	25		18	26
G	23	23	23	20	36		19
H	18	18	20	16	31	6	

The serological distances are given in the upper triangular matrix and the biochemical in the lower triangular matrix.

It has been the usual practice to draw dendograms or evolutionary trees based on a distance (or a dissimilarity) matrix between populations. Such diagrams provide a hierarchical clustering and a broad picture of the configuration of the populations in the space of gene frequencies, but they obscure the links between clusters which may explain some of the complex relationships between populations. The author has advocated the method of complete subgraphs to obtain overlapping clusters at least at some lower threshold values of distances in addition to drawing dendrograms and evolutionary trees (Rao, 1971, 1977).

The following tables provide the overlapping clusters based on different distance measures and types of data.

Table 5. Complete subgraphs based on $(\delta_{\alpha\beta}^{(1)})$ matrix

Serological data		Biochemical data	
Threshold	Subgraphs	Threshold	Subgraphs
.329-.413	FEG,BE	.116-.133	CE,GH
.413-.500	FEG,BEGH,AG	.133-.323	BCE,ABH,GH
.500-.610	BEFGH,AEFG	.323-.412	ABCEH,BEF,AGH
.610-.713	ABEFGH,BC	.412-.633	ABCEFH,ABCEGH
>.713	ABCEFGH	>.633	ABCEFGH

Table 6. Complete subgraphs based on $(\theta_{\alpha\beta}^{(5)})$ matrix

Serological data		Biochemical data	
Threshold	Subgraphs	Threshold	Subgraphs
15-20	EGH,FG,BE,BH	6-10	GH
20-25	AFG,BEGH,EFG	10-15	GH,BCE
25-30	AEFG,BEGH,BC	15-21	ABH,GHE,BF,BCEH
30-35	ABEFGH,BC	21-25	ABCEGH,BEF
>35	ABCEFGH	>25	ABCEFGH

It is seen from Tables 5 and 6 that the clusters of populations provided by the serological and biochemical data are somewhat different for each distance measure, while for each type of data the results for the two distance measures are consistent. Part of this discrepancy may be due to large sampling errors in the estimation of the low gene frequencies for the biochemical characteristics, which make the distance matrix somewhat unreliable. This shows the need for collection of large samples and covering a wide variety of gametic material to obtain a meaningful study of differences between populations. In general, for given data, the results are likely to be consistent for different choices of distance measures. A diagramatic representation of the configuration of the populations is given in Figures 1 and 2.

4. CONCLUDING NOTE

The object of the paper as mentioned in the introduction is to develop systematic methods for studying diversities within and dissimilarities between populations in the gene or genotypic frequencies. The data on serological and biochemical characteristics of the Makiritare Indians was chosen only for illustrating the methods. It must be noted that in general the results depend on the nature of the diversity and dissimilarity measures chosen and

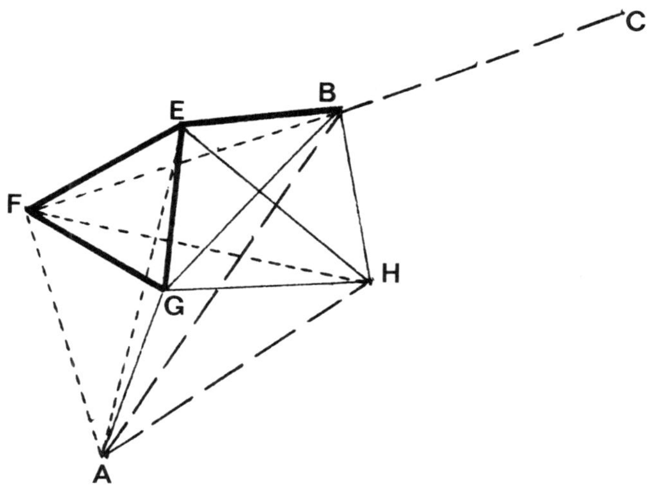

Figure 1. Configuration of the Makiritare Indians based on serological data

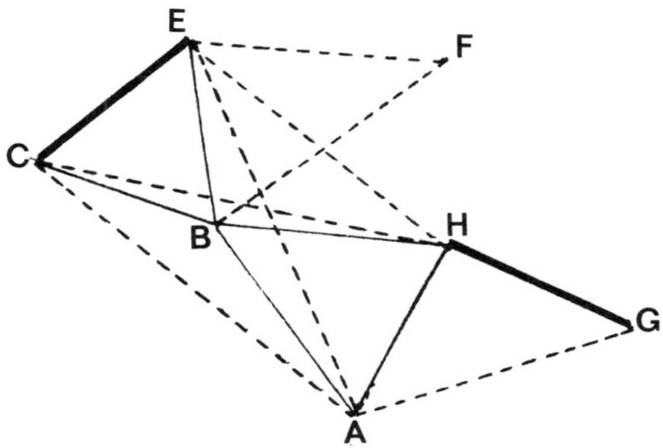

Figure 2. Configuration of Makiritare Indians based on biochemical data

▬▬, ——, ----, — — indicate connections at the 1st, 2nd, 3rd and 4th threshold intervals

the number and nature of characteristics included in the study.

It is better to examine given data using different diversity and dissimilarity measures and to compare the results. Inconsistencies in results may suggest further investigations or reexamination of data to extract more information.

It is generally recommended that all available genetic data should be used in diversity and dissimilarity analyses, which involves pooling of information over a number of genetic systems. While this should be done, it will be more informative to do the analyses separately for different subsystems (formed according to some criterion) and examine consistency of results. Suitable explanations may be sought if the results are inconsistent.

The methods developed in the paper are applicable to any qualitative data, such as those that arise in sociological studies.

The limitations of data used such as inadequacies of sample sizes, improper sampling, unrepresentativeness of samples and possible differences between investigators and techniques used in measurements should be clearly mentioned. Further, original data should be scrutinized before and as the analysis proceeds to find out any peculiar features of the data which need careful examination (see Rao and Boudreau, 1982 for illustration).

References

Bhattacharya, A.: 1946, A measure of divergence between two multinomial populations. Sankhyā, 7, 401.

Burbea, J. and Rao, C. Radhakrishna: 1982a, On the convexity of some divergence measures based on entropy functions. IEEE Trans. Information Theory, 28, 489-495.

Burbea, J. and Rao, C. Radhakrishna: 1982b, On the convexity of higher order Jensen differences based on entropy functions. IEEE Trans. Information Theory, 28, 961-963.

Cavalli-Sforza, L.L. and Edwards, A.W.: 1967, Phylogenetic analysis models and estimation procedures. Amer.J.Hum.Genet., 19, 234-257.

Chakraborty, R.: 1974, A note on Nei's measure of gene diversity in a substructural population. Humangenetik, 21, 85-88.

Csiszár, I.: 1967, A class of measures of informativity of observation channels. Periodica Math. Hungar., 2, 191-213.

Gershowitz, H. et.al.: 1980, Gene frequencies and microdifferentiation among Makiritare Indians (Parts I-IV). Amer.J.Hum.Genet., 22, 515-561.

Jorde, L.B.: 1980, The genetic structure of subdivided populations. A review, pp. 135-208, in Current Developments in Anthropological Genetics, Plenum Press, Edited by J.H. Mielke and M.H. Crawford.

Karlin, S., Kenett, R. and Bonne-Tamir, B.: 1979, Analysis of biochemical genetic data on Jewish populations. 11. Results and interpretations of heterogeneity indices and distance measures with respect to standard. Amer.J.Hum.Genet., 31, 341-365.

Lalouel, Jean-Marc: 1980, Distance analysis and multidimensional scaling, pp. 209-250, in: Current Developments in Anthropological Genetics, Plenum Press, Ed. J.H. Mielke and M.H. Crawford.

Latter, B.D.H.: 1973, Measures of genetic distance between individuals and populations, in Genetic Structure of Populations, pp. 27-39, edited by N.E. Morton, Univ. of Hawaii Press.

Lau, Ka-Sing: 1983, Characterization of Rao's quadratic entropy. Sankhyā (in press).

Lewontin, R.C.: 1972, The apportionment of human diversity. Evolutionary Biology, 6, 381-398.

Matusita, K.: 1957, Decision rule based on the distance for the classification problem. Ann.Inst.Stat.Math., 8, 67-77.

Morton, N.E.: 1973, Editor, Genetic Structure of Populations. Honolulu: University of Hawaii Press.

Nei, M.: 1973, Analysis of gene diversity in subdivided populations. Proc.Nat.Acad.Sci., 70, 3321-3323.

Nei, M.: 1978, The theory of genetic distance and evolution of human races. Japan J. Human Genet., 23, 341-369.

Rao, C. Radhakrishna: 1971, Taxonomy in anthropology, in: Mathematics in Archaelogical and Historical Sciences, pp. 19-20. Edin. Univ. Press.

Rao, C. Radhakrishna: 1977, Cluster analysis applied to a study of race mixture in human populations, in: Proceedings, Michigan Univ. Symposium, pp. 175-197.

Rao, C. Radhakrishna: 1982a, Diversity and dissimilarity coefficients: a unified approach. Theoret. Pop. Biol., 21, 24-43.

Rao, C. Radhakrishna: 1982b, Diversity: Its measurement, decomposition, apportionment and analysis. Sankhyā 44A, 1-22.

Rao, C. Radhakrishna: 1982c, Gini-Simpson index of diversity: a characterization, generalization and applications. Utilitas Mathematica, 21B, 273-282.

Rao, C. Radhakrishna and Boudreau, R.: 1982, Diversity and cluster analyses of blood group data on some human populations. Tech. Rept. 82-35, Center for Multivariate Analysis, University of Pittsburgh.

APPENDIX

Table 7. Gene frequencies of serological characteristics of Makiritare Indians in seven villages

Serological characteristics	Villages						
	A	B	C	E	F	G	H
MN	67 33	80 20	87 13	69 31	68 32	64 36	66 34
Ss	41 59	53 47	77 23	32 68	40 60	36 64	49 51
P	51 49	46 54	47 53	43 57	30 70	35 65	52 48
Duffy	64 36	78 22	76 24	72 28	68 32	80 20	77 23
Kidd	36 64	31 69	49 51	28 72	40 60	30 70	21 79
Diego	04 96	32 68	52 48	22 78	02 98	10 80	15 85
Lewis	77 23	50 50	22 78	47 53	42 58	55 45	33 67
Cc	42 58	44 56	42 58	36 64	37 63	45 55	47 53
Ee	53 47	54 46	56 44	62 38	64 36	53 47	52 48

Source: Gershowitz et.al. (1980).

Table 8. Gene frequencies of biochemical characteristics of Makiritare Indians in seven villages

Biochemical characteristics	Villages						
	A	B	C	E	F	G	H
Hp	27	48	51	53	65	24	29
	73	52	49	47	35	76	71
Gc	81	80	86	89	64	88	86
	19	20	14	11	36	12	14
Alb	100	100	96	97	100	98	99
	0	0	04	03	0	02	01
Ap	05	05	13	05	0	06	05
	95	95	87	95	100	94	95
PGM	96	88	85	86	91	71	77
	04	12	15	14	09	29	23
6 PDG	100	98	97	100	100	100	100
	0	2	3	0	0	0	0

Source: Gershowitz et.al. (1980).

MULTIVARIATE DISTANCES AND MULTIVARIATE CLASSIFICATION SYSTEMS USING NON-METRIC TRAITS IN BIOLOGICAL STUDIES

M. Finnegan, Osteology Laboratory,
R.M. Rubison, Department of Statistics,
Kansas State University

INTRODUCTION

Multivariate statistical measures of biological distance utilizing non-metric traits on the human cranium was first explored by Berry and Berry (1967) utilizing the multivariate statistics developed by C.A.B. Smith and known as the Grewal-Smith statistic (Grewal, 1962). By 1975 attempts were being made in the classification of human crania by means of non-metric traits in order to assign one or some few individuals into a parent or larger group (Sjøvold, 1975; Finnegan, 1975). However, these first attempts at classification utilized very cumbersome statistics (hypergeometric probabilities), and the methods had a high rate of misclassification. The "tally" method (Anderson, 1968) was also utilized, but did not provide the accuracy in classification that Anderson had originally achieved.

The possibility of developing a statistic where the individual cranium or infracranial skeletal remains could be correctly allocated to one race, or more specifically to a subracial group, and the implications of correct placement for archaeological materials and recent forensic cases brought a new wave of studies (Finnegan, 1977). By 1979 a number of papers had been published presenting theoretical refinements and empirical testing of the Grewal-Smith statistic for use in archaeological populations (Green and Suchey, 1976; Finnegan and Cooprider, 1978), and active classification of individual skeletons or crania into ethnic or isolate population groups was accomplished (Finnegan and McGuire, 1979).

G. N. van Vark and W. W. Howells (eds.), Multivariate Statistical Methods in Physical Anthropology, 69–80.

A NON-METRIC EXAMINATION

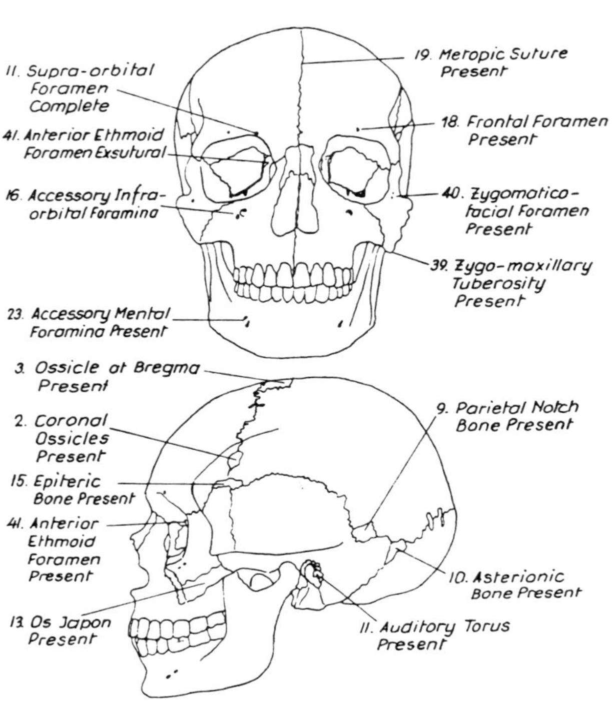

Figure 1. Normal frontal and normal lateral views of the cranium depicting the location of non-metric traits used in this study.

MATERIALS AND METHODS

Figure 1 shows a number of cranial traits utilized in population comparisons. These traits are considered non-metric or discontinuous in that for each skeleton or cranium a trait is scored as either present or absent. It is also believed that these traits have a very high genetic component, although no simple Mendelian segregation ratios have been found, and that the environmental effects on these traits are minimal as compared with metric traits. The expression of these traits are probably "threshold" in nature and the traits have been picked for their lower production of intertrait correlation. Most of the traits are bilaterally expressed with four states: bilateral absence, present left absent right, absent left present right, or bilaterally present. Some few traits are in the midsagittal line (or central planes) and are scored absent or present.

Numerous studies have been published dealing with the symmetry and asymmetry of non-metric traits, with results suggesting that relatively high correlations exist in those traits which are bilateral in nature, showing possible dependency between sides for a given trait (Trinkaus, 1978; Green, et.al., 1979; Molto, 1979). Most researchers use the bilaterality of traits in the production of biological distances among populations with the theory that while few significant differences in frequency between sides exist, these differences are greater than zero, and the use of only one side of the cranium or skeleton would result in a loss of biological information. Of course, by pooling sides for those traits which have a possibility of bilateral expression, we are to some extent increasing the sample size with a resultant reduction in the variance displayed within a particular population sample.

RESULTS

Appendix 1 shows the statistical derivation for the Bayes procedure with proportional priors (Duda and Hart, 1973), the weight of evidence procedure (Osteyee and Good, 1974), Fisher's linear discriminate function (LDF) (Gilbert, 1968; Moore, 1973; Bahadur, 1961), the "tally" method (see Cooprider, 1978), and a bilateral procedure developed by Rubison (1979). Table 1 is a summary of the percent misclassification rates averaged over all pairwise infracranial sample comparisons utilizing as sample populations, Coast Eskimo and Yukon Eskimo, Aleut, American Blacks and American Whites, all housed at the Smithsonian Institution. In this table, Bayes' procedure and Fisher's LDF utilize both equal priors and proportional priors, the "tally" method after Anderson (1968) and Rubison's procedure are listed. As can be seen by the low mean error production, Rubison's bilateral procedure is by far superior in classifying an infracranial skeleton into a larger population.

Table 2 shows the error production rate for cranial samples of Coast Eskimo, Yukon Eskimo, St. Lawrence Island Eskimo, Canaveral Point Indians, Aleut and pre-Aleut populations from the Smithsonian collections. Again, Bayes' procedure utilizes both the equal priors and proportional priors, and Fisher's LDF utilizes equal priors only, as sample sizes are so nearly equal. The "tally" method of Anderson is dropped for its high misclassification rate, but Rubison's bilateral procedure is presented. As with the infracranial skeleton, the Rubison bilateral procedure has the lowest error production in the "hit-miss" classification rates when averaged over all pairwise cranial samples.

These results are given in order to show our current direction in classifying biological populations and assigning crania or skeletons into larger population samples. The same procedure could also be useful to the biologists in looking at species or subspeciation and indeed, a number of studies have been done utilizing non-metric traits on animal populations including white-tailed deer (Rees, 1969a, 1969b), the Swedish red fox (Sjøvold, 1977), many non-human primates (Berry, 1974, Berry and Berry, 1971), and

Table 1. Summary of percent misclassification rates averaged over all pairwise infracranial sample comparisons. NP denotes that sufficient numbers of complete observations were not available to obtain useful estimates of the parameters

Population Pair	Sample Size	Bayes' Procedure				Fisher's LDF				Tally Method		Rubison's Procedure
		[1] Equal Priors		[2] Prop Priors		[3] Equal Priors		[4] Prop Priors		[5]		[6]
	n_1/n_2	Left	Right	Left	Right	Left	Right	Left	Right	Left	Right	
Coast Eskimo - Yukon Eskimo	62/53	26.09	25.22	25.22	26.09	33.91	36.52	33.91	36.52	33.91	28.70	15.65
Coast Eskimo - Aleut	62/51	22.12	25.66	22.12	24.78	46.90	NP	46.90	NP	42.48	36.28	14.16
Coast Eskimo - Black	62/100	20.37	16.05	18.52	16.67	29.63	25.93	29.63	25.93	26.54	19.14	9.26
Coast Eskimo - White	62/96	15.82	12.03	14.56	12.66	15.19	20.89	16.46	22.15	22.15	19.62	4.43
Yukon Eskimo - Aleut	53/51	22.12	20.19	23.08	20.19	35.58	37.50	35.58	37.50	30.77	27.88	11.54
Yukon Eskimo - Black	53/100	20.92	15.03	18.30	13.73	19.61	19.61	20.26	19.61	30.07	28.76	6.54
Yukon Eskimo - White	53/96	10.07	14.09	12.08	14.09	9.40	13.42	9.40	13.42	34.90	18.12	6.71
Aleut - Black	51/100	15.89	7.95	17.22	9.27	18.54	14.57	20.53	13.91	21.85	17.22	5.30
Aleut - White	51/96	10.88	7.48	11.56	5.44	12.24	14.97	12.24	16.33	17.01	16.33	5.44
Black - White	100/96	21.43	21.94	20.92	22.45	18.37	18.37	18.37	17.86	33.16	35.20	13.78
$\bar{X}$		18.57	16.56	18.36	16.54	23.94	22.42	24.32	22.58	24.73	29.28	9.23
$\bar{X}$ Left and Right		17.57		17.45		23.22		23.50		27.00		9.23

Table 2. Summary of percent misclassification rates averaged over all pairwise cranial sample comparisons. NP denotes that sufficient numbers of complete observations were not available to obtain useful estimates of the parameters. Proportional priors in Fisher's LDF were not considered because of nearly equal sample sizes (see Bayes' theorem)

Population Pair	Sample Size	Bayes' Procedure				Fisher's LDF		Rubison's Procedure
		[1]		[2]		[3]		[6]
		Equal Priors		Prop Priors		Equal Priors		
	n_1/n_2	Left	Right	Left	Right	Left	Right	
Coast Eskimo - Yukon Eskimo	50/50	25.00	31.00	25.00	31.00	21.00	26.00	15.00
Coast Eskimo - St.Law.Is.Esk.	50/50	28.00	19.00	28.00	19.00	43.00	43.00	16.00
Coast Eskimo - Canaveral Pt.	50/48	17.35	12.24	17.35	13.26	20.41	12.24	4.08
Coast Eskimo - Aleut	50/50	16.00	24.00	16.00	24.00	14.00	21.00	8.00
Coast Eskimo - Pre Aleut	50/47	18.56	19.59	17.53	19.59	13.40	15.46	12.37
Yukon Eskimo - St.Law.Is.Esk.	50/50	17.00	20.00	17.00	20.00	32.00	38.00	11.00
Yukon Eskimo - Canaveral Pt.	50/48	16.33	21.43	16.33	19.39	13.27	24.49	8.16
Yukon Eskimo - Aleut	50/50	17.00	15.00	17.00	15.00	16.00	17.00	13.00
Yukon Eskimo - Pre Aleut	50/47	13.40	17.53	13.40	15.46	8.25	9.28	10.31
St.Law.Esk.-CanaveralPt.	50/48	18.37	8.16	18.37	8.16	31.63	NP	8.16
St.Law.Is.Esk.-Aleut	50/50	19.00	21.00	19.00	21.00	22.00	28.00	12.00
St.Law.Is.Esk.-PreAleut	50/47	16.49	17.53	15.46	14.43	30.93	29.90	11.34
Canaveral Pt. - Aleut	48/50	13.27	12.24	14.29	11.22	18.37	22.45	7.14
Canaveral Pt - Pre Aleut	48/47	18.95	12.63	18.95	13.68	21.05	17.89	11.58
Aleut-Pre Aleut	50/47	17.53	29.90	18.56	28.87	17.53	23.71	17.53
$\bar{X}$		18.15	18.75	18.15	18.27	21.52	23.46	11.04
$\bar{X}$ Left and Right		18.45		18.21		22.46		11.04

many genetic purebred strains of mice and wild field populations of mice (see Berry 1979, for reference list).

DISCUSSION

We have found that at present there are three major applications of non-metric data in classifying skeletal remains. 1) The classifying of one cranium or infracranial skeleton in an archaeological setting. 2) Classifying one cranium or infracranial skeleton in recent forensic cases in order to aid identification of the skeletal remains. 3) Specific or subspecific classification of animals in a given geographical region and/or biological distances among free ranging animal populations.

There are, however, some problems inherent in classification of this nature. The first type of problem (class I) is the data base. For human populations we have, at the present time, 15 or 20 local populations which have been recorded for non-metric traits. On the finding of one skeleton recent enough to be a forensic case, or one lone skeleton of prehistoric interest without associated artifacts, it is difficult to classify this individual into its correct population when data on the correct base population may not be available. Subsets of this problem are the exact usefulness of each trait in the non-metric trait list scored per particular skeleton; dummy traits, which are defined as those traits producing the same frequency, usually low or zero, in each population tested; and, a trait may act as a one-way trait, where, in the case of the trait Os Japonicum, it is present in Japanese populations at about a 3% level and mainly absent in all other world populations where non-metric traits have been scored. Therefore, the absence of this trait does not indicate the skeleton could not be Japanese, but the presence of the trait would indicate quite highly a Japanese heritage.

In order to solve the problems here, we must increase the data base of populations represented, and at the same time, continue testing our non-metric traits in order to search for as many one-way traits as possible. The basic traits used have been tested for both inter- and intra-observer error. Molto (1979) found that scoring consistency in intra-observer error was high for most traits, although most errors were random rather than systematic in nature. Finnegan (1978) found that overall recording errors amounted to 2.76% and that these errors were also random. The only interobserver error study to date (Finnegan and Rubison, 1980), shows that the highest correspondence between any two observers on the same population is about 84%. This correspondence is much too low to allow the mixing of data scored by different observers in comparisons among populations.

Problems of the second type (class II) relate to the vehicle used in the placement of one or some few skeletons into larger population groups. On the basis of research completed to date, we suggest the use of Rubison's bilateral classification rule in assigning an individual skull or skeleton to a larger population based on the lower misclassification rate in utilizing pairwise population tests (See Finnegan and McGuire, 1979).

We are hopeful, however, that continued research on the behaviour of non-metric variation, on the part of the anthropologists, will help solve some of the problems of the traits used, the method of scoring these traits, and the reduction of inter- and intra-observer error. And, that continued research in the theoretical and applied uses of new multivariate statistical techniques will improve our success in analyzing recent forensic cases and earlier human populations.

References

Anderson, J.E.: 1968, Skeletal "Anomalies" as Genetic indicators. In: The Skeletal Biology of Earlier Human Populations. D.R. Brothwell, ed. Pergamon Press, London.

Bahadur, R.R.: 1961, A representation of the joint distribution of responses in n dichotomous variables. In: Studies in Item Analysis and Prediction. H. Solomon, ed. Palo Alto, Stanford University Press.

Berry, A.C.: 1974, Non-metrical variation in the prosimian skull. Prosimian Biology (Ed.) Martin, R.D., G.A. Doyle, & A.C. Walker, University of Pittsburg Press.

Berry, A.C. and R.J. Berry: 1967, Epigenetic variation in the human cranium. Journal of Anatomy. Vol. 101, No. 2, pp. 361-379.

Berry, A.C. and R.J. Berry: 1971, Epigenetic polymorphism in the primate skeleton. In: Comparative Genetics in Monkeys, Apes and Man. In D. Brothwell and A.B. Chiarelli (eds.), Academic Press, London.

Berry, R.J.: 1979, Section I. Genes and skeletons, ancient and modern. J. Human Evolution. Vol. 8, No. 7, pp. 669-677.

Buikstra, J.E. and T.L. Pistono: 1978, Non-metric traits in forensic research: A case study. Paper presented at the annual meeting of the American Academy of Forensic Sciences. 23 February 1978, St. Louis.

Cooprider, K.B.: 1978, Racial Classification based on non-metric skeletal traits. M.A. Report, Kansas State University.

Cooprider, K.B., R.M. Rubison, and M. Finnegan: 1980, Racial classification based on non-metric skeletal traits. Homo. Vol. 31, No. 1, pp. 1-21.

Duda, R.O., and P.E. Hart: 1973, Pattern Classification and Scene Analysis. New York: John Wiley and Sons.

Finnegan, M.: 1975, The use of non-metric traits in assigning individuals to a parent or larger group. Paper presented at the Plains Conference, 7 November 1975, Lincoln, Nebraska.

Finnegan, M.: 1977, Non-metric traits and forensic identification. Paper presented at the 29th annual meeting of the American Academy of Forensic Sciences, February 17, San Diego, California.

Finnegan, M.: 1978, An analysis of intra-observer error in non-metric studies. Abstract: American Journal of Physical Anthropology, Vol. 48, No. 3, p. 393.

Finnegan, M. and K. Cooprider: 1978, Empirical comparison of distance equations using discrete traits. American Journal of Physical Anthropology. Vol. 49, No. 1, pp. 39-46.

Finnegan, M. and S.A. McGuire: 1979, Classification systems for discrete variables used in forensic anthropology. American Journal of Physical Anthropology. Vol. 51, No. 4, pp. 547-553.

Finnegan, M. and R.M. Rubison: 1980, The assessment and analysis of interobserver error in non-metric cranial studies. American Journal of Physical Anthropology, Vol. 52, No. 2, p. 226.

Gilbert, E.S.: 1968, On discrimination using qualitative variables. Journal of the American Statistical Association, Vol. 63, pp. 116-122.

Green, R.F. and J.M. Suchey: 1976, The use of inverse sine transformations in the analysis of non-metric cranial data. American Journal of Physical Anthropology. Vol. 45, pp. 61-68.

Green, R.F., J.M. Suchey, and D.V. Gokhale: 1979, The statistical treatment of correlated bilateral traits in the analysis of cranial material. American Journal of Physical Anthropology, Vol. 50, No. 4, pp. 629-634.

Grewal, M.S.: 1962, The rate of genetic divergence of sublines in the C57BL strain of mice. Genetics Research, Vol. 3, pp. 226-237.

Molto, J.E.: 1979, The assessment and meaning of intraobserver error in population studies based on discontinuous cranial traits. American Journal of Physical Anthropology, Vol. 51, No. 3, pp. 333-344.

Moore, D.H., II: 1973, Evaluation of five discrimination procedures for binary variables. Journal of the American Statistical Association, Vol. 68, pp. 399-404.

Osteyee, D.B. and J.G. Good: 1974, Information, weight of evidence, the singularity between probability measures and signal detection. Lectures Notes in Mathematics, No. 376. Berlin: Springer Verlag.

Rees, J.W.: 1969a, Morphologic variation in the cranium and mandible of the white-tailed deer (Odocoileus virginianus): a comparative study of geographical and four biological distances. Journal of Morphology, Vol. 128, No. 1, pp. 95-112.

Rees, J.W.: 1969b, Morphologic variation in the mandible of the white-tailed deer (Odocoileus virginianus): a study of populational skeletal variation by principal component and canonical analysis. Journal of Morphology. Vol. 128, No. 1, pp. 113-130.

Rubison, R.M.: 1979, Classification based on correlated bilateral non-metric traits. Technical Report # 38, Department of Statistics and Statistical Laboratory, Kansas State University.

Russell, F.: 1900, Studies on cranial variation. American Naturalist. Vol. 34, pp. 737-745.

Sjøvold, T.: 1975, Allocation of single or some few individuals to one of two or more groups by means of non-metrical variants in the skeleton. OSSA, Vol. 2, pp. 41-46.

Sjøvold, T.: 1977, Non-metrical Divergence Between Skeletal Populations. OSSA, Vol. 4, Supplement 1, 133 p.

Suchey, Judy: 1976, Use of non-metric skeletal traits in forensic cases. Paper presented at the 28th annual meeting of the American Academy of Forensic Sciences, Washington, D.C., February, 1976.

Trinkaus, E.: 1978, Bilateral asymmetry of human skeletal non-metric traits. American Journal of Physical Anthropology. Vol. 49, No. 3, pp. 315-318.

APPENDIX 1

Five classification procedures are summarized below. They are labeled as the Tally Method, Bayes' Theorem, Weight of Evidence, the Linear Discriminant Function (LDF) and the Bilateral Classification Rule.

We define the following variables and notations:

$\underline{x}$ = $(x_1, x_2, x_3, \ldots, x_k)$ = trait sequence for an unknown individual

k = number of traits

where $x_i = \begin{cases} 0 & \text{if i-th trait is absent} \\ 1 & \text{if i-th trait is present} \\ & \text{if i-th trait is missing} \end{cases}$

s = number of populations in the data set

m_{ij} = number of tallies for trait $\underline{i}$ in population $\underline{j}$

n_{ij} = number of observed individuals for trait $\underline{i}$ in population $\underline{j}$

n_j = number of individuals in sample from population $\underline{j}$ $[n_{ij}<n_j]$

π_{ij} = $P(x_i = 1|w_j)$ = probability that the $\underline{i}$-th trait is present in population $\underline{j}$

$1-\pi_{ij}$ = $P(x_i = 0|w_j)$ = probability that the $\underline{i}$-th trait is absent in population $\underline{j}$

$P_{ij} = \frac{m_{ij}}{n_{ij}}$ = sample estimate of π_{ij}.

$1-P_{ij} = \frac{n_{ij}-m_{ij}}{n_{ij}}$ = sample estimate of $1-\pi_{ij}$

$P(w_j)$ = prior probability of an individual belonging to population $\underline{j}$.

The Tally Method (Anderson 1968)

Select an individual for classification, $\underline{x} = (x_1, x_2, x_3, \ldots, x_k)$.

If x_i is a missing observation, proceed to the next trait.

If x_i is a 1, place a tally or a unit of support for the population $\underline{j}$ which has the largest proportion P_{ij}.

If x_i is a 0, place a tally or a unit of support for the population $\underline{j}$ which has the smallest proportion P_{ij}.

When all k traits have been examined, classify the individual to the population which has accumulated the largest number of traits. If a tie occurs between populations with the same accumulated supports, classify to the population with the largest sample size.

Bayes' Theorem

Define $P(w_j|x)$ as the probability that an individual with trait sequence x belongs to population $\underline{j}$. By Bayes' rule,

$$p(w_j\ x) = \frac{P(x|w_j)P(w_j)}{\sum_{t=1}^{s} P(x|w_t)P(w_t)}$$

Since the denominator remains constant for any given discrimination procedure regardless of the chosen population w_j, the classification rule is based on $P(x|w_j)P(w_j)$. The individual will be classified to the population j for which the probability of membership is highest.

The classification rule is then to classify an individual to the population j which has the largest values $g(w_j)$:

$$g(w_j) = \sum_{i=1}^{k} [x_i \log P_{ij} + (1-x_i) \log(1-P_{ij})] + \log P(w_j)$$

Estimates for the priors $P(w_j)$ have been based on sample sizes for this study, i.e.,

$$P(w_j) = \frac{n_j}{\sum_{t=1}^{s} n_t}$$

Weight of Evidence

This classification procedure is identical to the Bayes' Theorem procedure with the adjustment term due to priors excluded.

The procedure classifies an individual to the population j which gives the largest value of $G(w_j)$:

$$G(w_j) = \sum_{i=1}^{k} [x_i \log P_{ij} + (1-x_i) \log(1-P_{ij})]$$

Linear Discriminant Function (LDF)

In the case of only two populations (s=2), Fisher's linear discriminant function is calculated as:

$$g(\underset{\sim}{x}) = \underset{\sim}{x}' \Sigma (\underset{\sim}{p}_1 - \underset{\sim}{p}_2) - \tfrac{1}{2}(\underset{\sim}{p}_1 + \underset{\sim}{p}_2)' \Sigma^{-1} (\underset{\sim}{p}_1 - \underset{\sim}{p}_2)$$

where $\underset{\sim}{p}_i = (p_{1i}, p_{2i}, \ldots, p_{ki})$ i=1, 2 and Σ is the pooled variance-covariance matrix associated with the random variables $\underset{\sim}{x}$.

The individual $\underset{\sim}{x}$ is classified to population 1 if $g(\underset{\sim}{x}) \geq \ln (P(w_2) / P(w_1))$.

Bilateral Classification Rule

The final classification method proposed by Rubison (1979) is a procedure which utilizes the side-to-side correlation of bilateral traits in classifying an individual into one of two or more populations.

Given a population w_j, let $\underset{\sim}{x} = (x_1,\ldots,x_k)$ and $\underset{\sim}{y} = (y_1,\ldots,y_k)$ represent the left and right trait sequences for one member of that population.

Define

$$P_{i,qr,j} = P\{(x_i = q,\ y_i = r)|w_j\}$$

$$i = 1,\ 2,\ \ldots,\ k$$
$$q = 0,\ 1$$
$$r = 0,\ 1$$
$$j = 1,\ \ldots,\ s(s = \text{number of populations}).$$

Then assuming independence of traits,

$$P(\underline{x},\ \underline{y}|w_j) = \prod_{i=1}^{k} P_{i,00,j}^{(1-x_i)(1-y_i)}\ P_{i,01,j}^{(1-x_i)y_i}\ P_{i,10,j}^{x_i(1-y_i)}\ P_{i,11,j}^{x_i y_i}$$

The decision rule is to classify to the population $\underline{j}$ which has the largest value $g(\underline{x},\ \underline{y}|w_j)$ where:

$$g(\underline{x},\ \underline{y}|w_j) = \log P(\underline{x},\ \underline{y}|w_j) + \log P(w_j)\ .$$

GENERALIZED DISTANCE IN FAMILIAL STUDIES OF ANTHROPOMETRICAL CHARACTERS

C. Susanne

Laboratory of Human Genetics, Free University of Brussels (Belgium)

INTRODUCTION

In the analysis of multiple continuous variables in man and in the study of the heritable component of variation, methods have been published to partition the variation and covariation among a group of characters into its environmental and genetical components (Martin et.al., 1977, Eaves et.al., 1977). This methodology has been influenced by the fundamental works on analysis of covariances of Jøreskog (1973). This method has the advantage to determine by statistical test the adequacy of the models.

We tried to use a descriptive methodology which cannot estimate or distinct additive genetical variation from common specific environmental effects, we hope however that this method will help less mathematical minded biologists to describe in a first approach familial material in terms of genetic, environmental, and assortative mating factors.

We used a methodology based on the concept of generalized distances and described by Defrise (1967, 1968). Subjects of different age and sex can be pooled, we are thus able to compare the distances observed between different relatives. The differences between groups of relatives (twins, sibs, parent offspring, spouses) are expressed quantitatively in number of standard deviations between observed and expected mean distances. The method does not take into account sampling effects but, on large samples, such as we used in this study, these effects are negligible.

MATERIAL AND METHODS

For p normally distributed characters and a covariance matrix Σ, the generalized distance (GD) between two subjects A and B

G. N. van Vark and W. W. Howells (eds.), Multivariate Statistical Methods in Physical Anthropology, 81–88.

characterized respectively by the vectors $a = (a_1,\ldots,a_p)$ and $b = (b_1,\ldots,b_p)$ is defined by Defrise (1967):

$$\Delta(A,B) = [(a-b)'\Sigma^{-1}(a-b)]^{\frac{1}{2}}$$

It was shown (Defrise, 1955) that these GD have common properties with the classical GD of Mahalanobis.
The p variables of a subject can be interpreted as the coordinates on p orthogonal axes of a hyperspace. These measurements are transformed, by a translation of the origin of the axes to the center of the hyperellipsoids (hypersurfaces of equally probable points), then by a rotation of the axes with the effect that the original axes coincide with the axes of the hyperellipsoids and finally by a reduction of the units making the variances of the new variables equal to 1. In this new hyperspace the Euclidean distance between the new points A' and B' representing the subjects is equal to the GD between A and B

$$A'B' = \Delta\ (A,B)$$

If the individuals A and B are chosen at random in a normal p variate population,

$$\Delta^2(A,B)/2$$

is distributed as χ^2 with p degrees of freedom. This property can be applied to n pairs of subjects belonging to a normal p variate population. The half sum of the Δ^2 between n pairs chosen at random is distributed as χ^2 with np degrees of freedom:

$$\Sigma\ \Delta^2\ (A,B)/2 = \chi^2_{np}.$$

Moreover, in these conditions, when $np \geq 100$,

$$\sqrt{\Sigma\Delta^2\ (A,B)}$$

is normally distributed about mean $\sqrt{2np-1}$ with variance 1.
It is also worthwhile to take into account the distance of each subject to the centre of the population M: indeed, if one of the members of a pair, say A, lies far from the center M, Δ^2 (A,B) has a greater chance of being large than when A lies nearer the center. M. Defrise (1968) showed that when A is given and B chosen at random, the distribution of $\Delta^2(A,B)$ is a non-central χ^2 with p degrees of freedom (number of variables) and $\Delta^2(A,M)$ as non-centrality parameter.
The probability P that the distance in the new hyperspace between a random subject and given A is smaller than the distance between A and his related partner (e.g. cotwin) B can then be calculated by a non-central χ^2 test. A small value of P indicates that the

twins A and B are very similar. In order to make all the twin pairs comparable, B denotes always the co-twin nearer to the center of the population than partner A. For instance, when p = 2, P is the probability of a random point falling inside an offset circle of radius r_d with center at a distance $\Delta(A,M)$ from the center M, taking $\Delta^2(A,M) \geq \Delta^2(B,M)$, $r_d = \Delta(A,B)$(distance to be tested for smallness). This method can easily be applied to the biometrical analysis of same-sexed twin pairs (Defrise, 1968, 1970). Therefore each same-sexed twin pair must be considered in its own population of identical age and sex. A large sample of 16.000 children (Twiesselmann, 1969) has been used to apply this standardisation, groups of half year with more than 300 individuals are available in this study, sampling effects are thus negligible. In the analysis of partners, parent/child or sibs, the two subjects of a studied pair do not belong to the same age and sex population. We must standardize the p studied measurements, not only for sex but also for age. Indeed, even with adults, most of the measurements still change with increasing age in relation with both secular change and individual changes connected with senescence (Susanne, 1967, 1974a, 1977a). The same large samples of children and unpublished sample of 500 adults has been used as reference population.

Under the assumption that the correlation matrix is identical in the different groups, all subjects can be studied as if belonging to the same population. The described method is used in a biometrical study of sibs (Susanne, 1972, 1978) and of husband and wife (Susanne, 1974b). Our material has been collected mainly in Brussels and covers 123 marriages. The subjects are of Belgian origin, husband and wife are 38-65 years old. In each family, two or three children have been studied. This sample is rather homogeneous for socio-economical status (Susanne, 1975). The correlation matrix we used is a common matrix calculated on standardized measurements of masculine and feminine subjects. We calculated the generalized distances for the following groups of measurements:

a) four head measurements, head length, head breadth, frontal breadth and head height,

b) five measurements of the body: stature, arm length, sitting height, biacromial diameter and biiliac diameter,

c) eleven measurements of the face: frontal breadth, internal biocular breadth, bizygomatic breadth, bigonial breadth, nasion gnathion height, nasion stomion height, nose height, nose breadth, lips height, mouth breadth and nose depth.

All these measurements were taken by the author, excluding systematic measuring errors among technicians.

We compared these results for assortative mating and parent-children relationships with distances calculated in the same population between sibs (Susanne, 1972) and between twin pairs (Defrise, 1970).

RESULTS

Four head measurements

As already described in the methods, if the different pairs are chosen at random the $\sqrt{\Sigma\Delta^2}$ is distributed normally about the mean $m = \sqrt{2np-1}$ with a variance equal to 1. Thus the differences $\sqrt{\Sigma\Delta^2} - \sqrt{2np-1}$ demonstrates the similarity of the two partners from the different pairs when $\sqrt{\Sigma\Delta^2}$ is significantly smaller than the average value m.

As far as this measure of similarity is concerned table 1 shows the intermediate position of the following relatives (non-identical twin pairs, sibs, parent-child), it is evident that identical twin pairs are found much more similar and that the husband-wife similarity is not significant.

Table 1. Generalized distances for head measurements (p = 4)

		$m-\sqrt{\Sigma\Delta^2}(A,B)$	$\bar{\Delta}^2(A,B)$	$\bar{P}$	% which P<.05
Identical twin-pairs (n=30)	(1)	9.292	1.268	.034	70.0
Non-identical twin pairs (n=33)	(1)	6.633	2.784	.127	51.3
Sibs (n=196)	(2)	5.696	4.604	.215	20.9
Parent-child (n=552)		6.422	5.209	.251	14.3
Husband-wife (n-123)		1.919	7.040	.371	6.6

m = mean = $\sqrt{2np-1}$, P : mean of probab. distance between random subject and partner A smaller than distance between partner A and B.

(1) Defrise, E., 1970
(2) Susanne, C., 1972.

A non-significant tendency to assortative mating is thus observed. However in the intermediate position of the relatives with a coefficient of relationship of 1/2 the similarity seems greater in non-identical twin pairs than in sibs, and in sibs greater than in parent-child: these results are indeed observed in the comparison of the means of distances between relatives. The use of the non-central χ^2 method confirms all these results, the differences between non-identical twin pairs, sibs and parent-child are very clear for the mean probability $\bar{P}$ and for the frequency of P less than .05.

Five body measurements

With the use of the square of the distances (A,B) the intermediate position of the relatives with a coefficient of relationship of 1/2 is again shown in table 2. However it is perhaps less evident than for the head measurements: the difference $m - \sqrt{\Sigma\Delta^2}$ takes for instance an aberrant position for the non-identical twins.
With the method of non-central χ^2, results equivalent to table 1 are however again observed: t.i. intermediate position of non-identical twin pairs, sibs, parent-child and greater similarity of the non-identical twins than of the face an assortative mating is also clearly observed for the five measurements used in table 2.

Table 2. Generalized distances for body measurements (p = 5)

		$m-\sqrt{\Sigma\Delta^2(A,B)}$	$\bar{\Delta}^2(A,B)$	$\bar{P}$	% where $P < 0.05$
Identical twin pairs (n=30)	(1)	7.933	2.920	.080	63.0
Non-identical twin pairs (n=33)	(1)	3.247	6.720	.164	42.0
Sibs (n=196)	(2)	5.568	6.298	.218	23.7
Parent-child (n-552)		7.105	6.555	.234	23.7
Husband-wife (n=123)		3.880	7.894	.296	15.6

m = mean = $\sqrt{2np-1}$, $\bar{P}$: mean of probab. distance between random subject and partner A smaller than distance between partners A and B.

(1) Defrise, E., 1970
(2) Susanne, C., 1972.

Eleven facial measurements

Table 3 shows the existence of an assortative mating and the intermediate similarity of non-identical twins, sibs and parent-child in comparison with the identical twin pairs and the husband-wife pairs.
The similarity of non-identical twins is however very great as shown by the $\Delta^2(A,B)$ but especially by the method of non-central χ^2 (average of P values and frequencies of P values lower than .05). It is again with the method of non-central χ^2 that the rank position of non-identical pairs, sibs and parent-child is clearly observed.

Table 3. Generalized distances for Facial measurements (p = 11)

		$m-\sqrt{\Sigma\Delta^2(A,B)}$	$\bar{\Delta}^2(A,B)$	$\bar{P}$	% where P <.05
Identical twin pairs (n=30)	(1)	11.329	6.856	.019	90.0
Non-identical twin pairs (n=33)	(1)	7.581	11.340	.056	60.6
Sibs (n=196)	(2)	7.611	14.626	.170	36.6
Parent-child (n=552)		8.762	15.561	.185	33.4
Husband-wife (n=122)		3.581	19.058	.282	14.6

m = mean = $\sqrt{2np-1}$, $\bar{P}$: mean of probab. distance between random subject and partner A smaller than distance between partners A and B.

(1) Defrise, E., 1970
(2) Susanne, C., 1972.

DISCUSSION

It is possible to make an interpretation of the results of the tables 1, 2 and 3 in terms of influence of genetical factors, of environmental factors and of assortative mating. The influence of all these factors is well known in the study of the multifactorial characters (Susanne, 1975, 1977b). However, generalized distances have the advantage of permitting comparisons within, as well as between, different groups of relatives of not only the absolute measurements but also of the correlations and proportions between the different characters. Generalized distances will be more frequently small when a subject lies near the centre of the population than when the subject lies far from the centre: we used therefore a more accurate method (Defrise, 1968) based on a non-central χ^2 and taking into account the generalized distance of the relatives from the centre of the population.
The results suggest clearly the existence of genetic factors in the determination of head size and shape (head length, breadth, frontal breadth and head height), of body dimensions (stature, arm length, sitting height, biacromial and biiliac diameter), and of facial size and proportion (frontal breadth, internal biocular breadth, bizygomatic breadth, bigonial breadth, nasion-gnathion height, nasion stomion height, nose height, nose breadth, lips height, mouth breadth and nose height): indeed the four different

measures of similarity used in this study showed an order of mean values expected for multifactorial characters, t.i. concordant twins, discordant, sibs, parent-children, husband-wife.
Typical of our results is also the decreased order of the mean similarity discordant twins > sibs > parent-children: however all these relatives have the same coefficient of relationship. This decreasing order is especially clear with the non-central χ^2 method.
This observation results from the influence of environmental factors: indeed, the environments of relatives are often positively correlated and these correlated environments are higher between twins than between sibs and still higher than between parents and children. This implies an overestimation of the genetical factors between twins and sibs compared to the parent-child estimations.
Assortative mating is relatively well known for anthropometric characters (Susanne, 1967, 1974b): the use of a multivariate analysis is however interesting to situate this assortative mating for the shape and the proportions of segments of the body. The generalized distances showed that for the shape of the body and the dimensions and proportions of the face an assortative mating is very evident. On the contrary in respect of the shape of the head the assortative mating is less evident.
If the method we used is not able to determine partitioned environmental and genetical components, we were able to measure the similarity between relatives and we test whether this similarity is greater than that between random pairs. We showed that these similarities for the different relatives follow the rank order expected if there was a genetical component of variation.

References

Defrise, E.: 1955, Mesure de divergence Λ^2 entre un sujet déterminé et une population multivariée normale. Bulletin Institut Royal des Sciences Naturelles de Belgique, 31, 1-15.
Defrise, E.: 1967, Generalized distance in genetic studies. Acta Genetica, 17, 275-288.
Defrise, E.: 1968, Non-central chi-square in twin studies. Acta Genetica, 18, 170-179.
Defrise, E.: 1970, Multivariate analysis in twins. Acta Geneticae Medicae et Gemellologiae, 19, 150-154.
Eaves, L.J., Martin, N.G., and Eysenck, S.B.: 1977, An application of the analysis of covariance structures to the psychogenetical study of impulsiveness. Brit. J. Mathem Stat. Psych, 30, 185.
Joreskog, K.G.: 1973, Analysis of covariance structures. In multivariate analysis III, ed. PP. Krishnaiah. Academic Press, 410 pp.
Martin, N.G. and Eaves, L.J.: 1977, The genetical analysis of covariance structure. Heredity, 38, 77-95.

Susanne, C.: 1967, Contribution à l'étude de l'assortiment matrimonial dans un échantillon de la population belge. Bulletin Société Royale belge d'Anthropologie et de Préhistoire, 78, 147-196.

Susanne, C.: 1972, Multivariate analysis in genetic studies. Acta Geneticae Medicae et Gemellologiae, 21, 204-210.

Susanne, C.: 1974a, Les changements morphologiques liés à l'âge. L'Anthropologie, 78, 693-718.

Susanne, C.: 1974b, Analyse multivariée dans une étude de l'assortiment matrimonial. Homo, 25 (3), 166-171.

Susanne, C.: 1975, Genetic and environmental influences on morphological characteristics. Annals of Human Biology, 3, 279-287.

Susanne, C.: 1977a, Individual age changes of the morphological characteristics. J. Hum. Evolution, 6, 181-189.

Susanne, C.: 1977b, Heritability of anthropological characters. Human Biology, 49, 573-580.

Susanne, C. and P. Dash Sharma: 1978, Multivariate analysis of head measurements in Punjabi families. Annals of Human Biology, 5, 179-183.

Twiesselmann, F.: 1969, Développement biométrique de l'enfant. Presses Univ. Bruxelles

GENERALIZED DISTANCE BETWEEN DIFFERENT THIGH-BONES AND A REFERENCE POPULATION

E. Defrise-Gussenhoven and R. Orban-Segebarth

Brussels

Summary

A generalized distance Λ is defined measuring how far an individual femur is from a reference population of 416 belgian femora; a "rate of remoteness" is introduced. The sampling distribution of the sample value L^2 of Λ^2 is derived.
An unbiased and consistent estimate of Λ^2
$\hat{\Lambda}^2 = \frac{(n-p-2)L_p^2}{n} - \frac{p}{n}$, with $E(\hat{\Lambda}^2) = \Lambda^2$ is obtained. The theory is applied to a particular case with a sample of n=416 elements from recent belgian femora found in the same cemetery.

1. GENERALIZED DISTANCE. POPULATION VALUES

If Σ is the covariance matrix of one or more populations represented in an Euclidian space R^p, its inverse Σ^{-1} provides the space with the following metric. Let A and B be any two points in R^p with coordinates $a_1,\ldots,a_p$ and $b_1,\ldots,b_p$ and a and b the corresponding column vectors such that

$$a' = (a_1,\ldots,a_p) \text{ and } b' = (b_1,\ldots,b_p)$$

then

$$\Delta_p(A,B) = \{(a-b)'\Sigma^{-1}(a-b)\}^{\frac{1}{2}} \tag{1}$$

is the generalized distance (G.D.) between the points A and B.

G. N. van Vark and W. W. Howells (eds.), Multivariate Statistical Methods in Physical Anthropology, 89–99.

1.1. Non-redundancy of the G.D., a property not shared by other distances

An interesting property of $\Delta_p(A,B)$ is its non-redundancy, obvious when the p measurements are introduced stepwise in $\Delta_p(A,B)$; denoting by

$$d_j = (a_j - b_j)/\sigma_j \tag{2}$$

the standardized difference for the jth variable $(j=1,\ldots,p)$, we find

$$\Delta_1^2(A,B) = d_1^2 \tag{3}$$

$$\Delta_2^2(A,B) = d_1^2 + \left\{\frac{d_2 - \rho_{12} d_1}{\sqrt{1-\rho_{12}^2}}\right\}^2 \tag{4}$$

ρ_{12} being the correlation coefficient between the two first variables. The second term on the right side of (4) is independent of d_1^2; it is the contribution of the second measurement and it adds only the new information given by this variable.
This is so because the part of d_2, predictable by d_1 and due to the correlation between the two measurements, has been substracted from d_2.
Finally, $\Delta_p^2(A,B)$ is the sum of p independent terms, each one being the square of the standardized residual part of the newly added measurement non-predictable by the preceding ones.
Therefore, $\Delta_p(A,B)$ is an Euclidean distance in a transformed space.

1.2. Mahalanobis Δ_p^2

From 1927 till 1936 Mahalanobis invented and deepened the concept of G.D. which he introduced to measure divergence between the mean points of several populations sharing a common covariance matrix.
μ_α and μ_β being the mean vectors of the populations P_α and P_β,

$$\Delta_p(M_\alpha, M_\beta) = \{(\mu_\alpha - \mu_\beta)'\Sigma^{-1}(\mu_\alpha - \mu_\beta)\}^{\frac{1}{2}} \tag{5}$$

is the generalized distance between the mean points M_α and M_β of the populations P_α and P_β.

1.3. Generalized distance in twin and family studies and in cytogenetics

The author used the same metric in genetic twin studies (Defrise-Gussenhoven 1967, 1968).

$$\Delta_p(A,B) = \{(a-b)'\Sigma^{-1}(a-b)\}^{\frac{1}{2}} \tag{6}$$

is the G.D. between twins A and B belonging to a population of same age and sex with covariance matrix Σ. The advantage of the use of G.D. lies in the fact that resemblance between members of different pairs can be compared, each pair being situated in its own age and sex group.
In genetic family studies, Susanne (1972) uses the same distance as the G.D. for twins.
In cytogenetics, the G.D. between centromeres of two chromosomes can also be calculated, the metaphasis with the 46 chromosomes serving as reference population (Hens, L. et.al., 1975).

1.4. Generalized distance Λ between a single femur and a reference population

In 1955 the first author was working in the laboratory of physical anthropology directed by Prof. Twiesselmann who at the time was studying the incomplete Neanderthal femur of Fond-de-Forêt in Belgium. A sample of 416 femora of a modern population was available and the author, knowing Mahalanobis' G.D., introduced the generalized distance Λ between a single femur and this population.
Let μ and Σ be the mean vector and the covariance matrix of the p-uple population and d a fixed (non random) vector with components $d_1,\ldots,d_p$ equal to the measurements of the femur D. By

$$\Lambda_p(D,M) = \{(d-\mu)'\Sigma^{-1}(d-\mu)\}^{\frac{1}{2}} \tag{7}$$

is then defined the generalized distance between D and the mean point M of the population.

1.5. Equiprobable hyperellipsoids in the case of a normal population

If ξ is a random vector belonging to the population supposed normal $N_p(\mu,\Sigma)$, then $(\xi-\mu)'\Sigma^{-1}(\xi-\mu)$ is distributed as χ^2_p.
Choosing α (for instance $\alpha=.05$) and denoting by $\chi^2_p(\alpha)$ the χ^2_p value exceeded in α x 100% of the cases, the equation

$$(\xi-\mu)'\Sigma^{-1}(\xi-\mu) = \chi^2_p(\alpha), \tag{8}$$

when ξ varies in the normal population, represents a hyperellipsoid outside of which lie α x 100% of the points of the population. The points lying on the hyperellipsoid all have the same G.D. to the center M. The hypersurface may be called equiprobable.

2. GENERALIZED DISTANCE Λ_p. SAMPLE VALUE L_p AND ESTIMATE $\hat{\Lambda}^2_p$ OF Λ^2_p IN THE NORMAL CASE

Let $\bar{x}$ and S be the mean vector and the covariance matrix of the sample of n=416 femora, with $\bar{x}' = (\bar{x}_1,\ldots,\bar{x}_p)$;

$\bar{x}_j = \frac{1}{n}\sum_{i=1}^{n} x_{ij}$; i=1,...,n ; j,k=1,...,p.

$$S = (s_{jk}) \; , \; s_{jk} = \frac{1}{n}\sum_{i=1}^{n}(x_{ij}-\bar{x}_j)(x_{ik}-\bar{x}_k).$$

We denote by $L_p^2 = (d-\bar{x})'S^{-1}(d-\bar{x})$ (9)

the sample value of Λ_p^2, d being as before the fixed vector representing the measurement of the femur D. When all the possible random samples of size n of the population are considered, the corresponding values $\bar{x}$, S and L_p^2 vary and become random variables.

2.1. Sampling distribution of L_p^2

The first author (1955) found the exact sampling distribution of L_p^2 with the help of a beautiful article, based on geometrical reasoning, by Bose and Roy (1938) who calculated the sampling distribution of Mahalanobis G.D. But it was difficult to write down this geometrical reasoning and finally the author demonstrated her result with nice theorems exposed in "Advanced Statistical Methods in Biometric Research" (Rao, 1970).

$$\frac{\Gamma(\frac{n}{2})e^{-n\Lambda_p^2/2}\,(L_p^2)^{\frac{p-2}{2}}}{\Gamma(\frac{p}{2})\,\Gamma(\frac{n-p}{2})\,(1+L_p^2)^{\frac{n}{2}}}\;{}_1F_1\left(\frac{n}{2},\frac{p}{2},\frac{n\Lambda_p^2L_p^2}{2(1+L_p^2)}\right)dL_p^2 \qquad (10)$$

with $${}_1F_1(\alpha,\rho,z) = 1 + \frac{\alpha}{1!\rho}z + \frac{\alpha(\alpha+1)}{2!\rho(\rho+1)}z^2 + \ldots \qquad (11)$$

is the sampling distribution of L_p^2 in the normal case.

2.2. Estimate of Λ_p^2

$$\hat{\Lambda}_p^2 = \frac{(n-p-2)L_p^2}{n} - \frac{p}{n} \qquad (12)$$

is an unbiased and consistent estimate of Λ_p^2. The formula (12) shows that $\hat{\Lambda}_p^2 < L_p^2$ and therefore L_p^2 overestimates Λ_p^2. The same happens with the D^2 of Mahalanobis.

2.3. Variance $V(\hat{\Lambda}_p^2)$ and its estimate $V(\hat{\Lambda}_p^2)$

Formula 10 yields:

$$V(L_p^2) = \frac{2n^2\Lambda_p^4 + 2(n-2)(p+2n\Lambda_p^2)}{(n-p-2)^2(n-p-4)} ; \tag{13}$$

$$V(\hat{\Lambda}_p^2) = \frac{2n^2\Lambda_p^4 + 2(n-2)(p+2n\Lambda_p^2)}{n^2(n-p-4)} \tag{14}$$

is the variance of $\hat{\Lambda}_p^2$. Putting $\hat{\Lambda}_p$ instead of Λ_p in (14) we find an estimate of the variance of $\hat{\Lambda}_p^2$.

2.4. Estimate of the "rate of remoteness" of the femur D

$\hat{\Lambda}_p^2$ being calculated by (9) and (12), the table of χ_p^2 yields a value of α such that

$$\chi_p^2(\alpha) = \hat{\Lambda}_p^2 \tag{15}$$

$(1-\alpha)$ 100% might be denoted by "rate of remoteness" of femur D from the reference population. It means that $(1-\alpha)\times 100\%$ points of the population lie nearer its mean point M than the thigh bone D.

2.5. Special cases

When $\Lambda_p^2=0$, the point D is confounded with the mean point M of the population. In this case, the sample value L_p^2 multiplied by $\frac{n-p}{p}$ has a F distribution with p and n-p degrees of freedom.
The null hypothesis $\mu=d$ can therefore be tested with the F test.
When p=1 with $\Lambda^2\neq 0$, the distribution of $L^2(n-1)$ is that of the non-central t^2 with non-centrality parameter equal to $n\Lambda^2$ (Rao, p. 48).

3. APPLICATION OF THE G.D. TO THE COMPARISON OF FEMORA

3.1. Femora used in the study

The reference population is represented by a sample of n=416 right and left recent belgian femora.
The separate femora that are compared are:
1) Recent Zaïre;
2) Four fossil Palestina femora: Skhùl IV right and left, Skhùl V left and Tabùn I;
3) Neanderthal right and left and the Neanderthal thigh-bone of Spy (Belgium);
4) Pithecanthropus;
5) As control, one of the 416 femora, chosen because of its mean length.

3.2. Measurements; mean values and st. dev. of the sample (n=416)

Table 1. Mean and st. deviations

	mean	st.dev.
1. Maximum length of femur in anatomical position	434.9	28.9
2. Anteroposterior diameter of the medial condyle	60.0	4.6
3. Bicondylar width	77.5	5.5
4. Anteroposterior diameter of midshaft	27.9	2.8
5. Transverse diameter of midshaft	27.4	2.2
6. Transverse diameter of proximal shaft	31.5	2.7
7. Anteroposterior diameter of proximal shaft	26.1	2.5
8. Vertical diameter of the neck	33.0	3.4
9. Anteroposterior diameter of the neck	25.8	2.7
10. Vertical diameter of the head	45.7	3.7
11. Anteroposterior diameter of the head	45.1	3.7

3.3. Univariate $\hat{\Lambda}_1$

Table II. Situation of each studied femur above (+) or under (-) the mean of the 416 femora. $\hat{\Lambda}_1$ and "rate of remoteness" of exceptional measurements

	Belg.	Zair.	Sk.IV L	SK.IV R	Tab.I	Nea.L.	Spy	Pith.
max. length	0	+	+ 1.90*	+ 1.75*	-	+	-	+
A-P D. cond.	+	+	+	+	- 2.28**	+	+	-
bicon. width	+	0	+	+	-	+ 1.71*	+ 2.62**	+
mids. A-P D.	+	-	+ 2.02**	+ 1.91*	- 1.70*	+ v	+ v	+ v
mids. tr.D.	-	-	-	-	-	+	+	+
prox.s. tr.D.	+	-	-	-	-	+	+ 1.68*	+ 2.43**
prox.s. A-P D.	-	- 1.82*	-	-	-	+	-	+
neck vert.	+	- 1.79*	+	+	- 1.79*	+	+	-
neck A-P D.	+	-	+	+	-	+ 2.99***	+ 4.08****	-
head vert.	+	-	+	+	-	+	+ 2.21**	-
head A-P D.	+	-	+	+	-	+ 1.83*	+ 2.36**	-

"rate of remoteness": *>90%; **>95%; ***>99%; ****>99,9%

The table II shows that Tabùn and Spy are short as compared to the reference group. Neanderthal and Spy are robust, Skhùl IV left and right have a slender shaft, Zaïre is slender with small head and neck. The value of univariate $\hat{\Lambda}_1$ and the rate of remoteness complete the univariate description.

3.4. Multivariate $\hat{\Lambda}_p$

Table III. The measurements are entered stepwise

		Belg.	Zair.	Sk.IV L	Sk.IV R	Sk.V L	Tab.I	Nea.L	Nea.R	Spy	Pith
$\hat{\Lambda}_1$	max. length	0	1.0	1.9	1.8	2.8**	0.7	0.1	0.1	0.4	0.7
$\hat{\Lambda}_2$	A-P D. cond.	0.6	1.3	1.9	1.8	2.8*	2.7*	1.8	2.1	2.9*	1.6
$\hat{\Lambda}_3$	bicon. width	0.7	1.5	1.9	1.8	3.0*	2.8*	2.3	2.5	4.4***	1.8
$\hat{\Lambda}_4$	mids. A-P D.	0.7	2.0	2.0	1.9	3.4*	2.9	2.4	2.5	4.4***	1.8
$\hat{\Lambda}_5$	mids. tr.D.	1.5	2.7	3.4*	3.2	4.3**	3.3	2.5	2.5	4.4**	1.8
$\hat{\Lambda}_6$	prox.s. tr.D.	1.6	2.7	3.4	3.2	4.7**	3.3	2.6	2.5	4.4**	3.6*
$\hat{\Lambda}_7$	prox.s. A-P D.	1.9	4.1*	4.6**	4.3**	4.7**	3.3	2.7	2.7	4.8**	3.7
$\hat{\Lambda}_8$	neck vert.	2.1	5.1**	4.7**	4.4*		3.5	2.7		4.9**	4.1*
$\hat{\Lambda}_9$	neck A-P D.	2.2	5.1**	4.9**	4.5*		3.5	3.8		6.6***	4.6*
$\hat{\Lambda}_{10}$	head vert.	2.3	5.3**	5.4**	4.7*		3.5	3.9		6.6***	4.6*
$\hat{\Lambda}_{11}$	head A-P D.	2.4	5.3**	5.6***	5.0*		5.1**	4.3		6.6***	4.6*

"rate of remoteness": *>95%; **>99%; ***>99,9%

The table III gives the G.D. between each separate femur and the mean of the reference population. The measurements are introduced stepwise. Only the Belgian control femur and Neanderthal left have all the rates of remoteness smaller than 95%, although this last femur is near 90% for the eleven measurements taken together. The most extreme case is Spy, with a mean shaft but very large head and broad condyles. A contrast is presented by Tabùn, small with slender condyles, but not very divergent considering its size.
It is worthwhile to observe in table III the fact that the distances increase as more measurements are considered. The rate

of remoteness also tends to increase, but there are exceptions, for instance for Tabùn, with rates of remoteness exceeding 95% for $\hat{\Lambda}_2^2$ and $\hat{\Lambda}_3^2$, but inferior to 95% when the anteroposterior diameter of midshaft is added. This measurement is very small, but in harmony with Tabùn's three first measurements. This example shows how the G.D. takes shape and correlation into account.
The Zaïre femur is rather elegant, a little above the Belgian mean for length, but with slender shaft and head. The three Skhùl are much alike, long, robust, but with small transversal diameter for the shaft. On the contrary, neck and head are strong.
Finally the Pithecanthropus diverges only because of its large transversal diameter of proximal shaft where its well known exostosis is localized; otherwise this femur does not diverge much.

3.5. Increase of $\hat{\Lambda}_j^2$ and sign of the residual value of the regression of x_j in function of $x_1,\ldots,x_{j-1}$ $(j=1,2,\ldots,p=11)$

Table IV indicates the increase $\hat{\Lambda}_j^2 - \hat{\Lambda}_{j-1}^2$ for each added measurement. As this difference has practically a χ_1^2 distribution, one, two or three asterisks indicate whether $\hat{\Lambda}_j^2 - \hat{\Lambda}_{j-1}^2$ exceeds 3.841, 6.635 or 10.827, the 5%, 1%, and 0.1% values of χ_1^2. On the other hand, the linear regression function of $x_j/x_1,x_2,\ldots,x_{j-1}$ is calculated for each $j = 2,3,\ldots,11$ in the sample of 416 femora.
For each separate thigh-bone with measurements $d_1,\ldots,d_{j-1}$ an expected value d_j^* is calculated with the regression function.
In the columns of table IV, the sign (+) means that $d_j > d_j^*$ and (-) means $d_j < d_j^*$, d_j being the real measurement number j of the femur.
Table IV reveals which newly added measurement adds original information about the shape of a studied femur. For instance, measurements 7 (anteroposterior diameter of proximal shaft) and 8 (vertical diameter of the neck) add new significant information to the divergence of the Zaïre thigh-bone. The sign (-) indicates that the real measurements d_7 and d_8 of the Zaire femur are smaller than the expected values d_7^* and d_8^* obtained with the regression function.
For Skhùl, both right and left, the measurements 5 (transv. diam. of midshaft) and 7 (anteroposterior diam. of prox. shaft) add crucial divergence, both measurements being exceedingly small when the preceding ones are taken into account. Skhùl V shares the same fate, except that it starts with large length.
For Tabùn I, the extreme smallness of the anteroposterior diameter of the medial condyle (2) leads to divergent increase of $\hat{\Lambda}^2$. The effect of the eleventh measurement is striking and is caused by the fact that the vertical and anteroposterior diameters of the head are 40 and 43 mm, whereas the corresponding mean values in the reference sample are 45.7 and 45.1 mm, a reversed order. So

Table IV. $\hat{\Lambda}^2_j - \hat{\Lambda}^2_{j-1}$; sign of $d_j - d^*_j$

	Belg.	Zair.	Sk.IV L	Sk.IV R	Sk.V L	Tab.I	Nea.L	Nea.R	Spy	Pith
$\hat{\Lambda}^1_2$	0	+ 1.1	+ 3.6	+ 3.1	+ 7.6**	− 0.5	0	0	− 0.2	+ 0.5
$\hat{\Lambda}^2_2 - \hat{\Lambda}^2_1$	+ 0.4	− 0.7	0	0	− 0.4	− 7.0**	+ 3.1	+ 4.5*	+ 8.0**	− 2.0
$\hat{\Lambda}^2_3 - \hat{\Lambda}^2_2$	+ 0.1	− 0.6	0	0	+ 1.0	+ 0.4	+ 2.4	+ 1.6	+ 10.9***	+ 0.6
$\hat{\Lambda}^2_4 - \hat{\Lambda}^2_3$	0	− 1.9	+ 0.5	+ 0.6	+ 2.3	− 0.6	+ 0.2	+ 0.2	− 0.1	+ 0
$\hat{\Lambda}^2_5 - \hat{\Lambda}^2_4$	− 1.7	− 3.0	− 7.3**	− 6.3*	− 7.1**	+ 2.1	+ 0.8	0	0	+ 0.1
$\hat{\Lambda}^2_6 - \hat{\Lambda}^2_5$	+ 0.3	+ 0.3	0	0	− 3.3	0	+ 0.1	0	+ 0.4	+ 9.5**
$\hat{\Lambda}^2_7 - \hat{\Lambda}^2_6$	− 1.2	− 9.2**	− 10.2**	− 8.8**	0	0	+ 0.6	+ 0.7	− 3.8	+ 1.2
$\hat{\Lambda}^2_8 - \hat{\Lambda}^2_7$	+ 0.9	− 9.1**	− 0.7	+ 0.7		+ 1.7	+ 0.2		+ 0.1	− 2.7
$\hat{\Lambda}^2_9 - \hat{\Lambda}^2_8$	− 0.2	− 0.5	+ 2.2	+ 0.9		0	+ 7.3**		+ 20.0***	− 4.8*
$\hat{\Lambda}^2_{10} - \hat{\Lambda}^2_9$	+ 0.5	+ 1.3	− 4.7*	− 2.0		0	− 0.5		+ 0.5	− 0.2
$\hat{\Lambda}^2_{11} - \hat{\Lambda}^2_{10}$	+ 0.7	− 0.9	− 2.7	− 2.3		+ 13.5***	+ 3.6		+ 0.2	0

Significance of the difference $\hat{\Lambda}^2_j - \hat{\Lambda}^2_{j-1}$: *<5%; **<1%; ***<0.1%.

it is seen how sensitive a G.D. is, since a difference of 3 mm in the wrong direction causes the final rejection of the Tabùn thigh-bone.

Neanderthal left and right are rather alike for the seven measurements that could be taken on both thigh-bones. Neanderthal left has a specially large anteroposterior diameter of the neck (9) causing divergent increase of the G.D.

Spy shows jumps in G.D. for the very broad condyles (2 and 3) and also, like Neanderthal left, for the ninth measurement. Spy might be described as an exaggerated Neanderthal form.

Pithecanthropus diverges with 6, the transverse diameter of proximal shaft, but this is due to its exostosis discussed in 3.4.

The few preceding examples suggest how thoroughly a morphological description could be completed with the generalized distance.

Still more enlightening would be the case where several reference samples are used, each with its own covariance matrix. One could then compare the G.D. of a fossil to each population and indicate to which one the fossil is nearest.

4. SOME OTHER STUDIES IN HUMAN PALEONTOLOGY MAKING USE OF $\hat{\Lambda}$

Thoma (1969) compared several fossil occipitalia - among others the Mindel-Interstadial Vertesszöllös - with a reference series of crania of Hungarian Gräberfeldes. In another study (1975), the same author analyses the occipital of Riss/Würm age of the cave Bourgeois-Delauney in France.
Orban (1979) compared Os coxae of three Pygmies and of Australopithecus Sterkfontein (STS 14) with a recent Belgian population.

References.

Bose, R.C., and Roy, S.N.: 1938, The distribution of the Studentized D^2 statistic. Sankhyā, Calcutta, 4 part 1: pp. 19-37.

Defrise-Gussenhoven, E.: 1955, Mesure de divergence Λ^2 entre un sujet déterminé et une population multivariée normale. Sa distribution d'échantillonnage. Bull. Inst. roy. Sc. nat. Belgique. 31, 56, pp. 1-16.

Defrise-Gussenhoven, E.: 1957, Mesure de divergence entre quelques fémurs fossiles et un ensemble de fémurs belges récents. Etude biométrique. Bull. Inst. Roy. Sc. nat. Belgique, 33, 37, pp. 1-13.

Defrise-Gussenhoven, E.: 1967, Generalized distance in genetic studies. Acta Genetica, 17, pp. 275-288.

Defrise-Gussenhoven, E.: 1968, Non-central chi-square in twin-studies. Acta Genetica, 18, pp. 170-179.

Hens, L., Kirsch-Volders, M., Susanne, C., Galpérin-Lemaitre, H.: 1975, Relative position of Trypsin banded Homologous chromosomes in Human (♀) Metaphase figures. Humangenetik, 28, pp. 303-311.

Mahalanobis, P.C.: 1930, On tests and measures of group divergence. Jn. and Proc. Asiat. Soc. of Bengal, 26, new-series, pp. 541-588.

Mahalanobis, P.C.: 1936, On the generalized distance in Statistics. Proc. Nat. Inst. of Sc. of India, 2,1, pp. 49-55.

Orban-Segebarth, R.: 1979, Ostéométrie comparée de l'os coxal chez les anthropomorphes. Thèse de doctorat U.L.B.

Orban-Segebarth, R.: 1980, A biometrical comparitive study of the os coxae of Hominidae, Pongidae and Australopithecus STS 14. Proceedings II[d] Congress European Anthropological Association Brno (in press)

Rao, C.R.: 1970, Advanced Statistical Methods in Biometric Research. Hafner publishing company, 390 pp.

Susanne, C.: 1972, Multivariate analysis in genetic studies. Acta Geneticae Medicae et Gemellologiae, 21, pp. 204-210.

Thoma, A.: 1969, Biometrische Studie über das Occipitale von Vértesszöllös. Z. Morph. Anthrop. 60, 3, pp. 229-241.

Thoma, A.: 1975, L'occipital de la grotte Bourgeois-Delauney (La Chaise, Charente). Etude biométrique. C.R. Acad. Sc. Paris, 281, série D, pp. 1821-1824.

Twiesselmann, F.: 1961, Le fémur néanderthalien de Fond-de Forêt (Province de Liège). Inst. roy. Sc. nat. Belgique, Mémoire 148, 164 pp., 65 fig., 2 pl.

THE USE OF MULTIVARIATE DISTANCES FOR NON-CLASSIFICATORY PURPOSES IN ANTHROPOBIOLOGY

Jean Hiernaux

Equipe d'Ecologie Humaine
Université de Paris
Place Jussieu, 75230 Paris, France

By reviewing some of the author's work, this publication attempts to show how much multivariate distances may be useful tools for answering questions of a non-classificatory nature. It focuses on the potential usefulness of following a number of lines of logical analysis of distance matrices. It does not attempt to discuss how much information is lost or distorted when the most accurate statistics must be renounced, as has been the case for anthropometric distances in the author's work: Mahalanobis' D^2 was used whenever all raw data were available, C_H^2 when only means and SD's were available, and Δ^2 was used whenever all raw data were available, and Δg (Hiernaux, 1965) when only means were available on all populations. The apparent fruitfulness of the analysis of even the rawest statistics seems to justify giving up more sophisticated ones when it permits the inclusion of key populations in the set, e.g., in a study of the populations of the African rainforest, the Mbuti Pygmies whose anthropometry is known by their arithmetic means only.

The work reviewed began in an area including the boundaries between Zaire, Rwanda and Burundi in central Africa. All populations of the area ((I to XV on Figure 1) were sampled. In Zaire, the populations were tribes; in Rwanda and Burundi they were castes: Tutsi, Hutu and Twa. The Twa of the two countries were grouped together, because the national boundary was of little significance to them. The matrix of D^2 distances was computed among the 15 populations for a set of 9 morphological variables (stature, height of the upper iliac spine, upper limb length, lip thickness, head height, radiohumeral index, cephalic index, facial index and nasal index).

Three stocks are known to have contributed to the present

G. N. van Vark and W. W. Howells (eds.), Multivariate Statistical Methods in Physical Anthropology, 101–114.

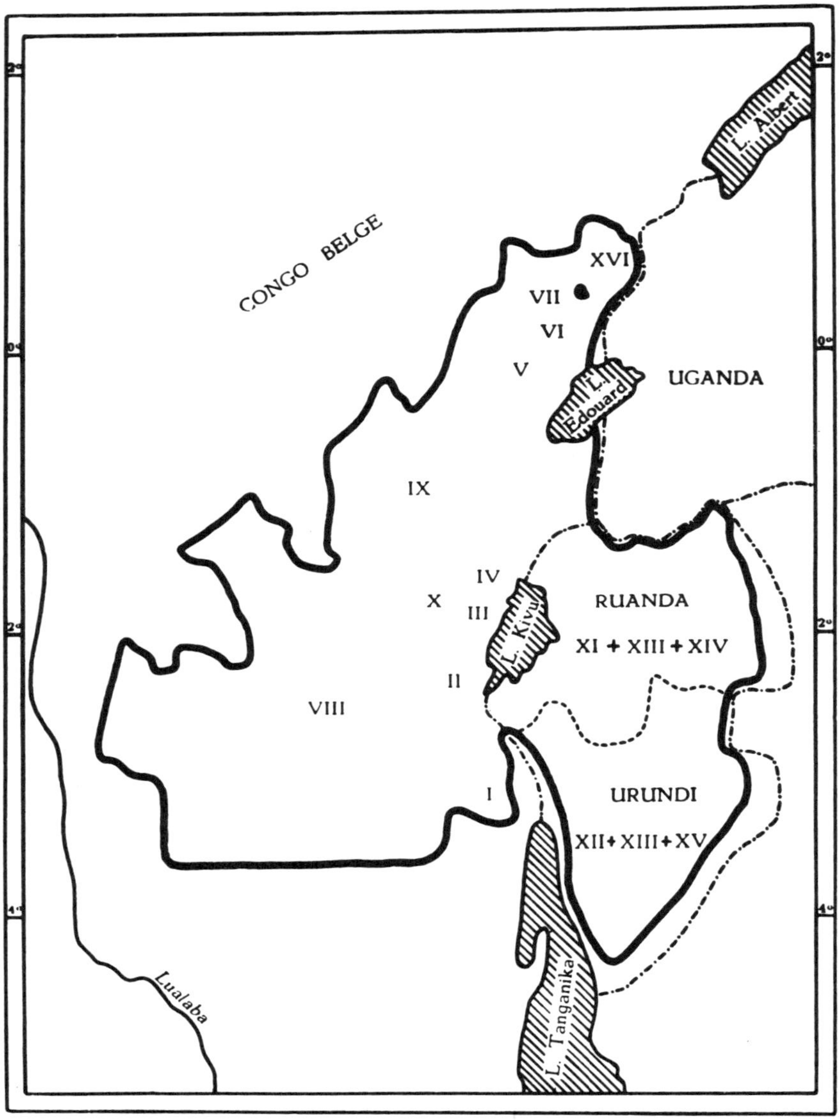

Figure 1. Map of the 16 populations studied in Hiernaux, 1956. Reproduced from Hiernaux, 1956. Congo Belge, Ruanda and Urundi are now Zaire, Rwanda and Burundi, respectively. I: Fulero, II: Shi, III: Havu, IV: Hunde, V: Swaga, VI: Shu, VII: Mbuba, VIII: Rega, IX: Nyanga, X: Tembo, XI: Hutu of Rwanda, XII: Hutu of Burundi, XIII:Twa of Rwanda and Burundi, XIV: Tutsi of Rwanda, XV: Tutsi of Burundi, XVI: Humu

peopling of the area: an anciently settled stock of hunter-gatherers ancestral to the Twa (XIII), a stock of Bantu-speaking agriculturalists who started spreading over the area some seventeen centuries ago and of whom the populations I to XII are the descendants, and a stock of cattle pastoralists, the Tutsi (XIV and XV) who settled much more recently in Rwanda and Burundi. Sociological and historical data indicate some intermating between Tutsi and Hutu in both countries, and ancient military and political contacts of Rwanda with the Shi of Zaire (II). As shown in Figure 2, all other Zairian tribes have the same sequence of increasing D^2 with the Shi, Hutu, and Tutsi: the sequence Shi-Burundi Hutu- Rwanda Hutu- Burundi Tutsi- Rwanda Tutsi. For all of them, D^2 increases moderately along the first three steps, and by far the sharpest step is the following one, from Hutu to Tutsi. The explanation of this constant sequence is apparent: it expresses gene flow from the highly different Tutsi, marking the Hutu to a greater extent than the Shi and failing to differentiate the other Zairian tribes, Tutsi genetic influence being either absent or nearly equal among them.

This explanation could be refined, in terms of genetic exchange between Hutu and Tutsi, by logically analysing the 4x4 submatrix of D^2 between Tutsi and Hutu. It is compatible with only one set of propositions: -gene flow from Hutu to Tutsi has been stronger in Burundi than in Rwanda; -gene flow from Tutsi to Hutu has been stronger in Rwanda than in Burundi. This again agrees with historical and sociological data. Tutsi settled in Rwanda much earlier than in Burundi; their genetic influence on the Hutu could thus operate during a larger number of generations in the first country. In both countries, the Tutsi became the ruling caste under a Tutsi king. Whereas in Rwanda they maintained an ideal of genetic purity, the royal family of Burundi proclaimed the presence of Hutu blood in its lineage, thus indicating a less negative attitude of the Tutsi of Burundi toward accepting Hutu mates (Hiernaux, 1956).

Reducing the 9-dimensional space of the D^2 to a plane with Gower's (1966) principal coordinates analysis (Hiernaux, 1972) subsequently generated the statements which had resulted from the logical analysis of the matrix of D^2 (Figure 3). Figure 4 illustrates them as plainly, although it involves a larger set of African populations (32, including the 16 of the 1956 study), uses much cruder statistics (Δg), and computes it from a different set of 10 variables (stature, shoulder width, hip width, head length and breadth, face height and width, nose height and breadth, and lip thickness). On this figure, the plots representing the Rwanda Tutsi (30), Burundi Tutsi (31), Rwanda Hutu (20), Burundi Hutu (21), and Shi (18) are almost linearly disposed (Hiernaux, 1973).

In the Zairian part of the area, the main barriers to gene migration are the tribal boundaries. In this subarea peopled by ten tribes, all offshoots of the Bantu expansion, D^2 showed a

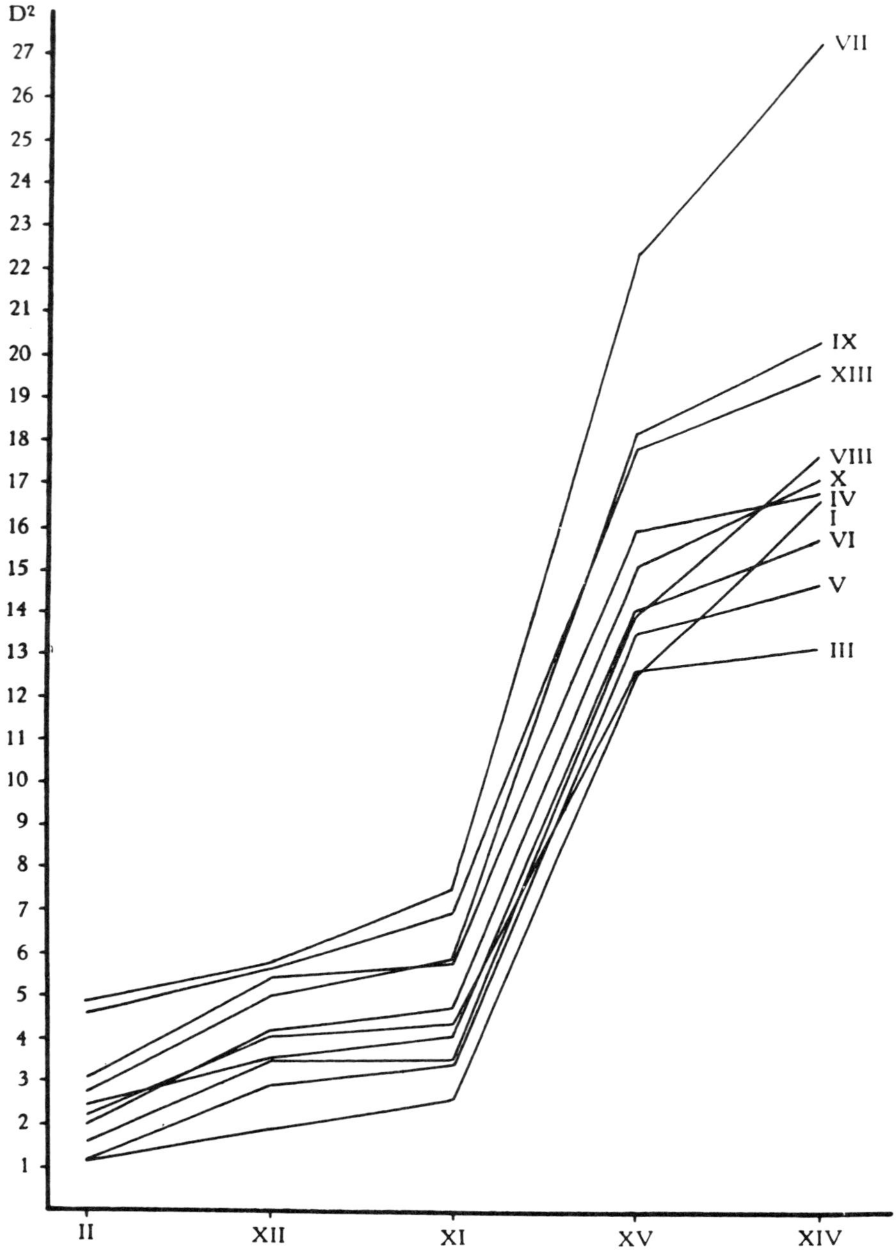

Figure 2. D^2 distances from populations II, XII, XV, and XIV to each of the other ones. See legend of figure 1 for the names of the populations.

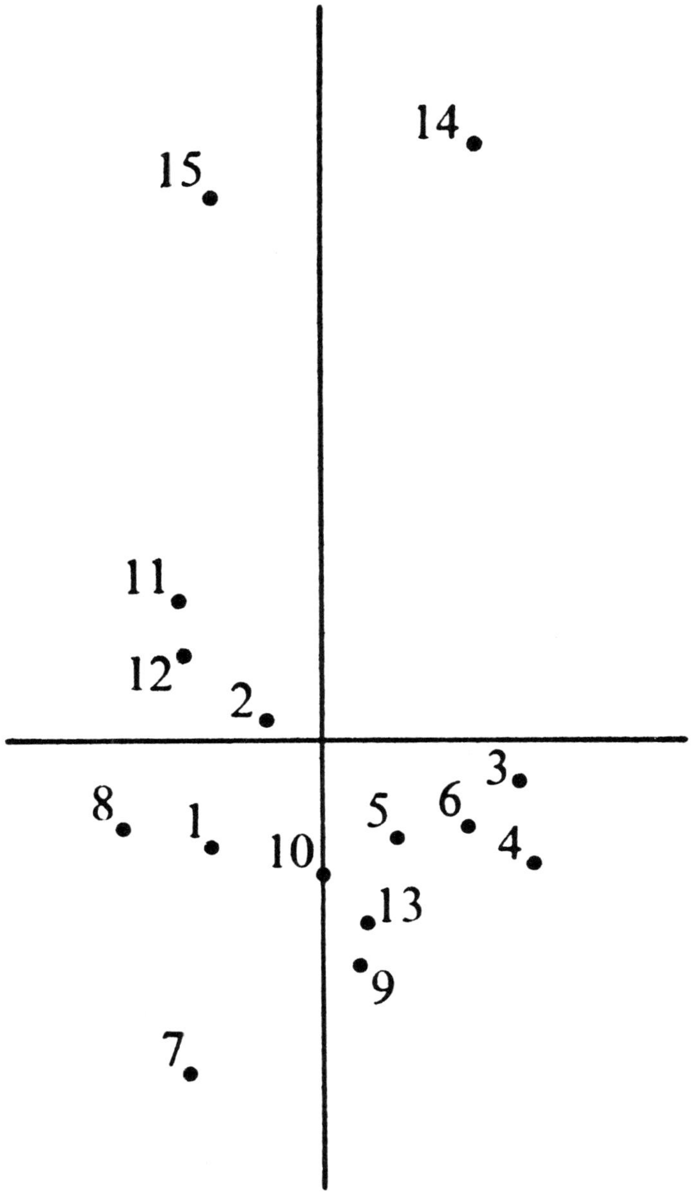

Figure 3. Two-dimensional plotting of populations I to XV of figure 1 (here in arabic ciphers) from a matrix of 9-dimensional D^2 distances by Gower's principal coordinates analysis. The axes are the first two latent vectors.

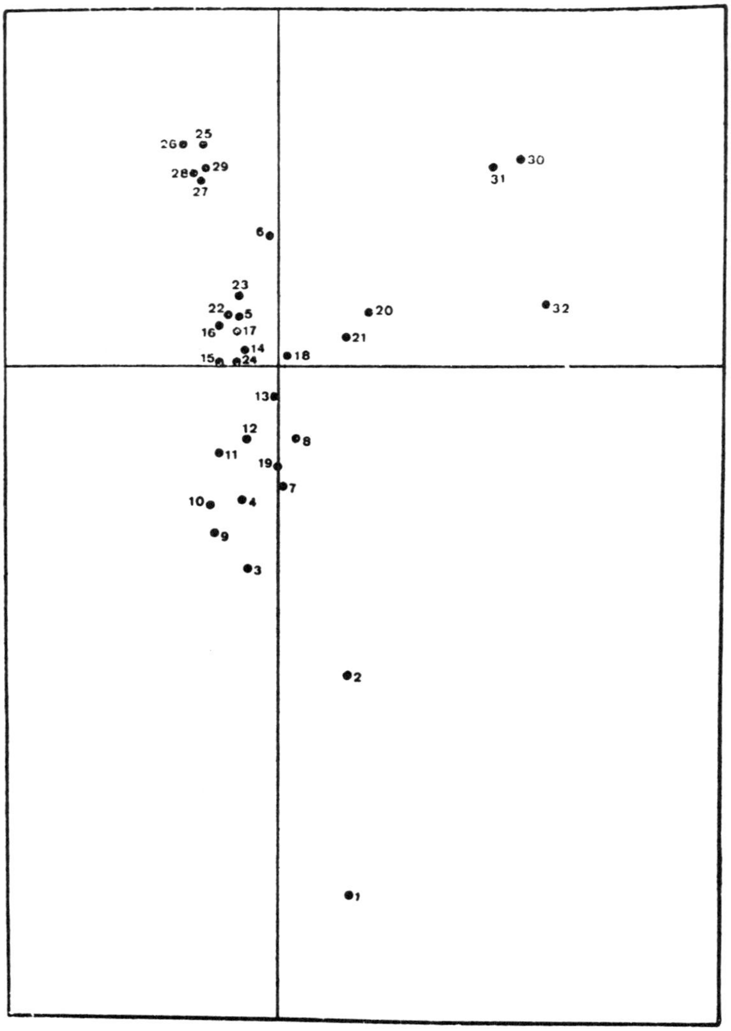

Figure 4. Two-dimensional plotting of 32 African populations from a matrix of Δg distances for 10 anthropometric measurements by principal coordinates analysis. 1: Mbuti pygmies, 2: Twa of Rwanda and Burundi, 3: Twa of the Kuba, 4: Twa of the Ekonda, 5: Ekonda (Oto), 6:Bushong, 7:forest Bira, 8: savanna Bira, 9: Humu-Amba, 10: Mvuba, 11: Nyanga 12: Tembo, 13: Lega, 14: Shu, 15: Swaga, 16: Hunde, 17: Havu, 18: Shi, 19: Fuliru, 20: Rwanda Hutu, 21: Burundi Hutu, 22: Songye, 23: Luba Kasai, 24: Luba Katanga, 25: Majingay, 26: Mbay, 27: Day, 28: Ngama, 29: Kaba, 30: Rwanda Tutsi, 31: Burundi Tutsi, 32: Sab Somali.

significant correlation (r= +0.33) with the number of such boundaries between the populations. This expresses a correlation between morphological differentiation and genetic isolation. This correlation was nearly equal to that (r = 0.34) between the number of tribal boundaries and crude statistics of genetic dissimilarity: the sum of χ^2 for the ABO and MN systems and the $\underline{Hb^S}$ allele, all samples being of equal size (300). D^2 and these statistics were correlated with r = +0.63 (Hiernaux, 1956).

In the west, the Zairian subarea is covered with an equatorial forest, home of the Rega (or Lega), Tembo, Nyanga, and Mbuba (or Mvuba). In the east, it is covered with savanna, where the other six tribes dwell. The mean within- and between-biomes distances were computed. D^2 has a mean of 1.65 between savanna-dwelling populations, 2.85 between forest-dwellers, and 3.58 between one population in the savanna and the other in the forest. This sequence does not result from an increasing mean number of tribal boundaries, which follows an inverse sequence: 2.6 between savanna-dwellers, 2.5 between forest-dwellers, and 2.3 between populations living in different biomes. The lower mean D^2 in the savanna than in the forest is coherent with linguistic and historical data, which point to a relatively recent common origin of the savanna tribes in western Uganda and a much larger heterogeneity of origin of the forest populations, but this category of data does not suggest heterogeneity of origin as an explanation for the higher inter-biome mean D^2. This line of analysis thus appears to show the differentiating influence of the biome on human morphology.

Geographically and morphologically, the set of savanna-dwellers is closer to the Tutsi, whose ancestral line lived from time immemorial in a savanna or hot steppe environment, and the set of forest-dwellers is nearer to the Mbuti Pygmies, the most ancient known inhabitants of the eastern Zaire equatorial forest, who live some distance to the north. Figure 7 strikingly shows this. Genetic influence of the Tutsi in the savanna, and of the Pygmies in the forest, may therefore be conceived of as the mechanism of the morphological difference between the inhabitants of the two biomes, even in the absence of geographical contact. However, this hypothesis is not supported by the pattern of ABO allele frequencies, which are powerful for detecting such influences. Tutsi and Mbuti largely differ for the ABO system and the ten Zairian populations are intermediate; the sets of savanna- and forest-dwellers show similar distances to Tutsi and Mbuti (Figure 5). For the MN system, the Tutsi also stand largely apart from the savanna-dwellers, and are not closer to them than to the forest-dwellers: the frequencies of the $\underline{M}$ allele are 0.606 in the Tutsi, 0.499 in the forest, and 0.481 in the savanna. No data on MN in the Mbuti had been published in 1956. A frequency of $\underline{M}$ of 0.504 in a small sample of Mbuti was later published by Fraser et al. (1966), indicating that the MN system offers no means of detecting a differential genetic influence of the Mbuti in the Zairian tribes considered here.

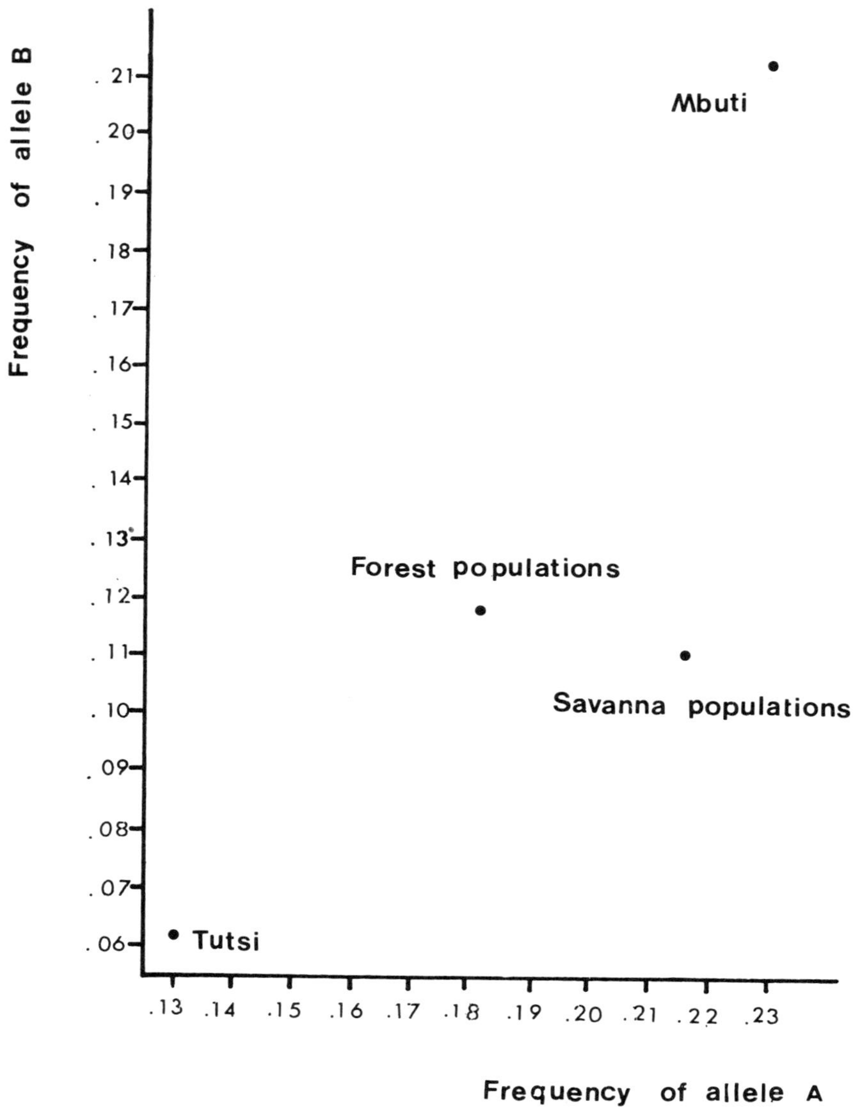

Figure 5. Plotting of Tutsi, Mbuti pygmies, and populations I to X of Figure 1 grouped by their biome on a frequency of allele A/ frequency of allele B (ABO system) graph.

If the hypothesis of differential mixture may be ruled out, what has been shown by this analysis, there is a direct influence of the biome on morphology. Since ultimately the cradle of all Bantu agriculturalists was in the savanna, it is more precisely a shift of the morphology of the groups relatively recently migrated into the equatorial forest toward the morphology of the Pygmies, the much more ancient inhabitants of this biome (Hiernaux, 1956).

The study of the Bira strengthened this conclusion (Hiernaux, 1979). This group of Bantu agriculturalists lives to the north of the previously studied area (Figure 6). One section dwells in the equatorial forest, in contact with the Mbuti, the other in the savanna east of its forest kin. Although, as all Bantu groups, the Bira have their ultimate origin in the savanna, local tradition sees the savanna Bira as a recent offshoot of the forest group. A matrix of Δg distances for the same 10 measurements as in the 1973 study was computed among 16 populations: the two Bira sections, the Mbuti, and 13 out of the populations studied in 1956, including the Tutsi of Rwanda. A biometric map was derived from this matrix with a principal component analysis (Figure 7). The ordering of the populations according to biome is striking: the savanna-dweller Pygmoids forcedly relocated in the savanna during the with the Tutsi and Mbuti standing at extreme opposites. Only one savanna population is among the forest-dwellers: the savanna Bira. They deviate however from the forest Bira in the direction of the other savanna-dwellers. This is the location expected on the basis of an influence of the biome on morphology such as that described before and of a recent migration of the savanna Bira from the forest. In this case again, the hypothesis of a stronger genetic influence of the Mbuti on the forest Bira is not supported by the ABO data: contrary to the morphological picture, the forest Bira are more distant from the Mbuti and closer to the Tutsi than the savanna Bira, without the two Bira sections significantly differing from each other. This does not imply the absence of a gene flow in the forest from the Mbuti to the Bira who, contrary to the Rega, Tembo and Nyanga, are symbiotic with the Mbuti. Forest Bira males marrying Mbuti girls is a recorded occurrence; children born from such matings are identified as Bira. What it suggests is only that the recent withdrawal of the savanna Bira from the genetic influence of the Mbuti has not been a significant factor in their differentiation from their forest kin.

The last study to be reported here (Hiernaux <u>et al</u>., 1976) revealed the morphological impact of such genetic contacts between African agriculturalists and Pymies or Pygmoids in different geographical and sociological settings. Four pairs of populations were compared, each one composed of an agriculturalist group and its Pygmy or Pygmoid vassal population: forest Bira with Mbuti Pygmies, the Mbimu in the equatorial forest of the Central African Republic with their vassal Binga Pygmoids, the Oto caste of the Konda in the forested Lake Tumba area of western Zaire with the vassal Twa Pygmoid caste, and the Bushong, the hegemonic tribe of

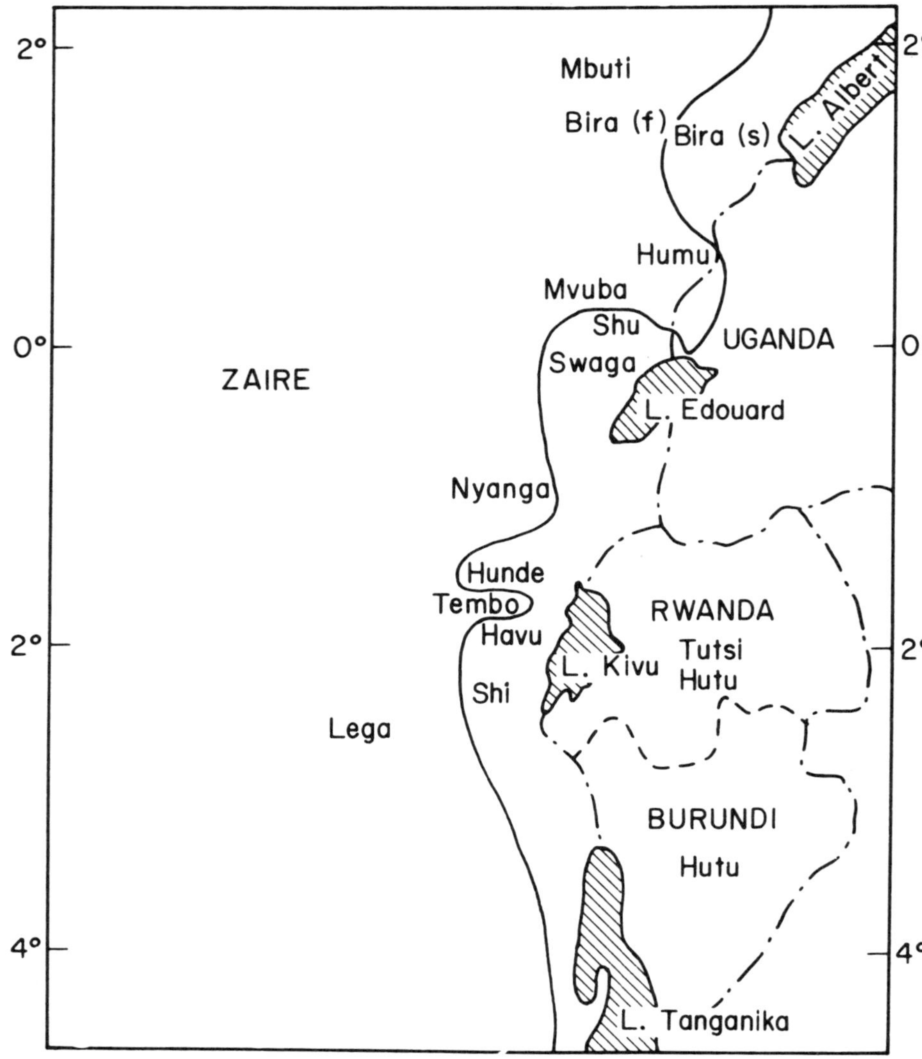

Figure 6. Location of the 16 populations of Figure 7. Heavy line: forest-savanna boundary.

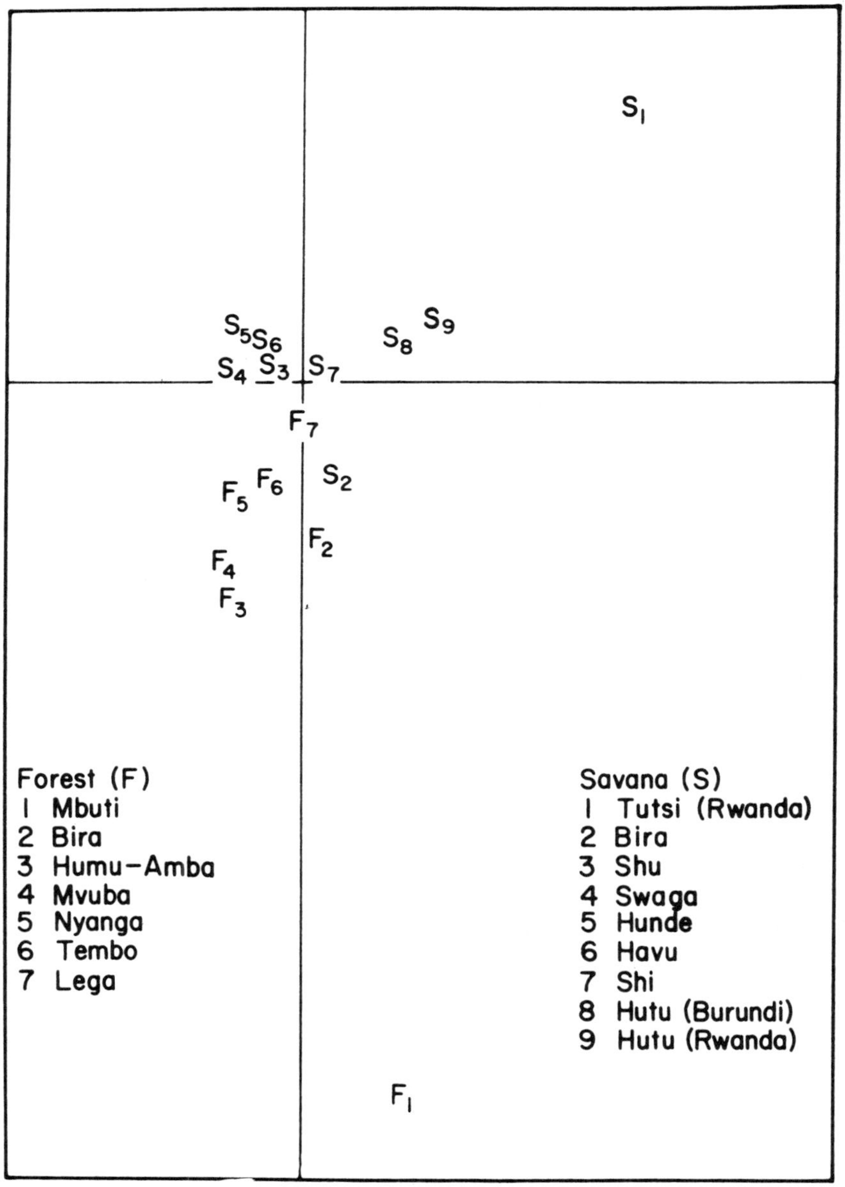

Figure 7. Two-dimensional plotting of 16 populations from a matrix of Δg distances for 10 anthropometric by principal coordinates analysis.

the Kuba kingdom living in Zaire just south of the forest, with the Cwa tribe of the same kingdom, a group of formerly forest-weller Pygmoids forcedly relocated in the savanna during the present century. Apparently very little gene exchange has occurred between the Bushong and Cwa Kuba, but the latter possibly had genetic contacts with other landlords while in the forest. Between Mbimu and Binga, the mating system apparently generates, as between Bira and Mbuti, a unidirectional gene flow from Pygmies or Pygmoids to their suzerains. Sociological data assert an inverse unidirectional gene flow in the Konda, among whom the children born from Oto-Twa matings are deemed as Twa.

A matrix of Δg distances between these eight populations was computed for the same set of 10 measurements as that used in the studies published in 1973 and 1979. Reduction to a plane (Figure 8) was obtained by the non-parametric multidimensional scaling method as described by Lalouel (1973). The pattern shown by this figure is consistent with that of gene migration as implied by sociological data. The Bushong, who were the least exposed to a gene flow from Pygmies or Pygmoids, stand the farthest away from them. Their position is also coherent with the influence of the biome as previously shown, since they are the only one of the four suzerain groups to dwell in the savanna. Of the three forest suzerain groups, the Oto Konda are the farthest away from the Mbuti and the closest to the Bushong, as are the Twa Konda among the Pygmoids. This conforms with the pattern of at least predominantly unidirectional gene flows resulting from the mating systems, which generate a minimal Pygmoid input in the Oto Konda and a maximal suzerain input in the Twa Konda.

Two conclusions on the analyses of matrices of morphological multivariate distances emerge from this review. First, they repeatedly show the influence of the biome, equatorial forest or savanna, on morphology. However, the data were not such as to enable to discriminate the relative role of different phenotypic expression and directional selection in the resulting differentiation. Secondly, such analyses have shown that morphological differentiation truly reflects the occurrence of gene flows and of barriers, geographical or social, to mating. In this, they directly support Lewontin's (1974, p. 171) assertion that "morphological similarity is, if not an infailible guide, at least a reliable indicator of genetic similarity".

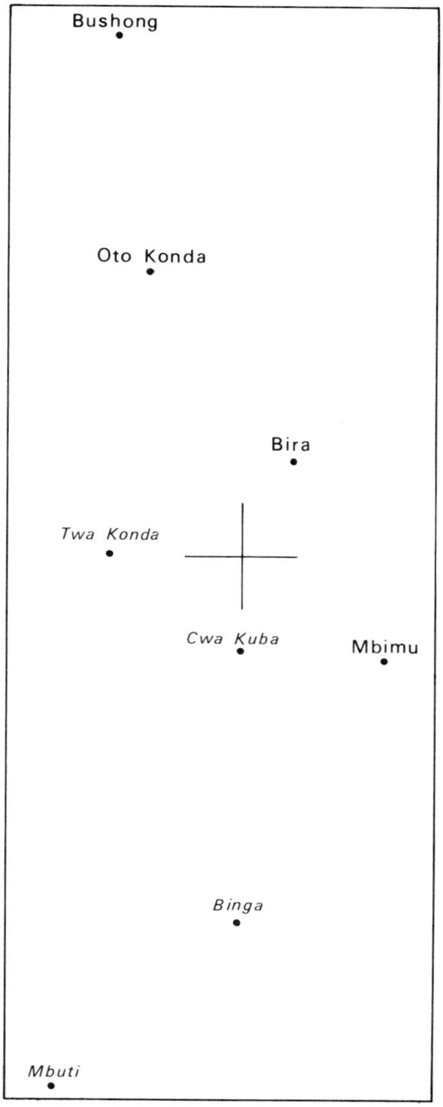

Figure 8. Two-dimensional plotting of four populations of Bantu agriculturalists (forest Bira, Mbimu, Bushong and Oto Konda), and, in italics, the corresponding symbiotic groups of Pygmies or Pygmoids (Mbuti, Binga, Cwa Kuba and Twa Konda) from a matrix of Δg distances for the same set of 10 measurements as for Figure 4 and 7. Reduction to a plane by non-parametric multidimensional scaling as elaborated by Lalouel (1973).

References

Fraser, G.R., E.R. Giblett, A.G. Motulsky: 1966, Population genetic studies in the Congo. III. Blood groups (ABO, MNSs, Rh, Js^a). American Journal of Human Genetics 18: 546-552.

Gower, J.C.: 1966, Multivariate analysis and multidimensional geometry. Statistician 17: 13-28.

Hiernaux, J.: 1956, Analyse de la variation des caractères physiques humains en une région de l'Afrique centrale: Ruanda-Urundi et Kivu. Tervuren: Annales du Musée Royal du Congo Belge, Sciences de l'Homme, Anthropologie, vol. 3.

Hiernaux, J.: 1965, Une nouvelle mesure de distance anthropologique entre populations, utilisant simultanément des fréquences géniques, des pourcentages de traits descriptifs et des moyennes métriques. Comptes-Rendus de l'Académie des Sciences de Paris 260: 1748-1750.

Hiernaux, J.: 1972, "The analysis of multivariate biological distances between human populations: principles, and application to sub-Saharan Africa" in The assessment of population affinities in man. Edited by J.S. Weiner and J. Huizinga, 96-114. Oxford: Clarendon Press.

Hiernaux, J.: 1973, "Numerical taxonomy of man: an application to a set of thirty-two African populations" in Physical anthropology and its extending horizons. S.S. Sarkar memorial volume. Edited by A. Basu et al., 151-161. Calcutta: Orient Longman.

Hiernaux, J.: 1979, "Adaptation of morphology to climate in central Africa" in Physiological and morphological adaptation and evolution. Edited by W.A. Stini, 89-96. The Hague: Mouton.

Hiernaux, J., E. Vincke, D. Commelin: 1976, Les Oto et les Twa des Konda. L'Anthropologie 80: 449-464.

Lalouel, J.M.: 1973, "Topology of population structure" in Genetic Structure of Populations. Edited by N.E. Morton, 139-152. Honolulu: University Press of Hawaii.

Lewontin, R.C.: 1974, The genetic basis of evolutionary change. New York: Columbia University Press.

INTERVAL ESTIMATES FOR POSTERIOR PROBABILITIES, APPLICATIONS TO BORDER CAVE

A.W. Ambergen* & W. Schaafsma**

* Mathematisch Centrum, Postbox 4079, 1009 AB Amsterdam
** Rijksuniversiteit Groningen, Postbox 800, 9700 AV Groningen

ABSTRACT

In this paper asymptotic variances of estimators for the posterior probability that an individual belongs to one of $k \geq 2$ populations are presented. It is assumed that a set of k prior probabilities and a $p \geq 1$ dimensional vector of scores of the individual are given. In the model the populations are represented by multivariate normal distributions. The case in which the dispersion matrices are assumed to be homogeneous as well as the case without this assumption are treated. To illustrate the theory an example from physical anthropology is given.

KEY WORDS & PHRASES: estimating posterior probabilities, confidence intervals for posterior probabilities, classification, discriminant analysis.

SUMMARY

During discussions with research workers, it became clear to the authors that there are many diagnostic situations where the involvement of the statistician should be restricted to the computation of point and interval estimates for posterior probabilities. Specification of prior probabilities should be left to the genuine decision maker who will also have to bear responsibility for the interpretation, in terms of future actions and decisions, of the generated confidence intervals for the posterior probabilities.

Let $x \in \mathbb{R}^p$ denote the vector of scores of the individual under investigation and let $\rho_1, \ldots, \rho_k$ denote the corresponding prior probabilities. Posterior probabilities $\rho_{1|x}, \ldots, \rho_{k|x}$ have to be

G. N. van Vark and W. W. Howells (eds.), Multivariate Statistical Methods in Physical Anthropology, 115–134.

estimated on the basis of training samples from k populations. We assume that X has a multivariate normal distribution for each of the k populations. In addition to this we shall sometimes postulate that homogeneity of dispersions holds. We use maximum likelihood estimators $R_{1|x},\ldots,R_{k|x}$ for estimating the posterior probabilities. The limit distribution (as $n \to \infty$) of $n^{\frac{1}{2}}(R_{t|x}-\rho_{t|x})$ is used to characterize the accuracy of the estimator for the t-th posterior probability ($n = \sum_h n_h$, n_h is the size of the h-th training sample).

Section 2 considers the case "$k \geq 2$, $p \geq 1$, normality and homogeneity of dispersions", Section 3 "$k \geq 2$, $p \geq 1$, normality" and Section 4 compares the results in Sections 2 and 3. It is shown how much the accuracy decreases by dropping the assumption of homogeneity of the dispersion matrices. In Section 5 the example of Border Cave is given as an application of the exposed theory.

1. INTRODUCTION

Discriminant analysis originated from the practical needs of research workers. This is particularly obvious from R.A. Fisher's contributions. Its history, like that of mathematical statistics at large, has been influenced by controversies concerning the probability concept. In broad outline, three competing definitions exist: (1) the classical definition, which is very convenient if games of chance are studied; (2) the frequentist definition, where probabilities of events are mathematical idealizations of relative frequencies; and (3) the subjectivist definition, where probabilities of events and also of statements are largely intended to describe personal degrees of belief.

Emphasizing the frequentist definition, Fisher, Neyman, Wald, and many others created their procedure-oriented approach to theoretical statistics. In discriminant analysis and in the related theory of pattern recognition the concepts of allocation rules, pattern recognition systems, etc., obtained a central position. WALD (1944) provided already a clear formulation. On the basis of an observation vector X_0 for the individual under classification and independent random samples $X_{h1},\ldots,X_{hn_h}$ from population h, the individual has to be assigned to one of the populations. Wald, Anderson and many others proposed classification procedures and studied the corresponding misclassification probabilities as a function of the underlying unknown parameter, usually by giving asymptotic expansions. One of us has derived exact results for certain univariate cases (see SCHAAFSMA-VAN VARK (1977) for further references). The idea behind these objectivistic approaches is that any specification of prior probabilities and prior distributions is avoided. One works with "plug-in", "maximum likelihood", "minimax risk", and "invariance" considerations.

Emphasizing the subjectivist definition, L.J. Savage and many others created their opinion-oriented approach to theoretical statistics. In discriminant analysis and in medical diagnosis the concepts of prior and posterior probabilities, or plausibilities, deserve a central position. Bayes' Theorem is the basic tool of the subjectivist.

In practice the distinctions between the procedure-oriented and the opinion-oriented approach are smaller than one might expect. Research workers well understand that the conclusions may depend on the specification of prior probabilities $\rho_1,\ldots,\rho_k$ and that they, and not the statistician, should make this specification. They want to make an assignment on the basis of the posterior probabilities

$$\rho_{t|x} = \rho_t f_t(x) / \sum_{h=1}^{k} \rho_h f_h(x) \quad t=1,\ldots,k \tag{1.1}$$

These probabilities are unknown parameters because, through $f_1(x),\ldots,f_k(x)$, they depend on the unknown densities $f_1,\ldots,f_k$. The training samples $X_{h1},\ldots,X_{hn_h}$ $(h=1,\ldots,k)$ have to be used to estimate the unknown parameters.

The problem of estimating the posterior probabilities (1.1) has received considerable attention in the literature. A recent reference is MCLACHLAN (1979), who compares the bias of estimators based on the estimative method with those based on the predictive method. He notes that it would have been instructive also to consider the asymptotic mean square errors, but that their derivation is a much more difficult problem. Note that our paper presents solutions to this problem. The following discussion outlines situations where asymptotic variances, etc., are not only instructive but essential. We distinguish between (1) pattern recognition applications where a classification procedure has to serve as the basic ingredient in some automatic pattern recognition system, and (2) diagnostic applications where a possible classification is only one aspect of fundamental group-membership discussions.

Focussing on such diagnostic situations, the authors have reached the opinion that it would often be wise if attention were to be shifted from the application of a classification procedure to the estimation, by means of confidence intervals, of the posterior probabilities (1.1). In fact, the cooperation between the statistician and his client (medical doctor or physical-anthropologist) might be much improved if the statistician were to restrict his activities to the construction of such intervals, leaving the actual classification to the genuine decision maker. Since the construction of exact confidence intervals is extremely difficult, this paper is devoted to the formulation of limit theorems for the distribution of various estimators $(R_{1|x},\ldots,R_{k|x})$ for the vector $(\rho_{1|x},\ldots,\rho_{k|x})$ of posterior probabilities. The asymptotic variance of $R_{t|x}$ can then be used to construct an approximate confidence interval for $\rho_{t|x}$.

Sections 2 and 3 are based on the assumption that f_h is the p.d.f. of a multivariate normal distribution $N_p(\mu_h,\Sigma_h)$. Section 2 considers the case $\Sigma_1 = \ldots = \Sigma_k$, while Section 3 deals with the more general situation in which no homogeneity assumptions are made. This leads to larger asymptotic variances (see Section 4).

Physical anthropologists interested in basic concepts and applications are invited to continue with Section 5 before venturing into the details of Sections 2,...,4.

2. NORMALITY AND HOMOGENEITY OF DISPERSION MATRICES

With $x \in \mathbb{R}^p$ and $\rho_1,\ldots,\rho_k$ as prescribed constants, we wish to estimate the posterior probabilities $\rho_{1|x},\ldots,\rho_{k|x}$ on the basis of training samples from the p.d.f. f_h of a p-variate normal distribution $N_p(\mu_h,\Sigma)$. Hence $\theta = (\mu_1,\ldots,\mu_k,\Sigma)$ plays the part of the unknown parameter. Note that

$$\rho_{t|x} = \rho_t \exp(-\tfrac{1}{2}\Delta^2_{x;t,t})/\{\sum_{h=1}^{k} \rho_h \exp(-\tfrac{1}{2}\Delta^2_{x;h,h})\} \tag{2.1}$$

where

$$\Delta^2_{x;h,h} = (x-\mu_h)^T\Sigma^{-1}(x-\mu_h) \tag{2.2}$$

is the squared Mahalanobis distance between the observed vector and the h-th population mean. Let $X_{h1},\ldots,X_{hn_h}$ denote the h-th training sample; these random vectors are assumed to be independent with $N_p(\mu_h,\Sigma)$ distributions. Let $n = n_1 + \ldots + n_k$ denote the total sample size. Note that $\rho_{t|x}$ is a function of $\theta = (\mu_1,\ldots,\mu_k,\Sigma)$ which can be estimated, e.g. by means of maximum likelihood. (Various other methods were considered in AMBERGEN-SCHAAFSMA (1983); these methods will not be repeated here because they did not lead to a significant improvement over maximum likelihood.)
Using the notation

$$X_{h.} = n_h^{-1}\sum_{i=1}^{n_h} X_{hi} \text{ and } S = \sum_{h=1}^{k}\sum_{i=1}^{n_k} (X_{hi}-X_{h.})(X_{hi}-X_{h.})^T,$$

for the h-th sample mean and the pooled matrix of cross-products, we see that

$$\hat\theta = (X_{1.},\ldots,X_{k.},n^{-1}S)$$

is the maximum likelihood estimator for θ, where

$$R_{t|x} = \hat\rho_{t|x} = \rho_t \exp(-\tfrac{1}{2}\hat\Delta^2_{x;t,t})/\{\sum_{h=1}^{k} \rho_h \exp(-\tfrac{1}{2}\hat\Delta^2_{x;h,h})\}$$

is the maximum likelihood estimator for $\rho_{t|x}$, and

$$\hat{\Delta}^2_{x;h,h} = n(x-X_{h.})^T S^{-1}(x-X_{h.})$$

The distribution of the estimator $R_{t|x}$ can be studied by means of simulation experiments. These experiments suggest that approximations based on the following limit theorem are reliable if the sample sizes are not smaller than about 30; the theorem shows that $R_{t|x}$ is approximately normally distributed with expectation $\rho_{t|x}$ and variance $n^{-1}(\Psi\Gamma\Psi)_{tt}$. The reliability of the approximations follows from simulation experiments reported in AMBERGEN-SCHAAFSMA (1983).

2.1. Theorem

If $n \to \infty$ and $n_h n^{-1} \to b_h > 0$ $(h=1,\ldots,k)$ then

$$\mathcal{L}n^{\frac{1}{2}}(R_{.|x}-\rho_{.|x}) \to N_k(0,\Psi\Gamma\Psi) \tag{2.3}$$

where

$R_{.|x} = (R_{1|x},\ldots,R_{k|x})^T, \rho_{.|x}=(\rho_{1|x},\ldots,\rho_{k|x})^T$, the matrix Γ is given by

$$\begin{aligned}\Gamma_{h,h} &= 4b_h^{-1}\,\Delta^2_{x;h,h} + 2\Delta^4_{x;h,h}\\ \Gamma_{h,t} &= 2\{(x-\mu_h)^T\Sigma^{-1}(x-\mu_t)\}^2 \qquad (h \neq t)\end{aligned} \tag{2.4}$$

and Ψ by

$$\begin{aligned}\Psi_{h,h} &= \tfrac{1}{2}\rho_{h|x}(1-\rho_{h|x})\\ \Psi_{h,t} &= -\tfrac{1}{2}\rho_{h|x}\,\rho_{t|x} \qquad (h \neq t).\end{aligned} \tag{2.5}$$

Proof: See AMBERGEN-SCHAAFSMA (1983), which generalizes previous results in SCHAAFSMA-VAN VARK (1977) for "$k = 2$, $p = 1$, $\sigma_1^2 = \sigma_2^2$" and SCHAAFSMA-VAN VARK (1979) for "$k = 2$, $p \geq 1$, $\Sigma_1 = \Sigma_2$" to the case $k > 2$.

3. NORMALITY BUT NOT NECESSARILY HOMOGENEITY OF DISPERSION MATRICES

With $x \in \mathbb{R}^p$ and $\rho_1,\ldots,\rho_k$ as prescribed constants, we wish to estimate $\rho_{1|x},\ldots,\rho_{k|x}$ on the basis of training samples from the p.d.f. f_h of $N_p(\mu_h,\Sigma_h)$. Hence $\theta = (\mu_1,\ldots,\mu_k, \Sigma_1,\ldots,\Sigma_k)$ now plays the part of the unknown parameter and $\rho_{t|x}$ is determined by (1.1) where

$$f_h(x) = |2\pi\ \Sigma_h|^{-\frac{1}{2}} \exp(-\tfrac{1}{2}\ \Delta^2_{x;h}) \tag{3.1}$$

with

$$\Delta^2_{x;h} = (x-\mu_h)^T\ \Sigma_h^{-1}(x-\mu_h). \tag{3.2}$$

Again using maximum likelihood, we obtain

$$\hat{\theta} = (X_{1.},\ldots,X_{k.},n_1^{-1}S_1,\ldots,n_k^{-1}S_k)$$

where

$$X_{h.} = n_h^{-1} \sum_{i=1}^{n_h} X_{hi} \text{ and } S_h = \sum_{i=1}^{n_h} (X_{hi}-X_{h.})(X_{hi}-X_h)^T,$$

and hence the maximum likelihood estimators

$$\hat{\Delta}^2_{x;h} = n_h(x-X_{h.})^T S_h^{-1}(x-X_{h.}),$$

$$\hat{f}_h(x) = |2\ \pi\ n_h^{-1}\ \ S_h|^{-\frac{1}{2}} \exp(-\tfrac{1}{2}\hat{\Delta}^2_{x;h})$$

and

$$R_{t|x} = \hat{\rho}_{t|x} = \rho_t\ \hat{f}_t(x)\ /\ \{\sum_{h=1}^{k} \rho_h\hat{f}_h(x)\}. \tag{3.3}$$

3.1. Lemma

If $n_h \to \infty$, then

$$\mathcal{L}n_h^{\frac{1}{2}}\ (\hat{f}_h(x) - f_h(x)) \to N(0,\tfrac{1}{2}(\Delta^4_{x;h}+p)f_h^2(x)) \tag{3.4}$$

Proof: See AMBERGEN-SCHAAFSMA(1983).

3.2. Theorem

If $n \to \infty$ and $n_h n^{-1} \to b_h > 0 \quad (h=1,\ldots,k)$, then

$$\mathcal{L}n^{\frac{1}{2}}(R_{.|x}-\rho_{.|x}) \to N_k(0,\Psi\Theta\Psi) \tag{3.5}$$

where $R_{.|x},\rho_{.|x}$ and Ψ are as in Theorem 2.1 and

$$\left.\begin{aligned}\Theta_{h,h} &= 2\ b_h^{-1}(\Delta^4_{x;h}+p)\\ \Theta_{h,t} &= 0 \qquad\qquad (h \neq t).\end{aligned}\right\} \tag{3.6}$$

Proof: This is an immediate consequence of the lemma and (3.3).

4. WHAT DOES IT COST TO DROP THE HOMOGENEITY ASSUMPTION?

How many observations does "not knowing $\Sigma_1 = \ldots = \Sigma_k$" cost? What happens if the sample sizes are modified? What if the dimensionality p is increased, x is changed, or $\rho_1,\ldots,\rho_k$ are modified? In practice one will not know what the true values of the underlying unknown parameters are. One will have to estimate the approximate standard deviations. However, if one needs an intuitive feel for the magnitude of the standard deviations, then it helps to elaborate on a number of theoretical cases. The following cases were selected in order to suggest possible answers to the above-mentioned questions.

Case 1. Suppose $p = 2$, $k = 2$, $n_1 = n_2$, $\mu_1 = (1,0)^T$, $\mu_2 = (-1,0)^T$, $\Sigma = I_2$, $x = (0,1)^T$ (the reader should draw a picture and notice that $x - \mu_1$ is perpendicular to $x - \mu_2$), $\rho_1 = \rho_2 = \frac{1}{2}$. Note that the consequences of drawing training samples are studied without actually drawing them. The above-mentioned specifications imply $\rho_{1|x} = \rho_{2|x} = \frac{1}{2}$, $b_1 = b_2 = \frac{1}{2}$, $\Delta^2_{x;h} = \Delta^2_{x;h,h} = 2$, $\Gamma_{hh} = 4x2x2+2x2^2 = 24 = 2x2x(2^2+2) = \Theta_{h,h}$, $\Gamma_{h,t} = \Theta_{h,t} = 0 (h \neq t)$. Hence $\Gamma = \Theta$ and for this very special situation both estimators are asymptotically equivalent. It follows from

$$\Psi \; \Gamma \; \Psi = \Psi \; \Theta \; \Psi = 64^{-1} \begin{bmatrix} 1 & -1 \\ -1 & 1 \end{bmatrix} \begin{bmatrix} 24 & 0 \\ 0 & 24 \end{bmatrix} \begin{bmatrix} 1 & -1 \\ -1 & 1 \end{bmatrix}$$

that the approximate standard deviations of all $R_{t|x}$'s are $n^{-\frac{1}{2}}(64^{-1}48)^{\frac{1}{2}} = .87n^{-\frac{1}{2}}$. If $n_1 = n_2 = 25$, then we obtain $.87(50)^{-\frac{1}{2}} = .12$ for the approximate standard deviation. This is considerable in comparison with the true values $\rho_{1|x} = \rho_{2|x} = .50$.

Case 2. Suppose $p = 2$, $k = 4$, $n_1 = n_2 = n_3 = n_4 = 4^{-1} n$, $\mu_1 = (1,0)^T$, $\mu_2 = (0,1)^T$, $\mu_3 = (-1,0)^T$, $\mu_4 = (0,-1)^T$, $\Sigma = I_2$ and $x = (0,0)^T$ (the reader should draw a picture and notice that everything has been arranged nicely around the origin to facilitate computations). With $\rho_1 = \ldots = \rho_4 = .25$ we obtain $\rho_{t|x} = .25$ for the posterior probabilities $(t=1,\ldots,4)$ because $\Delta^2_{x;h,h} = 1$. Note that $(x-\mu_h)^T \; \Sigma^{-1}(x-\mu_t)$ is equal to -1 if $(h,t) = (1,3)$ or (2.4), and equal to 0 if $(h,t) = (1,2)$, $(1,4)$, $(2,3)$ or $(3,4)$. Hence

$$\Psi = 32^{-1} \begin{bmatrix} 3 & -1 & -1 & -1 \\ -1 & 3 & -1 & -1 \\ -1 & -1 & 3 & -1 \\ -1 & -1 & -1 & -3 \end{bmatrix}, \Gamma = \begin{bmatrix} 18 & 0 & 2 & 0 \\ 0 & 18 & 0 & 2 \\ 2 & 0 & 18 & 0 \\ 0 & 2 & 0 & 18 \end{bmatrix}, \Theta = \begin{bmatrix} 24 & 0 & 0 & 0 \\ 0 & 24 & 0 & 0 \\ 0 & 0 & 24 & 0 \\ 0 & 0 & 0 & 24 \end{bmatrix}$$

Notice that $\Theta - \Gamma \geq 0$ and hence $\Psi \; (\Theta - \Gamma) \; \Psi \geq 0$. If $\Psi \; \Gamma \; \Psi$ and $\Psi \; \Theta \; \Psi$ are computed, then the approximate standard deviations

$$n^{-\frac{1}{2}}\{(\Psi\Gamma\Psi)_{t,t}\}^{\frac{1}{2}} = .45n^{-\frac{1}{2}}, \; n^{-\frac{1}{2}}\{\Psi\Theta\Psi)_{t,t}\}^{\frac{1}{2}} = .53n^{-\frac{1}{2}}$$

are obtained for the estimators $R_{t|x}$. Note that the second standard deviation is about 1.2 times the first one: it costs about 40% of the observations if the homogeneity assumption $\Sigma_1 = \ldots = \Sigma_4$ is removed (the sample sizes should be about 1.4 times as large if the same standard deviation is required). If $n_1 = \ldots = n_4 = 25$, then $n = 100$ and the respective approximate standard deviations .045 and .053 are considerable in comparison with the true values .25 of the posterior probabilities (a comparison with Case 1 requires that the same observation vector $x = (0,0)^T$ is considered).

Case 3. We modify Case 2 by taking $n_1{:}n_2{:}n_3{:}n_4 = 1 : 2 : 3 : 4$ ($b_1 = .10$, $b_2 = .20$, $b_3 = .30$, $b_4 = .40$) and we leave the other specifications unchanged. Hence $p = 2$, $k = 4$, $\mu_1 = (1,0)^T$, $\mu_2 = (0,1)^T$, $\mu_3 = (-1,0)^T$, $\mu_4 = (0,-1)^T$, $\Sigma = I_2$, $x = (0,0)^T$, $\rho_1 = \ldots = \rho_4 = .25$, Ψ as in Case 2. For Γ and Θ we obtain

$$\Gamma = \begin{bmatrix} 24 & 0 & 2 & 0 \\ 0 & 22 & 0 & 2 \\ 2 & 0 & 15.33 & 0 \\ 0 & 2 & 0 & 12 \end{bmatrix}, \quad \Theta = \begin{bmatrix} 60 & 0 & 0 & 0 \\ 0 & 30 & 0 & 0 \\ 0 & 0 & 20 & 0 \\ 0 & 0 & 0 & 15 \end{bmatrix}$$

Notice that $\Theta - \Gamma \geq 0$ and hence $\Psi(\Theta-\Gamma)\Psi \geq 0$. Elaborating on $\Psi\ \Gamma\ \Psi$ we obtain the approximate standard deviations

$$n^{-\frac{1}{2}}\{\Psi\Gamma\Psi)_{tt}\}^{\frac{1}{2}} = .64n^{-\frac{1}{2}},\ .51n^{-\frac{1}{2}},\ .45n^{-\frac{1}{2}},\ .42n^{-\frac{1}{2}}$$

for $t = 1, 2, 3, 4$ respectively. The average of these values is larger than the value $.45\, n^{-\frac{1}{2}}$ in Case 2. The smallest value $.42n^{-\frac{1}{2}}$ is smaller than in Case 2 and this is caused by the large value of b_4. Elaborating on $\Psi\ \Theta\ \Psi$ we obtain approximate standard deviations for the $R_{t|x}$'s which are again about 1.2 times as large as for the estimators based on the homogeneity assumptions.

Case 4. We modify Case 2 by increasing the dimensionality p. It is interesting to do this by introducing variables which contain no discrimination information. The true values of the posterior probabilities will remain unchanged but the variances of their estimators will become larger. We take $p = 4$, $k = 4$, $n_1 = n_2 = n_3 = n_4 = 4^{-1}n$, $\mu_1 = (1,0,0,0)^T$, $\mu_2 = (0,1,0,0)^T$, $\mu_3 = (-1,0,0,0)^T$, $\mu_4 = (0,-1,0,0)^T$, $\Sigma = I_4$, $x = (0,0,x_3,x_4)^T$ and $\rho_1 = \ldots = \rho_4 = .25$. Note that any specification of x_3, x_4 leads to $\Delta^2_{x;1} = \ldots = \Delta^2_{x;4}$ and $\rho_{1|x} = \ldots = \rho_{4|x} = .25$. If we take $x_3 = x_4 = 0$ then Ψ, Γ and Θ are as in Case 2. If we take $x_3 = x_4 = 1$, then the matrix of inner-products $(x-\mu_h)^T(x-\mu_t)(h,t=1,\ldots,4)$ becomes

$$\begin{bmatrix} -1 & 0 & 1 & 1 \\ 0 & -1 & 1 & 1 \\ 1 & 0 & 1 & 1 \\ 0 & 1 & 1 & 1 \end{bmatrix} \begin{bmatrix} -1 & 0 & 1 & 0 \\ 0 & -1 & 0 & 1 \\ 1 & 1 & 1 & 1 \\ 1 & 1 & 1 & 1 \end{bmatrix} = \begin{bmatrix} 3 & 2 & 1 & 2 \\ 2 & 3 & 2 & 1 \\ 1 & 2 & 3 & 2 \\ 2 & 1 & 2 & 3 \end{bmatrix}$$

Hence Ψ is still as in Case 2 but

$$\Gamma = \begin{bmatrix} 66 & 8 & 2 & 8 \\ 8 & 66 & 8 & 2 \\ 2 & 8 & 66 & 8 \\ 8 & 2 & 8 & 66 \end{bmatrix}; \; \Theta = \begin{bmatrix} 104 & 0 & 0 & 0 \\ 0 & 104 & 0 & 0 \\ 0 & 0 & 104 & 0 \\ 0 & 0 & 0 & 104 \end{bmatrix}$$

Notice again that $\Theta - \Gamma \geq 0$. Elaborating on $\Psi\, \Gamma\, \Psi$ we obtain the approximate standard deviations $.84\; n^{\frac{1}{2}}$ for the estimators $R_{t|x}$ of Section 2. Computing the diagonal elements of $\Psi\, \Theta\, \Psi$ we obtain the approximate standard deviations $1.10\; n^{-\frac{1}{2}}$ for the estimators $R_{t|x}$ of Section 3. For $n_1 = n_2 = n_3 = n_4 = 25$ we now obtain the respective approximate standard deviations .084 and .110 for the estimators $R_{t|x}$. The values are very large with respect to the estimated true values $\rho_{t|x} = .25$. Not knowing that $\Sigma_1 = \ldots = \Sigma_4$ now costs about $(110/84)^2 - 1 = 72\%$ of the observations. This is not unexpected because increasing the dimensionality without supplying relevant extra information causes extra confusion, especially if $\Sigma_1 = \ldots = \Sigma_4$ cannot be postulated.

Case 5. We modify Case 2 by breaking the symmetry such that the posterior probabilities differ from the prior ones. The specifications $p = 2$, $k = 4$, $n_1 = n_2 = n_3 = n_4 = 4^{-1}n$, $\mu_1 = (1,0)^T$, $\mu_2 = (0,1)^T$, $\mu_3 = (-1,0)^T$, $\mu_4 = (0,-1)^T$, and $\Sigma = I_2$ are left unchanged but $x = (1,0)^T$ is taken such that the first posterior probability $\rho_{1|x}$ (again $\rho_1 = \ldots = \rho_4 = .25$) is close to .50. The matrix of inner-products $\Delta^2_{x;h,t} = (x-\mu_h)^T(x-\mu_t)\,(h,t=1,\ldots,4)$ becomes

$$\begin{bmatrix} 0 & 0 \\ 1 & -1 \\ 2 & 0 \\ 1 & 1 \end{bmatrix} \begin{bmatrix} 0 & 1 & 2 & 1 \\ 0 & -1 & 0 & 1 \end{bmatrix} = \begin{bmatrix} 0 & 0 & 0 & 0 \\ 0 & 2 & 2 & 0 \\ 0 & 2 & 4 & 2 \\ 0 & 0 & 2 & 2 \end{bmatrix}$$

with the consequence that

$$\rho_{1|x} : \rho_{2|x} : \rho_{3|x} : \rho_{4|x} = e^0 : e^{-1} : e^{-2} : e^{-1}$$

and hence

$$\rho_{1|x} = .52; \; \rho_{2|x} = .20; \; \rho_{3|x} = .07; \; \rho_{4|x} = .20$$

The matrices Ψ, Γ and Θ become

$$\Psi = \tfrac{1}{2}\begin{bmatrix} .25 & -.11 & -.04 & -.11 \\ -.11 & .16 & -.01 & -.04 \\ -.04 & -.01 & .07 & -.01 \\ -.11 & -.04 & -.01 & .16 \end{bmatrix}, \Gamma = \begin{bmatrix} 0 & 0 & 0 & 0 \\ 0 & 40 & 8 & 0 \\ 0 & 8 & 96 & 8 \\ 0 & 0 & 8 & 40 \end{bmatrix}, \Theta = \begin{bmatrix} 16 & 0 & 0 & 0 \\ 0 & 48 & 0 & 0 \\ 0 & 0 & 144 & 0 \\ 0 & 0 & 0 & 48 \end{bmatrix}$$

Notice again that $\Theta - \Gamma \geq 0$. If we compute $\Psi \, \Gamma \, \Psi$ then we obtain the respective approximate standard deviations $.56\ n^{-\frac{1}{2}}$, $.51\ n^{-\frac{1}{2}}$, $.34\ n^{-\frac{1}{2}}$, and $.51\ n^{-\frac{1}{2}}$ for the estimators $R_{t|x}$ $(t = 1,\ldots,4)$ of Section 2 and based on the assumption of homogeneity of dispersion matrices. Notice that the largest posterior probability $\rho_{1|x} = .53$ has an estimator $R_{1|x}$ with approximate standard deviation .06 if $n_1 = n_2 = n_3 = n_4 = 25$; in our opinion, this accuracy is quite satisfactory. If we compute $\Psi \, \Theta \, \Psi$, then we obtain the approximate standard deviations $.77\ n^{-\frac{1}{2}}$, $.62\ n^{-\frac{1}{2}}$, $.43\ n^{-\frac{1}{2}}$, $.62\ n^{-\frac{1}{2}}$.

Remark 1. It would be interesting to combine various modifications of Case 2, and to see what happens if the prior probabilities are changed. Modifications of the latter kind have a considerable effect on the parameters to be estimated: "bias" which appears if wrong prior probabilities are used may be much larger than the approximate standard deviations of the estimators $R_{t|x}$ for $\rho_{t|x}$. Conclusions should always be formulated with respect to the introduced prior probabilities.

Remark 2. We have seen that if we drop the assumption of homogeneity of the dispersion matrices we have to pay for this with larger standard deviations for the estimators of the posterior probabilities. Or in other words more observations are needed to obtain the same accuracy. The same holds if we take a further step and also drop the normality and only assume a reasonable smoothness of the population densities and use a nonparametric approach: the greater generality must be payed for with larger inaccuracies in the posterior probabilities. For this we refer to AMBERGEN-SCHAAFSMA (1982).

5. APPLICATIONS TO PHYSICAL ANTHROPOLOGY

The idea that posterior probabilities should be equipped with standard errors evolved during discussions between the second author and G.N. van Vark on statistical problems in physical anthropology. Important applications of our theory will be found in the area of medical diagnosis. However, group membership discussions are also important in physical anthropology and we shall illustrate our theory using the following case, suggested to us by Van Vark.

In 1940, W.E. Horton, while digging for guano at Border Cave (near the boundary between Swaziland and Zululand, South-Africa), found fragments of a human cranium. More of the adult cranium was found in his dump during 1941-42. It was supposed on the basis of circumstantial evidence, e.g. artifacts, that the cranium belonged to a Middle Stone Age hominid. The fragments of the cranium have been set in a plaster reconstruction by A.R. Hughes. RIGHTMIRE (1979) took p=11 measurements on the original fossil and compared Border Cave with crania drawn from k = 8 recent African populations (Bushman males and females, Hottentot males, Zulu males and females, Sotho males and females and Venda males). When all discriminants are considered, Border Cave lies closest to the Hottentot centroid and is contained within the .05 limits of this distribution. This assignment should not be interpreted in a strict sense to exclude it from all Bushman populations. Rightmire's paper was followed by an interesting discussion. CAMPBELL (1980) put the statistical comments in perspective and made useful suggestions which led to a reanalysis of the data by RIGHTMIRE (1981). While we welcome the progress made by concentrating on typicality indices (from the F distribution) and posterior probabilities (based on multivariate Student densities as suggested by the semi-Bayesian approach), we are not completely satisfied because we prefer a classical statistical approach where these indices and probabilities are regarded as estimates for basic unknown parameters, estimates which should be equipped with standard deviations if not replaced by confidence intervals.

Our evaluation of Border Cave is based on a comparison with crania drawn from k = 8 recent African populations (see Table 1). We used samples from Van Vark's data bank. It is a pity that Hottentots were not available because Rightmire had concluded that Border Cave is closest to the Hottentot centroid. The figures in Table 1 were obtained by converting those in Table 2 of RIGHTMIRE (1979) into Howells' measurement system.

The reader is invited to make univariate comparisons by looking at Table 1 and the column of standard deviations in Table 2 (Student's two sample test is the appropriate tool). Rightmire's Table 2 shows that Hottentot males and Bushman males are very similar, so that we need not be too concerned about the missing Hottentots. It is clear from Table 1 that most Border Cave measurements are too large in relation to the reference samples. Border Cave is not very "typical" for any of the eight populations involved, in fact it looks rather "atypical". Hence prior and posterior probabilities should be regarded with suspicion because they are based on the assumption that Border Cave has randomly been drawn from one of the k populations involved. This complication had deepened our insight. Physical anthropology has always been rich in motivating statisticians. One of the basic reasons for this phenomenon might be that sample sizes are essentially limited, especially when dealing with non-recent populations. In other areas of application, especially those going under the heading "pattern recognition", one can increase sample sizes more easily.

The next step in evaluating Border Cave is by computing Mahalanobis distances and canonical variates, and by performing Hotelling tests instead of Student tests. Table 3 presents some of our results. Note that $\hat{\Delta}^2_{x;h,h}$ was defined in Section 2 and $\hat{\Delta}^2_{x;h}$ in Section 3. The null-hypothesis H_h that Border Cave is from the same population as the h'th sample can be tested by referring the Hotelling T^2 statistic

Table 1. Measurements of Border Cave compared with means for eight modern African populations

	Measurements	Border Cave	Bushmen		Zulu		Dogon		Teita	
			males $N_1=41$	females $N_2=49$	males $N_3=55$	females $N_4=46$	males $N_5=48$	females $N_6=53$	males $N_7=34$	females $N_8=49$
1	SOS, Supraorbital projection	10	6.73	5.69	6.18	5.24	5.40	4.08	6.44	4.94
2	FMB, Bifrontal breadth	112	97.27	93.90	101.98	97.74	99.54	94.34	100.06	95.43
3	NAS, Nasio-frontal subtense	15	15.41	16.20	17.84	16.48	16.46	15.45	18.79	17.12
4	NFA, Nasio-frontal angle	150	143.20	143.65	141.51	142.70	143.46	143.68	138.88	140.49
5	WMH, Check height	21	20.93	19.84	20.73	20.06	21.21	19.96	22.21	20.18
6	FRC, Nasion-bregma chord	116	109.17	105.10	111.69	109.39	110.00	105.66	108.71	105.76
7	FRS, Nasion-bregma subtense	32	28.46	28.22	27.71	27.70	26.69	25.64	26.62	27.02
8	FRF, Nasion-subtense fraction	51	47.59	45.08	47.16	46.04	47.88	44.62	48.82	47.37
9	FRA, Frontal angle	122	124.29	122.73	126.33	125.33	127.58	127.28	127.41	125.43
10	OBB, Orbit breadth, left	45	39.27	37.67	40.44	39.20	39.71	38.08	39.65	37.76
11	MDH, Mastoid height	26	25.24	21.61	28.42	25.61	29.06	25.21	29.09	24.18

Table 2. Standard deviations and correlation matrix for the eleven measurements in the eight populations for the case with homogeneity of dispersion matrices

		standard deviation	Correlation-matrix										
			1	2	3	4	5	6	7	8	9	10	11
1	SOS	1.18	1.00										
2	FMB	3.43	0.29	1.00									
3	NAS	2.21	0.25	0.33	1.00								
4	NFA	4.43	-.18	-0.07	-0.96	1.00							
5	WMH	2.17	0.06	0.26	0.03	0.04	1.00						
6	FRC	4.63	0.04	0.22	0.09	-0.04	0.22	1.00					
7	FRS	2.62	-0.01	0.06	-0.12	0.15	-0.02	0.60	1.00				
8	FRF	3.46	-0.03	0.10	0.02	0.01	0.22	0.53	0.18	1.00			
9	FRA	3.78	0.03	0.04	0.19	-0.19	0.14	-0.18	-0.89	0.18	1.00		
10	OBB	1.65	0.08	0.63	0.25	-0.09	0.04	0.11	-0.01	0.02	0.06	1.00	
11	MDH	3.14	0.11	0.18	0.04	0.01	0.16	0.12	0.03	0.00	0.02	0.12	1.00

Table 3. Mahalanobis distances, F-probabilities, posterior probabilities and standard deviations of posterior probabilities for the two cases with and without homogeneity of dispersion matrices

1		N_i	with homogeneity of dispersion matrices				without homogeneity of dispersion matrices			
			$\hat{\Delta}_{x;h,h}$	F-prob	posterior probability	standard dev. post-prob.	$\hat{\Delta}_{x,h}$	F-prob	posterior probability	standard dev. post.prob.
1	Bushmen males	41	6.48	0.025	0.749	0.283	6.50	0.044	0.056	0.311
2	Bushmen females	49	7.59	0.008	0.000	0.000	7.69	0.014	0.000	0.000
3	Zulu males	55	6.66	0.021	0.231	0.271	6.14	0.046	0.943	0.313
4	Zulu females	46	7.13	0.013	0.009	0.013	7.34	0.020	0.001	0.004
5	Dogon males	48	7.13	0.013	0.009	0.013	7.45	0.018	0.000	0.001
6	Dogon females	53	8.30	0.004	0.000	0.000	9.76	0.002	0.000	0.000
7	Teita males	34	7.38	0.009	0.001	0.002	9.87	0.004	0.000	0.000
8	Teita females	49	7.93	0.006	0.000	0.000	10.96	0.001	0.000	0.000

$$n_h(n-k-p+1)n^{-1}(n_h+1)^{-1}\ p^{-1}\ \hat{\Delta}^2_{x;h,h} \tag{5.1}$$

to the F(p,n-k-p+1) distribution if normality is postulated together with homogeneity of covariances (see Section 2 and RAO (1965) 8.b.2.XII), and by comparing

$$(n_h-p)(n_h+1)^{-1}\ p^{-1}\ \hat{\Delta}^2_{x;h} \tag{5.2}$$

with the $F(p,n_h-p)$ distribution if normality is postulated but no assumptions are made concerning the covariances (if one tries to verify that (5.1) and (5.2) coincide if $k = 2$ and $n = n_h + 1$ then one should notice that $n_h\ \hat{\Delta}^2_{x;h,h} = n\ \hat{\Delta}^2_{x;h}$: the maximum likelihood estimators of Sections 2 and 3 are different because the underlying models differ).

It is also informative to compute the probability that the observed outcome is exceeded. These F-probabilities can be found in Table 3 and are called "typicality probabilities" in the recent publications of Campbell and Rightmire. Note that all these F-probabilities are smaller than .05 which suggests that Border Cave is significantly atypical at the 5% level for all recent African populations considered. This is in perfect agreement with expectations based on the earlier univariate comparisons, but not with Campbell's Table 2 which suggests reasonable affinity of Bushman males with Border Cave (see Campbell's paper in these proceedings).

To explain various conceptual difficulties, we note that if anything should be called the typicality probability of Border Cave with respect to population h, then this should be the unknown parameter

$$\alpha_h(x) = P(G > (x-\mu_h)^T\ \Sigma_h^{-1}(x-\mu_h)) \tag{5.3}$$

where G has the $\chi^2(p)$ distribution. This involves constructing a <u>confidence interval</u> for the typicality probability $\alpha_h(x)$ of Border Cave with respect to population h. The required confidence interval is easily obtained by transforming the confidence interval for the unknown parameter (2.2), if $\Sigma_1 = \ldots = \Sigma_k$ is postulated, or (3.3) if homogeneity of covariances is not required. Now x is regarded as a prescribed constant and not as a random drawing as in the theory behind (5.1) and (5.2). An exact confidence interval for $\Delta^2_{x;h,h}$, under the assumptions of Section 2, follows from the distributional result that

$$n_h(n-k-p+1)\ n^{-1}\ p^{-1}\ \hat{\Delta}^2_{x;h,h} \tag{5.4}$$

has the noncentral F distribution with p and n - k - p + 1 d.f.'s and non-centrality parameter $n_h\ \Delta^2_{x;h,h}$ (see e.g. RAO (1965)

8.b.2.XII). In practice one might content oneself with approximate results based on the unbiasedness of

$$(n-k-p-1)\ n^{-1}\ \hat{\Delta}^2_{x;h,h} - n_h^{-1}\ p \tag{5.5}$$

as an estimator for $\Delta^2_{x;h,h}$ and the corresponding variance

$$(n-k-p-3)^{-1}\ \{2\ \Delta^4_{x;h,h} + 4(n-k-1)n_h^{-1}\ \Delta^2_{x;h,h} + 2p(n-k-1)n_h^{-2}\} \tag{5.6}$$

We applied this approach to Border Cave and Zulus. Starting from $\hat{\Delta}_{x;3,3} = 6.66$ we obtained the approximate outcomes 42 and 13 for (5.5) and (5.6) with the consequence that [35,49] is an approximate confidence interval for $\Delta^2_{x;3,3}$. Converting this by means of (5.3) delivers a confidence interval for $\alpha_3(x)$ "left of everything". This makes very clear that Border Cave is very atypical for the Zulu population, at least if $\Sigma_1 = \ldots = \Sigma_k$ is postulated.

If the assumptions of Section 2 are weakened to those of Section 3, then the uncertainty is increased considerably because Σ_h has to be estimated on the basis of sample h only. Instead of (5.4) we now obtain that

$$(n_h-p)\ p^{-1}\ \hat{\Delta}^2_{x;h} \tag{5.7}$$

has the noncentral $F(p;n_h-p;n_h\ \Delta^2_{x;h})$ distribution and that

$$(n_h-p-2)n_h^{-1}\ \hat{\Delta}^2_{x;h} - n_h^{-1}\ p \tag{5.8}$$

is an unbiased estimator for $\Delta^2_{x;h}$ with variance

$$(n_h-p-4)^{-1}\{2\Delta^4_{x;h}+4(n_h-2)n_h^{-1}\ \Delta^2_{x;h}+2p(n_h-2)n_h^{-2}\} \tag{5.9}$$

Applying this approach to Border Cave and Zulus we obtain from $\hat{\Delta}_{x;3} = 6.14$ in Table 3 and $n_h = 55$, $p = 11$ that (5.8) and (5.9) are approximately equal to 28.5 and 5.6. Hence [23.7,33.3] is an approximate confidence interval for $\Delta^2_{x;3}$. Conversion by means of χ^2_{11} delivers [.0005,.02] as approximate confidence interval for the typicality probability (5.3) of Border Cave with respect to the population of Zulus. The impression made by this result differs from that made by the F-probability in Table 3 though a common feature is that everything is less than .05; whichever way we turn, Border Cave is certainly not a random drawing from the population of Zulus. Hence our computations suggest that Border Cave is certainly not a random drawing from any of the populations involved. We already remarked that Campbell, using other training samples, obtained the F-probability .292 for Border Cave with respect to Bushman males. His computations were based on p=9 variables, a sample of size $n_h = 22$ and $\hat{\Delta}^2_{x;h} = (22/21)20.9$. He did not assume equality of covariance matrices in his Table 2. (His Table 1 shows that, with this assumption, all F-probabilities are $\leq .05$.) Is there a conflict between Campbell's F-probability .292 and our value .044 which was based on a sample of size 41 and p=11? A partial answer to this question is obtained by interpreting Campbell's finding as a confidence interval for the typicality probability (5.3). Application of (5.8) and (5.9) delivers the unbiased estimate 10.5 for $\Delta^2_{x;h}$ and the estimated variance 29. Hence [0,21.3] is an approximate confidence interval for $\Delta^2_{x;h}$ and [.01,1.00] for the typicality probability (5.3). This shows that Campbell's sample size is so small that his F-probability .292 is not in conflict with that of us.

The next step in evaluating Border Cave is to compute approximate confidence intervals for posterior probabilities, though we are convinced that Border Cave cannot be regarded as a random drawing from any of the populations involved. Results of the computations are presented in Table 3. The enormous standard deviations for Bushman males and Zulu males show that <u>if Border Cave were known to be Bushman, Zulu, Dogon or Teita</u>, then it will either be a Bushman male or a Zulu male. It is impossible to discriminate between these two possibilities.

If one accepts the idea that Border Cave is not a random drawing from one of the populations considered, then new interesting problems appear if one tries to compare Mahalanobis distances. We will assume that $\Sigma_1 = \ldots = \Sigma_k$ is postulated. Rightmire's conclusion that Border Cave lies closest to the Hottentot centroid referred to sample properties. We prefer an approach where hypotheses are formulated concerning the populations and tested on the basis of samples. The hypotheses can refer to (1) the distances $\Delta_{x;h,h}$ between the Border Cave specimen and population h, or (2) the distances

$$\Delta_h = \{(\mu-\mu_h)^T \Sigma^{-1}(\mu-\mu_h)\}^{\frac{1}{2}}$$

between the population from which Border Cave is drawn, and population h. Similar problems for comparing Mahalanobis distances appeared in Van Vark's research (but then with Σ known).

As an example of the approach based on (1), we consider testing the null-hypothesis

$$H : \Delta_{x;1,1} = \Delta_{x;7,7}$$

that "the triangle with apex x (= Border Cave) and other vertices μ_1 (= Bushman males ≈ Hottentot males) and μ_7 (= Teita males) has equal legs". We content ourselves with a crude approach. If approximate confidence intervals are available for $\Delta^2_{x;1,1}$ and $\Delta^2_{x;7,7}$ then the interpretation will be clear. Starting from

$\hat{\Delta}_{x;1,1} = 6.48$ ($\hat{\Delta}_{x;7,7} = 7.38$) we obtained the approximate outcomes 39 and 13 (52 and 20) for (5.5) and (5.6) with the consequence that [32,46] is an approximate confidence interval for $\Delta^2_{x;1,1}$ (and [43,61] for $\Delta^2_{x;7,7}$). These crude computations suggest that the difference between $\hat{\Delta}_{x;1,1} = 6.48$ and $\hat{\Delta}_{x;7,7} = 7.38$ is on the verge of being significant. The Border Cave specimen is closer to Bushman males than to Bushman females, Dogon females and Teita females. It is very likely that Border Cave is also closer to Bushman males than tot Teita males.

As an example of the approach based on (2), we consider testing the null-hypothesis

$$H : \Delta_1 = \Delta_6$$

that "the triangle with apex μ (= expectation of the measurement vector in the population from which the Border Cave specimen is drawn) and vertices μ_1 (= Bushman males) and μ_6 (= Dogon females), has equal legs". Note that (5.1) has the noncentral $F'(p,n-k-p+1), n_h(n_h+1)^{-1} \Delta_h^2)$ distribution. We are content with crude approximate results based on the corresponding consequence that

$$n^{-1}(n-k-p-1)\,\hat{\Delta}^2_{x;h,h} - n_h^{-1}(n_h+1)\,p \tag{5.10}$$

is an unbiased estimator for Δ_h^2 with variance

$$\begin{aligned} & n^{-2}\,p^{-2}(n-k-p-3)^{-1}(n-k-p+1)^2\{2p(n-k-1)\,+ \\ & +\,4\,n_h(n_h+1)^{-1}(n-k-1)\,\Delta_h^2 + 2\,n_h^2(n_h+1)^{-2}\,\Delta_h^4\} \end{aligned} \tag{5.11}$$

Starting from $\hat{\Delta}_{x;1,1} = 6.48$, $n_1 = 41$, $p = 11$ and $n = 375$ we obtain the outcomes 29 and 1.1 for (5.10) and (5.11). Hence [27,31] is an approximate confidence interval for Δ_1^2. Starting from $\hat{\Delta}_{x;6,6} = 8.30$ and $n_6 = 53$, we similarly obtain outcomes 54 and 2.2 for (5.10) and (5.11). Hence [49,59] is an approximate confidence interval for Δ_6^2. Our conclusion is that the Border Cave population is definitely different from that of recent Dogon females.

Starting from $\hat{\Delta}_{x;2,2} = 7.59$ we obtain also that the approximate confidence intervals show no overlap though they are very close to each other.

Conclusion

Border Cave specimen is closer to Bushman males than to Bushman females, Dogon females and Teita females and probably also to Teita males. About the same conclusion holds if Border Cave specimen is replaces by the population from which it comes.

ACKNOWLEDGEMENTS

We wish to thank Dr. G.N. van Vark for suggesting the Border Cave application and for the use of his data of the South African Negro crania. We thank Dr. G.P. Rightmire for his assistance in transforming the Border Cave data into Howells' measurement system. Many thanks are also due to Dr. N.A. Campbell and Dr. A.J. Stam for careful reading and suggesting many improvements.

References

Ambergen, A.W.: 1981, *Approximate confidence intervals for posterior probabilities*. Report TW-224. Depart. of Math., postbox 800, Groningen.

Ambergen, A.W., & W. Schaafsma: 1982, *The asymptotic variance of estimators for posterior probabilities*. Report SW 86/82. Mathematisch Centrum, postbox 4079, Amsterdam.

Ambergen, A.W. & W. Schaafsma: 1983, *Interval estimates for posterior probabilities*. Submitted for publication.

Anderson, T.W.: 1973, *An asymptotic expansion of the distribution of the studentized classification statistic* W. Ann. of Stat., Vol 1, No. 5, 964-972.

Bean, S.J. & C.P. Tsokos: 1980, *Developments in nonparametric density estimation*. Int. Stat. Review, 48, 267-287.

Bowker, A.H. & R. Sitgreaves: 1961, *An asymptotic expansion for the distribution function of the W-classification statistic*. In: Solomon, H.

Cacoullos, T.: 1966, *Estimation of a multivariate density*. Ann. of the Inst. of Stat. Math., 18, 178-189.

Cacoullos, T.: 1973, *Discriminant analysis and applications*, Ac. Press.

Campbell, N.A.: 1980, *On the study of the Border Cave remains: Statistical Comments*. Current Anthropology, Vol. 21, No. 4, August 1980, 532-535.

Fisher, R.A.: 1936, *The use of multiple measurements in taxonomic problems*. Ann. Eugenics 7, 179-188.

Ganesalingam, S. & G.L. McLachlan: 1979, *A case study of two clustering methods based on maximum likelihood*. Statistica Neerlandica, Vol. 33, No. 2.

Geisser, S.:1964, *Posterior odds for multivariate normal classifications*. Journal of the Royal Statistical Society, Series B, 26, 26-76.

Habbema, J.D.F. & J. Hermans: 1978, *Statistical methods for clinical decision making*. Thesis, Leiden University.

Habbema, J.D.F., J. Hermans & K. van der Broek: 1974, *A stepwise discriminant analysis program using density estimation*. In G. Bruckmann (Ed.), Compstat 1974, Proceedings in Computational Statistics. (Physica Verlag, Wien).

Howells, W.W.: 1973, *Cranial variation in man. A study by multivariate analysis of patterns of difference among recent human populations*. Papers of the Peabody Museum 67.

McLachlan, G.J.: 1979, A comparison of the estimate and predictive methods of estimating posterior probabilities. Commun. Statist. Theor. Meth. A8(9), 919-929.

Okamoto, M.: 1964, An asymptotic expansion for the distribution of the linear discriminant function. Ann. Math. Stat., 34, 1286-1301, correction 39, 1358-1359.

Parzen, E.: 1962, On estimation of a probability density function and mode. AMS, 33, 1065-1076.

Rosenblatt, M.: Remarks on some nonparametric estimates of a density function. AMS, 27, 832-837.

Rao, C.R.: 1965, Linear statistical inference and its applications, Wiley.

Rightmire, G.P.: 1979, Implications of Border Cave skeletal remains for Later Pleistocene Human Evolution. Current Anthropology, Vol. 20, Nr. 1, March 1979, 23-35.

Rightmire, G.P.: 1981, More on the study of the Border Cave remains. Current Anthropology, Vol. 22, No. 2, 199-200.

Schaafsma, W.: 1976, The asymptotic distribution of some statistics from discriminant analysis. Report TW-176, Dept. Math, Postbox 800, Groningen.

Schaafsma, W.: 1982, Selecting variables in discriminant analysis for improving upon classical procedures. To appear in: Kanal, L and Krishnaiah, P.R. (eds.), Handbook of Statistics 2. North Holland, Amsterdam.

Schaafsma, W. & T. Steerneman: 1981, Classification and discrimination procedures when the number of features is unbounded. IEEE Transac. SMC 11/2, 144-151.

Schaafsma, W. & G.N. van Vark: 1977, Classification and discrimination problems with applications, part I, Statistica Neerlandica 31, 25-45.

Schaafsma, W. & G.N. van Vark: 1979, Classification and discrimination problems with applications, part II[a], Statistica Neerlandica 33, 91-126.

Sitgreaves, R.: 1961, Some results on the distribution of the W-classification statistic. In : Solomon, H.

Solomon, H.: 1961, Studies in item analysis and prediction. Stanford Univ. Press.

Stein, Ch.: 1966, Multivariate analysis, (mimeographed notes recorded by M.L. Eaton). Dept. Stat., Stanford.

Van Vark, G.N.: 1970, Some statistical procedures for the investication of prehistoric human skeletal material. Thesis, Groningen University.

Van Vark, G.N. & P.G.M. Van der Sman: 1982, New discrimination and classification techniques in anthropological practice. Zeitschrift für Morphologie und Anthropologie, Vol. 73, No. 1, 21-36.

CLUSTER ANALYSIS, HISTORY, THEORY AND APPLICATIONS

F.W. Wilmink, M.A., M.D., and H.T. Uytterschaut, Ph. D.

Lab. of Anatomy and Embryology, University of Groningen,
The Netherlands

PART I
HISTORICAL AND THEORETICAL OUTLINE

"None of these strategies for evaluating the adequacy of a cluster solution as a classification has been extensively studied. Each of the strategies involves potential methodological problems. Nonetheless, given the fact that the literature on cluster analysis is still in its infancy and that a solution from a cluster analysis can vary from repartitioned noise to accurately discovering the populations underlying a mixture, future users should be skeptical of any results from this set of statistical methods".

R.K. Blashfield (1976)

1. INTRODUCTION

Cluster analysis is a generic name for a large set of statistical methods that all aim at the detection of groups in a sample of objects, these groups usually being called clusters. Essential to cluster analysis is that, in contrast to discriminant analysis, a group structure need not be known a priori. This makes cluster analysis attractive as an exploratory tool. However, the naive user of cluster analysis (defined as someone who has some data and by running an available computer program hopes to find a group structure underlying his data) may well be compared with one who sets off for a foreign country hoping to find Fortune: without some idea about what type of fortune in which country our hero may well end up his life on a desert island or relapse in

G. N. van Vark and W. W. Howells (eds.), Multivariate Statistical Methods in Physical Anthropology, 135–175.

melancholy. Similarly, some reflection on the nature of cluster analysis may result in more fruitful application than can be expected from our naive user. We shall follow him first and try to learn from his experiences.

Our naive user, by no means simple, first tries to find a group structure by visual inspection of his data. However, the data being multivariate this is not an easy task to accomplish and he proceeds to compute principal components and plots the first versus the second principal component. Unfortunately, no clear group structure emerges. In such cases, he knows, cluster analysis may be of value and again he sits down at his computer terminal. By global inspection of the voluminous manual to the available cluster analysis computer program he selects several program options. These options will usually relate to ordination and selection of variables, transformation of variables and the choice of a similarity measure and a cluster method (see sections 3-5). On his line printer a nice dendrogram (visual display of the clustering process) then appears (c.f. figure 1).

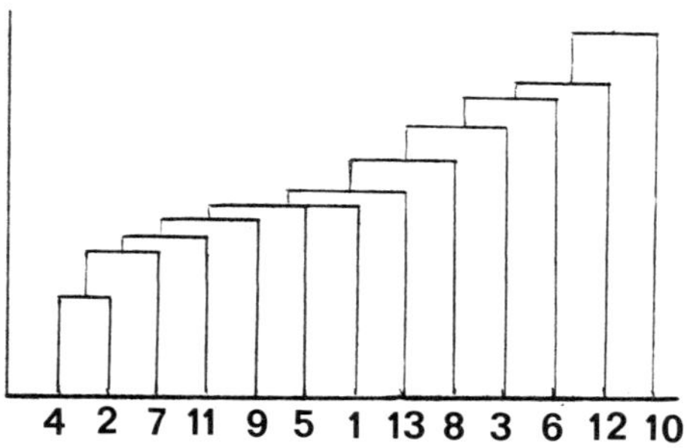

Figure 1. Single linkage dendrogram of 13 objects

Unfortunately, no clear-cut group structure can be detected. Typically, in this situation questions begin to arise. Is there really a structure underlying the data? And, if so, are the selected variables, transformation, similarity measure and cluster method likely to reveal this structure? (sections 6.1 - 6.2). A positive answer to these questions is a necessary but not sufficient condition for any emerging clusters to be valid. The implication is that our user will have to specify some prior beliefs about the expected number of clusters and about in what respect these clusters will differ from each other. From these specifications he will then be better able to decide which variables should be used and whether and how they should be transformed; also, which similarity or dissimilarity measure should be selected. Further issues are the amount of coverage (is it necessary to assign all objects to clusters) and whether overlap of clusters is admissible.

Finally, then, he will be able to choose a suitable method and to run the program again.

The reader may have noted that some circularity is introduced by tuning the cluster procedure to the expected results. As a consequence, the validity of the emerging clusters is questionable. This is one of the major methodological problems in cluster analysis. Although the validity problem may, strictly speaking, be unresolvable, along two lines can one try to get some guarantee that the clusters found do not represent "repartitioned noise".

The first line is to require that the resulting clusters show certain properties, e.g. stability under error perturbation and the like (sections 6.3 - 6.4).

The second line is to require that the cluster algorithm should allow the assessment of how successful it was with respect to its own purpose. Ward's method, e.g., tries to minimize intra-cluster variance, so by comparing intra-cluster variance to between-cluster variance one may assess how successful the method was. Clique-analysis, on the other hand, will produce all subsets of objects-to-be-analyzed that share some characteristic, and it will always succeed in this; consequently, there is no direct measure to evaluate the result and, too, the relation between clusters (cliques) remains obscure. Requiring that a cluster method should try to maximize some criterion therefore not only helps to clarify the nature of the method, but also allows comparison of methods by the analysis of the appropriateness of the criteria they try to maximize and by the degree to which they succeed in it. Moreover, inferences about a structure underlying the data can be evaluated by inspection of the criterion values, although two remarks should be made here. Firstly, it would seem logical to have the clustering criterion to take the number of clusters into account, thereby defining not only the optimal formation of clusters but also their optimal number. Surprisingly, only the information-theory based criteria show this feature explicitly. Although some suggestions have been made (e.g. Marriott, 1971) the issue seems not yet to have been thoroughly investigated. This is unfortunate since it renders most methods in a way indeterminate and therefore allows for much unwelcome subjectivity in the choice of the number of clusters. Secondly, a clustering algorithm will always produce clusters, whatever the nature of the data (excluding degenerate cases), and there are few ways to check the validity of the clusters found. Nevertheless the clusters thus produced are optimal given the data and the clustering criterion and when they are considered to be meaningless in some sense one must conclude that either the objects did not differ in that sense or that the criterion was inappropriate for the data (or both).

We have now given a short impression of what cluster analysis is all about and we have stipulated some methodological problems in cluster analysis. We shall discuss these problems in the next sections. Before doing so, we shall introduce some nomenclature and give a historical outline of the development of the field of cluster analysis.

1.1. Nomenclature

"Cluster analysis" probably is the most widely used name for the class of multivariate data analysis procedures under consideration. In graph theoretical contexts "unsupervised pattern recognition" is a more common name. Other names are numerical taxonomy, numerical analysis and, more specifically, classification, grouping, clumping, Q-analysis. We shall use the familiar term cluster analysis.

The input for a cluster analysis is multiple observations on a set of objects, the objects being called operational taxonomical units (OTU's) by Sokal and Sneath (1963). OTU's can be anything in principle, e.g., individuals, variables, ecological sites, etc. Very often, OTU's are individuals and the input is multivariate data on these individuals. In the latter case we are performing a Q-analysis. In some other cases OTU's are variables and the scores of individuals on these variables are the multiple observations; we then have the case of a R-analysis.

The input data are submitted to a computational algorithm. This is the clustering algorithm. Some clustering algorithms will yield N clusters in the first step (where N is the number of OTU's), N-1 clusters in the second step and so on until finally all OTU's are merged into one cluster. Some other algorithms work just the other way around, i.e., they start with one cluster of N OTU's and end up with N clusters of one OTU; and still others will relocate OTU's from one cluster to another where the number of clusters is specified by the user. These algorithms are called agglomerative, divisive and K-group methods, respectively. The first two have been subsumed under the heading "hierarchical cluster schemes" while K-group methods have also been called partition methods. Unfortunately, hierarchical methods have gradually become more or less synonymous with agglomerative methods, and K-group methods have sometimes been called divisive methods.

Cluster analysis results in clusters having been formed. Ironically, there appears to exist no precise definition of what a cluster is. More or less intuitively, clusters show internal cohesion and external isolation (Cormack, 1971). If the idea is accepted that cluster methods should maximize some criterion, clusters are operationally defined to be sets of OTU's which maximize the clustering criterion.

2. HISTORY

As early as 1939 Tryon published a book on cluster analysis. For other early applications of cluster analysis see Sneath and Sokal (1973). The advent of computers rapidly increased the number of potential users of cluster analysis but in the sixties the number of newly developed cluster algorithms may have exceeded the number of their users, and even in 1976 Blashfield notes that "there are as many programs for clustering as there are users of cluster analysis".

Especially in botany, ecology and biology interest has been greatest in those first years (e.g., association analysis, Lambert and Williams, 1962, 1966; Sheals, 1964; Watson, Williams and Lance, 1966; El. Gazzar et al., 1968; Crovello, 1968; Cole, 1969), but also from the social sciences there were contributors (Cattell, 1966; Johnson, 1967; Wilkins and McNaughton-Smith, 1964; Guttman et al., 1967; Stringer, 1967) while related techniques were developed as well (Principal Coordinate Analysis (Gower, 1966), Multidimensional Scaling (Shepard, 1962), Nonmetric Scaling (Kruskal, 1964), Latent Structure Analysis (Lazersfeld, 1950), Clique analysis (Harary and Ross, 1957)). Methods deriving from graph theory and pattern recognition were developed somewhat apart, leading to different jargon. Nevertheless, algorithms were stressed more than models (Bolshev, 1969) which led to critical sounds (e.g., Pritchard and Anderson, 1971 (p. 728): "Little attempt seems to have been made to assess the efficiency of the various available methods in any comparative sense ...", "It is generally accepted that each situation may require individually devised techniques ..."). In the re-edition (in 1973) of their very influential book of 1963 Sneath and Sokal took a more critical position than before, but Cormack (1971) probably was the first to give a very critical review of the field and he suggested to start comparative studies on cluster procedures and just stop formulating new ones. Moreover he paid attention to some methodological issues and conceptualized clusters as being "internally cohesive and externally isolated".

Indeed, in the seventies more attention was paid to consolidation of the field of cluster analysis rather than to the development of novel techniques. This was expressed in an increasing number of comparative studies (see section 4), the edition of reviews (see section 3), of computer packages (e.g., CLUSTAN, Wishart, 1969; MICKA, McRae, 1971), and of the formulation of a "probability theory of cluster analysis" (Rao, 1952; McQueen, 1967; Friedmann and Rubin, 1967; Bolshev, 1969; Day, 1969; Wolfe, 1970; Ling, 1972; Binder, 1978; Lee, 1979).

Possible future developments have been sketched by Blashfield and Aldenderfer (1978) and by Everitt (1980). One important line of research are comparative Monte Carlo studies, another line a formal statistical approach. Integration with maximum likelihood statistics is a recent development as is the integration of cluster analysis with research areas that have developed more or less apart, like pattern recognition and graph theory. It is our hope that this growing knowledge of the characteristics of cluster analysis methods will now and then be summarized in terms that are comprehensible to the interested (and not necessarily naive) user of cluster analysis.

3. SOME CLUSTER ANALYSIS METHODS

Good reviews of cluster analysis methods and implementations have been published in recent years (Sneath and Sokal, 1973; Anderberg, 1973; Everitt, 1974; Hartigan, 1975; Gnanadesikan, 1977; Späth, 1980) and the reader is referred to those reviews for more comprehensive information than will be given here. It is our purpose to mention only the most common procedures.

Cluster analysis methods can be devided into

- I Hierarchical methods
- II K-group methods
- III Rest group.

3.1. Hierarchical methods

"Humans can assimilate information from pictures better than from tables of numbers".

Dubes & Jain (1976).

Hierarchical methods have been very popular due perhaps to the easily visualized dendrogram they produce. This is a tree-like structure that summarizes the clustering process (see figure 1). Nevertheless, two problems are inherent to the nature of hierarchical methods. Considering agglomerative methods, in the first step one OTU is linked to one other, viz. the OTU to which it is most similar. This requires the specification of a similarity measure between individual OTU's. Numerous similarity (and dissimilarity) measures have been suggested (see, e.g., Cormack, 1971), from which a choice has to be made. Somewhere in the agglomerative process, however, OTU's will have to be linked to clusters, or clusters to other clusters, and this requires the additional specification of a similarity measure between clusters. Of course, for divisive methods the same holds true. Secondly, just as the best split of all N OTU's into two clusters, followed by the best split of one of the two resulting clusters, does not necessarily produce the best split of all N OTU's into three clusters, the best fusion of two OTU's followed by the best fusion of two of the resulting N-1 clusters does not necessarily lead to the best fusion of N OTU's into N-2 clusters. Hierarchical methods therefore maximize the clustering criterion in a suboptimal way. Nevertheless the classification found may be useful just as it stands, or it may serve as trial configuration for K-group methods.

Agglomerative procedures try to maximize intercluster distance by fusing at each step those two clusters that are least distant (= nearest). Possible definitions of the distance between two clusters are:

1. the minimum distance between one OTU from the first cluster and one OTU from the second: single link = nearest neighbour = minimum method = cutting minimal connected subgraphs;[1]
2. the maximum distance between one OTU from the first cluster and one OTU from the second: complete link = furthest neigbour = maximum method = cutting maximal connected subgraphs;[1]
3. the average of all $\binom{n_i+n_j}{2}$ distances between the n_i OTU's in the first cluster and the n_j OTU's in the second cluster: average link = group average = UPGMA;[1]
4. the distance between the cluster centroids: centroid method = UPGMC;[1]
5. the average of all $\binom{n_i+n_j}{2}$ distances between the n_i+n_j OTU's of both clusters after they have been fused: Ward's method = least squares methods = minimum variance method;[1]
6. as 4, but now distances are computed as if both clusters were of equal size: median method = WPGMC;[1]

The properties of these various methods are discussed in section 5.3.1.

Divisive methods have not been as popular as agglomerative methods. The reason for this might be sought in the heavy computational load they carry with them. To divide N OTU's over m clusters there are

$$\frac{1}{m!}\sum_{j=1}^{m} (-1)^{m-j} \binom{m}{j} j^N$$

different possibilities (Jensen, 1969; Anderberg, 1973), which for m=2 reduces to the familiar formula $2^{N-1}-1$, taking large values even for small values of N (e.g., exceeding 10^{10} for $N > 33$.) This approach was adopted by Edwards and Cavalli-Sforza (1965), while McNaughton Smith et al. (1964) consider only some of all possible splits. Both methods are polythetic, i.e. they use information on all p variables simultaneously. Monothetic methods use the information of one variable at a time, the variables being binary. Association analysis (Williams & Lambert, 1959) is the best known monothetic divisive method. Whether mono- or polythetic, divisive methods produce a dendrogram which reflects the successive splits.

Many of the hierarchical methods have been implemented in CLUSTAN (Wishart, 1969).

3.2. K-group methods

"There can be many different local minima
all of which seem to satisfy the client".

Ross.

As the number of possible splits of N OTU's into K clusters usually is far too large, K-group methods typically start from a trial configuration and then try to maximize the clustering criterion by successive relocation of OTU's. Of course, local optima[2] form a serious problem here and it is usually advisable to use several trial configurations in order to try to overcome the problem.

Before listing some K-group methods, the reader should be familiar with some basic concepts. If V denotes the matrix of variances and covariances of the p observations, then $N \times V \equiv T$ is the total SSCP matrix, i.e., the matrix that contains the Sums of Squares and Cross Products of the total sample of OTU's. We can write $T = W + B$ which means that the total sums of squares and cross products are due to differences between OTU's Within clusters as well as to differences Between the means of the clusters. By the trace of a matrix, e.g., tr(W), is meant the sum of its diagonal elements. The determinant of a matrix, e.g., det(W), can be interpreted as a kind of multivariate variance; consequently, the ratio det(B)/det(W) can be interpreted as a multivariate counterpart of the familiar F-ratio of between to within group variance. As there are several clusters, W itself is composed of the individual W-matrices of the clusters, denoted by W_g, so $W = \sum_{g=1}^{K} W_g$, where K is the number of clusters. Although by relocation of OTU's the clustering may change, the total SSCP matrix T obviously cannot change; consequently, det(T) is a constant, given the data. Finally, T^{-1} denotes the inverse of T.

We can now list some of the well-known clustering criteria:

1a - minimize the sum of all squared Euclidean distances
1b - minimize tr(W)
2 - minimize $tr(WT^{-1})$
3 - minimize det(W); this is of course equivalent to maximizing Wilks Λ, where Λ = det(T)/det(W), since det(T) is a constant. Variants of this criterion are minimizing $\sum_{g=1}^{K} n_g \ln \det(W_g)$ (Scott & Symons, 1971), where n_g is the number of OTU's in cluster g, and minimizing $K^2 \det(W)$ (Marriott, 1971)
4 - maximize trace $(W^{-1}B)$(or equivalently $tr(BW^{-1})$), where B is the between groups SSCP matrix
5 - minimize the information statistic (Wallace & Boulton, 1968)

Criterion 1a is identical to criterion 1b. It is also identical to Ward's minimum variance criterion, and its use for K-group methods recently has been advocated strongly by Späth (1980). Criterion 2 was mentioned by Korhonen (1977). When variables have been transformed to principal components and subsequently standardized to unit variance, it is identical to criterion 1. Criteria 1b, 3 and 4 were suggested by Friedman and Rubin (1967) and apply to the multivariate normal mixture model (Rao, 1952). They have been implemented in the MICKA program by McRae (1974). The fifth criterion was developed from an information theory background and has been implemented in the program SNOB. It should be noted that methods defined by criteria 2-4 take correlations between variables into account, whereas methods defined by criteria 1 and 5 do not.

In recent literature K-group methods enjoy increasing interest and the further investigation of their statistical properties will probably enhance their application to real data by researchers who until now have relied upon the classical hierarchical methods. Computer capacity and computer time may remain limiting factors to this development, however, since K-group methods are rather expensive as to the latter aspects.

It should be noted that K-group methods do not require the specification by the investigator of a distance measure: the distance measure is implied by the clustering criterion.

3.3. Rest group

In this section we mention only two methods, viz. mixture estimation (Wolfe, 1970) and mode analysis (Wishart, 1969). Of these, mode analysis does not maximize a clustering criterion. We have included this method here since it tries to combine the best of hierarchical and K-group methods, respectively, and because the method has gained some popularity. Mode analysis has been implemented in CLUSTAN, mixture estimation in the NORMAP/ NORMIX programs.

Mode analysis defines dense OTU's in the p-dimensional

observation space, and assigns other OTU's to these dense OTU's if at least they are within some specified distance of the dense OTU. Then the distance value is increased and the process repeated until most (or all) OTU's have been assigned to dense OTU's. Obviously, a distance measure has to be specified.

The mixture estimation method tries to maximize the likelihood of the distribution of the OTU's over K clusters (K is predefined), where this distribution is assumed to be a multivariate normal mixture. The parameters of this mixture, conditional to which the likelihood is maximized, are estimated from each cluster in turn. As a consequence, estimated mean vectors and covariance matrices are a result of this method. A distance measure need not be specified, but implicitly Mahalanobis distance is used. A disadvantage of the NORMIX procedure might be that N, the number of OTU's should fairly exceed $\frac{1}{2}p(p+3)$ (Boshev, 1969).

4. ORDINATION AND SELECTION

In the preceding section we have described some of the more common cluster analysis procedures. It now comes to the problems of which observations should be considered more important than others and which observations should be used. The former problem is usually referred to as ordination, the latter as selection. Both problems are, in our opinion, largely unresolvable, except for some special cases, and therefore one must make a choice out of several possibilities that have been suggested in the literature (e.g., Jain and Dubes, 1978; Becker, 1978).

Suppose we have two independent samples from some mixture of populations. If in one of these samples it is known which OTU stems from which population, we could use this sample to study which variables discriminate best between the populations. To do this in a proper way is an art in itself but for us discriminant analysis rather than cluster analysis is the method of choice in this situation.

If we do not possess prior information on group membership, it might seem a sound step to eliminate highly correlated observations. If, however, group differences are in one dimension, and observation A discriminates between clusters 1 and 4, but not between the intermediate 2 and 3, while observation B discriminates well between all four clusters, A and B probably will be highly correlated, but elimination of B would impair the results of the cluster analysis.

So in most cases where cluster analysis is considered, experience probably will be the base for the choice of the observations. In our view, observations that are logically or functionally related (Jardine and Sibson, 1971) can be eliminated.

The problem of selection is interwoven with the problem of ordination when the computation of Principal Components (PC) is considered. PC do appeal probably to some extent because many observations per OTU can be used (i.e., no selection problem as to raw variables), the procedure is strictly mathematical so responsibility is referred to matrix algebra and the computer, and notwithstanding this the first r PC can be chosen (ordination and selection) according to well-known and widely accepted criteria.

In cluster analysis PC have been computed or advocated by many authors (e.g., Lambert and Williams, 1962; Everitt et al., 1971; Paykel, 1971; Green et al., 1967; Bolshev, 1969; Spielman, 1973; Jain and Dubes, 1978) and the computation of PC prior to cluster analysis is more or less accepted as a first step (cf. CLUSTAN). Since the first PC has largest variance it will play an important role in cluster analysis if the distance measures used are scale dependent. This is sometimes considered an advantage if PC have been computed from a correlation matrix (Corruccini, 1975) while others (e.g., Paykel, 1971) first standardize PC scores to unit variance. Still others (e.g., Jolicoeur & Mosimann, 1960; Sneath and Sokal, 1973) have argued that the first PC represents mainly size differences and therefore should be discarded in studies of shape differences.

Several points should be considered before the decision is made to compute PC. Firstly if $p > N-1$ the covariance (or correlation) matrix will have a zero determinant which goes on to assure in that N should fairly exceed p. Otherwise as the rotation matrix is based on the sample to be cluster analyzed, chance effects may play an important role in the results if not all PC are selected. Dubes and Jain (1979) demand that $N/p > 5$ while Everitt (1975) feels that $N/p > 5$ is too low a requirement and suggests that the minimum value for this ratio has been lowered with the advent of computers. Nunnally (1967) advocates $N/p > 50$. Secondly, one usually only has the total covariance (or correlation) matrix to compute PC from as nothing more is known. Nevertheless it is by no means certain that the main axes of between variance (correlation) will coincide with the main axes of within covariance (correlation)(Cormack, 1971). Indeed, Uytterschaut and Wilmink (1982) did not observe any monotonic relation between eigenvalues and F-ratios of PC. As a consequence, by selection of the first few PC part of the between variance in the data may be lost. Van Vark and Van der Sman (1982) found similar results. Selection of all PC (e.g., Spielman, 1973), however, is counter to the very purpose of computing PC, viz. data reduction. Finally, weighted sums of multimodal distributions will tend to a unimodal distribution, so PC may help to obscure the picture (Garside & Roth, 1978).

Our position therefore will be to choose $p < N/20$ observations, say, which are judged or known to be relevant to the problem and to use all of these observations for cluster analysis. When estimates of the within standard deviation are available we would divide scores by these standard deviations (cf. section 5.1). From the above considerations, we would be inclined not to compute PC, unless p is large and N even much larger - such a situation, however, will not often occur in practice.

We would like to address one further point. The suggestion by Sneath and Sokal (1973) of adding new observations until the cluster structure is stable is, though interesting, unfeasible since it assumes that the number of attributes measuring the same thing is unlimited, or at least that this stability has occurred before determinants approach zero or we are running out of attributes. The procedure seems not to have been employed by themselves nor by any others.

5. TRANSFORMATION OF DATA AND CHOICE OF SIMILARITY MEASURE AND CLUSTER METHOD

Many papers of the literature have been devoted to the issues to be considered in this section. We can therefore only try to highlight the essentials.

To start with almost a triviality, there is no need to bother about which of two similarity measures to adopt when both yield the same cluster configuration. A table which depicts the consequences of data transformation, choice of similarity measure and cluster method for the cluster configuration would prove very useful. To mention only one example: if single linkage is the cluster method, all transformations and all similarity measures that produce only monotonic changes in the similarity matrix of OTU's will yield identical clusters.

As a second point, one should try to avoid incompatibilities like running Ward's method with any but the Euclidean distance or average linkage with a similarity measure of which the average is meaningless.

Thirdly we have not discussed the problem of weighing rare observations. This is a controversial issue (Cormack, 1971; Sneath and Sokal, 1973; Bolshev, 1969; Jardine and Sibson, 1971) arising in the case of categorical data. Probably the issue cannot be resolved by statistical argument alone. Since in physical anthropological practice usually only continuous data are being used, the issue is not further addressed here.

5.1. Transformation of data

As to transformation there are two main options: transformation by OTU or across OTU's. Transformation by OTU is usually standardization to unit variance and/or zero mean. Transformation to zero mean (per OTU) reduces the dimensionality of the measurement space from p to p-1 as is easy to show. According to Sneath and Sokal (1973) it is the first principal component that disappears, which may play a role in matters of size and shape. Also, average correlation between observations is reduced to about-zero level.

Transformation across OTU's has been treated by Sneath and Sokal (1973). Since they feel that a priori weighing of variables should be avoided they advocate transformations that make measurement scales about equal, like $(X - X_{min})/(X_{max} - X_{min})$ (Gower, 1971), standard scores or taking logarithms (which in some cases is similar to standardization (Burr, 1968) and seems to make matrices based on these scores "extremely congruent" (Schnell, 1970)). However, a problem with these transformations is that they remove not only scale differences but also intrinsic differences in variability (Cormack, 1971; Sneath and Sokal, 1973). Moreover, division by the total standard deviation tends to produce hyperspheres (Dubes & Jain, 1976) (or "hyperfootballs" (Needham, 1965)) which in turn may lead to the discovery of spurious hyperspheric clusters by methods that tend to find such clusters (e.g., Ward's method). Edelbrock (1979) in an experimental study found only slightly better results with standard scores over a variety of methods, while Milligan (1980) in a broad experimental study found slightly better results with raw scores. In conclusion, then, we would prefer to divide each observation by an estimate of the within group standard deviation; if such an estimate is not available, we would use the data just as they stand.

5.2. Similarity measures

As before we use the term similarity in a generic sense, i.e., covering both similarity and dissimilarity.

Many similarity measures have been proposed. Listings can be found e.g. in Sneath and Sokal (1973), Cormack (1971), the CLUSTAN manual (Wishart, 1969), etc. Of all measures the Euclidean distance and the correlation coefficient probably have been used most frequently. Formal properties of distance measures (with respect to, among others, level of measurement (van Meurs, 1978), triangle inequality, symmetry, etc. (Bock, 1975)) have been investigated but most applications do not really bother about them. Nevertheless, incompatibilities should again be avoided, like computing the Euclidean distance on ordinal or nominal data (van Meurs, 1978).

The choice of a distance measure between OTU's cannot be made on formal grounds alone. It also depends on the research question and on properties of the data. Two well-known objections to the Euclidean distance (Hall, 1969) are that it also measures scale differences and that it does not take correlation between observations into account. Division by the within-group standard deviation, or simply selection of observations using the same scale, will help to overcome the first of these objections, while in most applications correlations are positive but low (most values not exceeding .30) which means that axes are at angles between 70° en 90°, a distortion which is not likely to affect the final cluster configuration very much. Also, standardization by OTU will have the effect of making the average correlation between observationsabout zero. As an alternative, however, the generalized distance (Mahalanobis, 1936) might be computed but here other problems arise, e.g., the possibility of near-zero determinants

(Cormack, 1971) and the availability of Σ_{total} only, where Σ_W should be used. Moreover, it should have to be assumed that all clusters are of equal shape (equal covariance matrices). In spite of the above mentioned objections Wishart (1971) and Anderberg (1973) have suggested the use of the Euclidean distance, and Späth (1980) considers the sum of all squared Euclidean distances as a good criterion for choosing a final configuration (see section 6). Clearly, there is no agreement on the use of the Euclidean distance.

Similarly, use of the Q-correlation coefficient has been advocated e.g., by Cormack (1971) when comparable variables have been used, by Edelbrock (1979) and Strauss et al. (1973) and by Corruccini (1973)if shape is of interest, while others (e.g., Jardine and Sibson, 1971) reject this measure.

It will be clear, then, that the choice of a distance measure between OTU's will usually depend in part on subjectivity on behalf of the investigator. Therefore it may be wise to run several analyses with different distance measures and to check for correspondence in the resulting classifications. Validity problems arise here, and the reader is referred to section 6 for a discussion.

Sneath and Sokal (1973) advise to use as simple a measure as possible. This seems to undervalue the exercise of investigators who have constructed distances that measure special features of OTU's such as shape alone, or size alone. However, transformation of the raw scores and use of a simple similarity measure can usually achieve the same goals (e.g., elimination of size). We feel that the latter approach illuminates the essential of what is being measured much better than the use of a complicated similarity measure does.

5.3. Comparison of methods

Cluster methods can be compared in several ways. One may investigate whether a method recovers the clusters that are assumed or known in the data. Of course, some statistic expressing the degree of recovery has to be defined. Also, one must decide whether the evaluation will be at the 100% coverage level (i.e., all OTU's have to be assigned to a cluster) or that incomplete coverage is allowed. Results of such empirical studies are presented in tables 1 and 2. Another possibility is to investigate formal properties of cluster methods. In such studies, formal criteria of admissibility are formulated and methods are judged in the light of these criteria. The last possibility we mention here is to investigate statistical properties of cluster methods, e.g. what type of cluster a method, by its very nature, is likely to produce.

Watson, Williams and Lance (1966) were probably among the first to compare in a systematic empirical study two methods of cluster analysis, single linkage and centroid analysis, and five distance measures. Many comparative studies have been performed since then. In many of these studies some a priori classification or the experimenters judgment about the validity of the clusters found served as criterion for the feasibility of the cluster method(s) studied. Of course the danger here is that cluster methods producing classifications in deviation to traditional ones will be judged of no value without questioning the validity of the traditional classification (Cormack, 1971; Hirschfeld et.al., 1973). An exception is the application of information theory to cluster analysis (Wallace & Boulton, 1968) where the validity of the classification found is expressed in a statistic which is independent of the cluster method and distributed as a chi-square.

Friedman and Rubin (1967) were probably the first to use artificial data generated as random samples from known probability distributions. This line of investigation, known as Monte Carlo studies, has gained increasing interest and especially in recent years comprehensive studies have been published. One of the first of these was performed by Blashfield (1976) who studied four methods with six multivariate normal mixtures from each of which 50 samples were drawn. The criterion for judging the performance of cluster methods in this type of study usually is the percentage of correct re-classifications, or a variant thereof such as Rand's statistic (1971) or Cohen's kappa (1960). Recently, Milligan (1981) has given an excellent review of Monte Carlo studies in cluster analysis.

Formal properties of cluster methods were investigated in studies by Jardine and Sibson (1971), Fisher and van Ness (1971) and van Ness (1973). Dubes and Jain (1975), following a suggestion by Anderberg (1973), cluster analyzed cluster methods using formal properties of these methods as data.

The work of Binder (1978) is a good example of a study investigating statistical properties of cluster methods. He showed, among others, that the trace (W) criterion approximately maximizes some posterior probabilities when the prior distribution of cluster size is strongly peaked around equal cluster sizes.

From the many studies that have been reported in any of the above three categories two conclusions can be drawn:

1. there is no single outstanding cluster method; instead, the success of a cluster method depends on the type of data,
2. if the data are well-structured almost any cluster method will recover this structure. If, on the contrary, there is no structure in the data, one can almost go as far as to say that <u>any</u> configuration can be recovered by some cluster method. E.g., Milligan and Isaac (1980) arrived at three clusters, perfectly separated with discriminant analysis, from data that were drawn from a uniform distribution. Other examples are given by Everitt (1974). Cormack (1971) already noted that "the ability of procedures to find non-existent clusters is well-established" (p. 345).

We shall now present the results of some comparative studies for hierarchical methods, K-group methods and the rest group (cf.

section 3) and for each of the three types of comparison mentioned above (viz. experimental, formal and statistical). Interestingly, there are few studies comparing hierarchical and K-group methods (see section 5.3.4). This may be due to the rather different nature of both groups of methods.

Table 1. Empirical studies of hierarchical cluster methods

First author	Methods	Data	Conclusions
Watson (1966)	1 4	bin, plants	Both good
Hodson (1966)	1 2	m.s., archeology	AL > SL
True (1970)	2	bin, archeology	Good
Pritchard (1971)	1 2 3 4 5	bin, ecology	AL, CL and Ward > CEN and SL
Cunningham (1972)	1 2 3 4 5 6	cont, artif.	Ward best, but bad if data chain, AL > CL, CEN and M; SL worst.
Kuyper (1975)	1 2 3 4 5 6	cont, MC	Ward > CL > AL > M > CEN > SL
Dubes (1976)	1 3	m.s., handwritings	CL > SL
Blashfield (1976)	1 2 3 5	cont, MC	Ward > CL > AL > SL
Mojena (1971)	1 2 3 4 5 6	cont, MC	Ward > CL, AL > M > CEN > SL
Edelbrock (1979)	1 2 3 4 5	cont, MC	Ward > AL > CL, CEN > SL at 60% coverage (Euclidean distance) or 80% (r) all methods almost perfect. From there on: using Euclidean distance Ward remains good, SL, AL and CEN get bad, CL in between. Using r, AL remains best.
Milligan (1980)	1 2 3 5	cont, MC	AL > CL > Ward > SL. Critique to previous studies that these have adopted orthogonal variables. In this study data satisfy the ultra-metric inequality.
Milligan (1980)	1 2 3 4 5 6	cont, MC	AL > Ward, CL > CEN, M, SL.

Abbreviations - Methods: 1 = SL = single linkage; 2 = AL = average linkage (group average) 3 = CL = complete linkage; 4 = CEN = centroid analysis; 5 = Ward's method; 6 = M = median method. Data: bin = binary; m.s. = multistate; cont = continuous; artif. = artificial; MC = Monte Carlo; Conclusions: r = correlation coefficient; > = shows a better recovery state, to the present authors judgment, over a set of conditions.

5.3.1. Hierarchical methods

From table 1 it can be concluded that single linkage is not a good method. Everitt (1974) considers it to be obsolete. However, single linkage tends to detect outliers and may be used for that purpose: outliers should be removed from the data set since they disturb the cluster process (Kanal, 1972; cf. Milligan, 1980).

As to the other methods the reader should be aware that in table 1 (and table 2) the "conclusions" represent our judgment of overall recovery rate of these methods under various conditions, e.g., error perturbation of data, percentage of outliers present, dimensionality of attribute space, etc., with various distance measures and various underlying distributions. Nevertheless, from a variety of situations Ward's method and average linkage emerge as "best" methods. Ward's method maximizes an objective criterion, trace(W), given the number of clusters, but does so in an suboptimal way due to its hierarchical nature. On the other hand it tends to produce small, spheric clusters (Everitt, 1974), and it can only be used in combination with the Euclidean distance. From a practical point of view, average linkage might be preferred to the other methods since it usually performs well and it maximizes the cophenetic correlation coefficient (Farris, 1969). As a special feature, it tends to gather "nonconformists" into one cluster (Williams et al., 1971). Also, it appears to be rather stable under error perturbation of the data (Cunningham & Ogilvie, 1972; Milligan, 1980). Edelbrock (1979) notes that Ward's method (using Euclidean distance) retains values of Cohen's kappa above .7 at all levels of coverage, while other methods (single linkage, average linkage and centroid analysis) fall rapidly from kappa being about unity to about zero levels at a coverage level exceeding 60%. Using the correlation coefficient, however, average linkage performed best at coverage levels over 80%. Edelbrock therefore suggests that clusters that emerge both from average linkage using the correlation coefficient and from Ward's method (using Euclidean distance) can probably be considered robust.

From an axiomatic point of view Jardine & Sibson (1971) rejected all hierarchical methods but single linkage. Dubes and Jain (1976) used three groups of admissibility criteria and concluded that single linkage satisfied seven out of eight criteria, complete linkage six out of eight, while other methods (including K-group methods) satisfied less than six criteria. Similar scores are obtained when using criteria by Fisher and van Ness (1971) and van Ness (1973).

From a statistical point of view Degens and Federkiel (1978) showed that the distribution of the split point (when dividing a sample into two clusters) is asymptotically independent of the underlying univariate distribution for complete linkage and the median method. The authors conclude that these methods are inconsistent[3] in this respect. Single linkage shows a better asymptotic behaviour and Hartigan (1978) showed that Ward's method is asymptotically consistent.

5.3.2. K-group methods

From table 2 it can be seen that of the criteria tr(W), $tr(WT^{-1})$, Λ and $tr(W^{-1}B)$, Wilks Λ in general produced good results, while tr(W), by its tendency to produce hyperspheric clusters, sometimes forced this structure onto the data (Everitt, 1974). There is some evidence that criteria that allow for unequal cluster form, i.e., unequal covariance matrices, perform better if clusters are indeed differently shaped, while in the latter case criteria using the pooled within groups SSCP matrix show inferior results. If group differences are essentially in one dimension $tr(W^{-1}B)$ is the criterion of choice. It should be noted that the above mentioned good results are obtained using good trial configurations; otherwise K-group methods may suffer from local minima. The SNOB program seems to have enjoyed only temporary interest.

From a formal point of view, $tr(WT^{-1})$, Λ and $tr(W^{-1}B)$ have desirable invariance properties under linear transformations. They also take correlation between variables into account; this tr(W) (and hierarchical methods) do not and for this reason the latter methods are rejected by McRae (1973).

From a statistical viewpoint, minimizing tr(W) has been shown to lead to the production of hyperspheric clusters (Dubes and Jain, 1976; Korhonen, 1978), while $tr(WT^{-1})$ produces diffuse cluster structures and $tr(BW^{-1})$ concentrated ones (Korhonen, 1978). Binder (1978) showed that tr(W) and Λ will maximize certain posterior probabilities when priors are chosen that are highly peaked around equal cluster sizes. As a consequence these methods will tend to produce clusters of about equal size. As noted before, all methods suffer from local minima, except the SNOB program, which determines the number of clusters objectively and allows for significance tests on the clusterings.

5.3.3. Rest group

On mode analysis we have not found any comparative information. Everitt (1974) considers the method to be potentially very useful and reports that the two groups of data he constructed were essentially recovered.

The NORMAP/NORMIX programs have been used more often. Everitt et al. (1971) found results comparable to the tr(W) method (table 2). Everitt (1974) noted some convergence problems, but the NORMIX version (unequal covariances) recovered all structures. This was apparently not the case in Mezzich's (1978) study, with a mean value for the Rand statistic of 44.32. Symons (1981), however, found good results. The obvious problem with NORMIX is that far more than $\frac{1}{2}p(p+3)$ OTU's should be used in order to give the likelihood function some stability.

Table 2. Empirical studies of K-group methods

First author	Methods	Data	Conclusions
Friedman (1967)	1 3 4	cont, MC, bee data	tr(W) only with Euclidean distance if group differences are in one dimension, $tr(W^{-1}B)$ is best, otherwise $\Lambda > tr(W^{-1}B) > tr(W)$
Wallace (1968)	5	cont, seals	good
Pilowsky (1969)	5	m.s., psychiatry	in contrast to Pilowsky's conclusion: not very good
Everitt (1971)		m.s., psychiatry	some clear clusters, majority in vague clusters
Paykel (1971)	3	m.s., psychiatry	interpretable results
Everitt (1974)	1 3	cont, MC	$\Lambda > tr(W)$; the Scott and Symons version was better than Λ with unequal covariance-matrices.
Dubes (1976)	1	m.s., handwritings	good
Mezzich (1978)	1 3	cont, MC	about equally good
Korhonen (1978)	2 3 4	cont, MC	Λ good, $tr(W^{-1}B)$ slightly better if group differences are in one dimension, $tr(WT^{-1})$ slightly better with diffuse data sets
Milligan (1980)	1	cont, MC	good with good starting configuration, bad with random seeds
Symons (1981)	3	cont, diabetes	Λ and several variants were compared; the variants allowing for unequal covariance matrices did better than the variants using the pooled within matrices.

Abbreviations - Methods : 1 = tr(W); 2 = $tr(W^{-1})$; 3 = $\Lambda = |T|/|W|$, 4 = $tr(W^{-1}B)$; 5 = SNOB. Data: see table 1. Conclusions: see table 1. It should be noted that there exist many different programs that try to minimize tr(W). These programs may give different results on the same data. We have gathered these methods under the single heading tr(W). The same is done for the other criteria.

5.3.4. Hierarchical methods, K-group methods and restgroup compared

Almost no studies have been performed that compare hierarchical methods with K-group methods, mode analysis or NORMIX. Everitt (1974) used different data sets (though drawn from the same population) to compare some hierarchical and K-group methods and found that with spheric clusters all methods did well, with elliptical data only Λ (and single linkage!) did well while with differently shaped clusters Λ (and especially the Scott and Symons variant) was good. NORMIX was also applied to the data and did well. Dubes and Jain (1976) compared tr(W) methods with hierarchical ones and found the tr(W) methods to be superior. Mezzich (1978) compared tr(W) and Λ with complete and single linkage and found the K-group methods to be slightly superior to complete linkage and much better than single linkage. NORMAP did worst. Milligan (1980) compared several tr(W) programs with five hierarchical methods (table 1). The tr(W) programs gave better recovery rates than the hierarchical ones, but with random starts they performed worse. Symons (1981) found the NORMIX program to be as good as the rest of the K-group methods he studied (table 2).

In conclusion, then, it can be said that K-group methods seem to be superior to hierarchical methods if good start configurations are used. Especially K-group methods that allow for unequal cluster shape, such as the variant of Wilks Λ proposed by Scott and Symons, are promising. The latter procedure also seems to have less tendency to form clusters of about equal size, like some K-group methods. However, average linkage (using the correlation coefficient as similarity measure) when clusters do not overlap or Ward's method when clusters overlap may attain similar recovery rates, especially if one does not try to achieve 100% coverage. For instance, one might perform discriminant analysis (using new observations on the same OTU's) when about 70% coverage has been achieved in order to assign the remaining OTU's to the clusters already obtained. By their different nature hierarchical methods can serve as a check on clusterings obtained by K-group methods. Or, they may supply start configurations for K-group methods. Preliminary to any cluster analysis, we recommend to run single linkage in order to detect outliers. Using the configuration produced by some K-group method as start configuration for NORMIX, maximum likelihood estimates of normal parameters can be obtained.

6. VALIDITY OF CLUSTERS

In most practical situations the investigator, on using some cluster analysis method, will be confronted with the following four validity questions:

1. is there a structure underlying the data or have the data been drawn from one single population;
2. is the cluster method, in connection with the distance measure chosen, apt to recover the supposed structure;
3. what is the "true" number of clusters in the data set;
4. do the obtained clusters reflect the underlying structure.

6.1. Is there a structure underlying the data?

In experimental studies the underlying distribution is known, of course. Otherwise we have the problem that a clear cluster structure can arise while, e.g., the data have been drawn from a uniform distribution (section 5.3). Dubes and Jain (1979) in their excellent review of validity research in cluster analysis mention six methods to test for randomness of the data. As stated before, the investigator should do well to explicitely formulate his prior ideas about the structure possibly present in the data (Everitt, 1975; Blashfield, 1976).

6.2. Are distance measure and cluster method suitable for the data?

From section 5 it should be possible to decide on one or more methods. According to Milligan (1980) this choice and the selection of variables are more important than the choices of distance measure and transformation of data. Nevertheless, the method chosen should, if possible, be tried on similar data of known structure, but even good recovery here does not guarantee that the method will be good for the present data (Cormack, 1971; Hirschfeld et.al., 1973).

6.3. What is the "true" number of clusters?

In the information theory approach by Wallace and Boulton (1968) the information due to the number of clusters K is minimized along with other sources of information. Another approach is to compute estimates of K (Frank, 1978), but this approach seems unfeasible for large N. One might also split the dendrogram at the level where the cophenetic correlation coefficient is at maximum (where this maximum should exceed .80 (Dubes and Jain, 1979)). Since of all hierarchical methods average linkage maximizes the cophenetic correlation coefficient (Farris, 1969), the latter criterion would lead naturally to average linkage as the cluster method. Marriot (1971) has proposed to take K to be the value of g for which $g^2 \det(W)$ is minimal. Everitt (1974) showed that rather large cluster distances are needed for this method to decide on the correct number of clusters. An adjusted likelihood ratio test has been suggested by Wolfe (1971). Other tests have been reviewed by Dubes and Jain (1979). It is probably a good thing to have an independent investigator judge the classification found. - In spite of all these possibilities the issue is in fact still unresolved. We feel that incorporation of the number of clusters into cluster criteria would prove very useful; but this is a complicated matter.

6.4. Do the obtained clusters reflect the underlying distribution?

Again, only in experimental research is it possible to assess the degree to which a structure has been recovered. In practice, the structure has to be inferred from the data. Van Meurs et al. (1978) list some criteria for evaluating a classification. Dubes and Jain mention the work of Ling (1972, 1973) on tests for the validity of an individual cluster. Most common, however, has been the requirement that if there are several methods suitable for the data, these methods should yield roughly the same clusters. This introduces the problem of how much alike two classifications are. The Rand statistic is quite generally accepted to help in this evaluation. Nevertheless, stable clusters need not be valid clusters (Blashfield, 1976; Garside and Roth, 1978). Everitt (1975) suggests to check whether the clusters found are able to predict a new criterion observation. This approach has been adopted successfully by Paykel (1972), who demonstrated a difference in response to antidepressive treatment between clusters formed by cluster analysis of data on depression in psychiatric patients. Another possibility is to cross-validate the classification found by splitting the sample into two halves (Blashfield, 1976).

None of the above procedures, however, can guarantee the validity of the clusters produced by a cluster analysis. Nevertheless, if methods are known to be suitable for the data and if these (different) methods yield essentially the same classification that is in line with prior beliefs, then it seems reasonable to conclude that this classification is likely to reflect some structure underlying the data.

NOTES

1. For more synonyms see Blasfield and Aldenderfer (1978); UPGMA = Unweighted Pair Group Method using Arithmetic Average; UPGMC (WPGMC) = Unweighted (Weighted) Method using Centroid distance.
2. Procedures that seek a maximum may sometimes find so called local maxima (optima). By this is meant that minor changes in the clustering will decrease the criterion value, but that a major change might still increase the criterion value. Typically, by their iterative nature most maximum-seeking procedures will not make such major changes and the resulting clustering will be suboptimal. In contrast, a global maximum is reached if with the present data no other clustering results in a higher criterion value. Of course, a global maximum also is a local maximum.
3. If the distribution of the split point is asymptotically concentrated in one point, which is the split point of the real distribution of the population, then the clustering algorithm is said to be consistent.

PART II
APPLICATION OF CLUSTER ANALYSIS IN PHYSICAL ANTHROPOLOGY

1. SOME HISTORICAL APPLICATIONS OF CLUSTER ANALYSIS IN PHYSICAL ANTHROPOLOGY

In physical anthropology, clustering methods have not been used abundantly. Among the first to use cluster analysis was Boyce in 1964. Since then, the number of studies applying some form of cluster analysis has gradually increased. Some of these have been summarized in the appendix. From this table can be seen that many authors use different names for the same methods and sometimes badly specify which transformation, distance measure and cluster procedure has been used.

It cannot be stated that cluster analysis yet has become one of the standard tools of the physical anthropologist, but one can agree with Howells (1978) that multivariate distances and clustering methods have recently done much to simplify apparent relations among human populations, although forms of analysis are far from standard.

2. APPLICATION AND COMPARISON OF DIFFERENT "TRANSFORMATION - DISTANCE MEASURE - CLUSTER METHOD" - COMBINATIONS IN THE CLUSTER ANALYSIS OF CRANIOMETRIC DATA

Considering the conclusions of the literature study described in part I (section 5), we made a study of different "transformation - distance measure - cluster method" (TDC) - combinations in the cluster analysis of craniometric data.

We made use of two data sets; both data sets were measured and recorded by W.W. Howells. The first data set consisted of the Norse- (n=55), Teita- (n=34) and Hainan- (n=45) skulls, all male individuals. The second data set consisted of three Asiatic and Oceanic populations: the Mokapu (n=43), South Japan (n=42) and the Moriori (n=43), again only males. For the first data set the sex diagnosis was based on the visual sexing of Howells, which is described in his book "Cranial variation in man" (1973). For the second data set the sex diagnosis was based on discriminant analysis and the doubtful individuals were eliminated.

We used the cluster analysis program CLUSTAN IC, written by David Wishart (1978).

As mentioned in section 1.3, cluster analysis methods can be divided into hierarchical methods, K-group methods and a rest group. We first applied hierarchical methods to our data. Ward's method and average linkage (AL) were chosen because from the literature survey (see section 1.5.3.1) they emerged as "best" hierarchical methods. The results were, however, not satisfactory. K-group methods and also mode analysis, a method belonging to the rest group, gave better results. The superiority of K-group methods, if good start configurations are used, to hierarchical methods was already mentioned in section 1.5.3.4.

For each data set the comparison of K-group method options was done twice, i.e. once for 8 measurements and once for 46 measurements (see table 4). The comparison of mode analysis options was only based on 46 measurements (table 6). The 8 measurements were selected from the literature for their high discriminatory value. The 46 measurements were selected as follows. From the 70 variables measured by Howells we have excluded 13 angle-measures nrs. 58-70; six measures, nrs. 40, 41, 43, 44, 46, 47, for which in Howell's opinion the measurement accuracy is rather low; and five measures for which scores were missing frequently in our data, nrs. 12, 13, 19, 51, 57. For the enumeration of the measures belonging to the two variable combinations: see table 3.

Our decision as to whether an option gives good or less good results will be mainly determined by the number of correct classifications. Moreover, the number of clusters obtained and the enclosure ratio were taken into account. The enclosure ratio is a number between 0 and 1, indicating which part of the objects is already classified in the nuclei-classifications.

2.1. K-group methods, application of tr(W) (table 4)

The K-group method we have chosen uses tr(W) as the clustering criterion. We have chosen tr(W) for the following two reasons. First, because it is the only K-group method included in CLUSTAN IC. Second, because from table 2 it can be seen that tr(W) generally produces good results.

The choice and results of the different TDC-combinations we have used are summarized in table 4.

Table 3. Variable combinations used in the comparison of K-group method options (table 4) and in the comparison of mode analysis options (table 5)

n = 8 : (1,5,6,8,14,18,30,31)
n = 46: (1-11, 14-18, 20-39, 42, 45, 48-50, 52-56)

1	glabello-occipital length	20	mastoid length	36	supraorbital projection
2	nasion-occipitale length	21	mastoid width	37	glabella projection
3	basion-nasion length	22	bimaxillary chord	38	foramen magnum length
4	basion-bregma height	23	zygomaxillary subtense	39	nasion-bregma chord
5	maximum cranial breadth	24	bifrontal chord	42	bregma-lambda chord
6	maximum frontal breadth	25	nasio-frontal subtense	45	lambda-opisthion chord
7	bistephanic breadth	26	biorbital breadth	48	lambda-opisthion subtense
8	bizygomatic breadth	27	dacryon subtense	49	nasion radius
9	biauricular breadth	28	interorbital breadth	50	subspinale radius
10	minimum cranial breadth	29	naso-dacryal subtense	52	dacryon radius
11	biasterionic breadth	30	simotic chord	53	zygoorbitale radius
14	nasal height	31	simotic subtense	54	frontomalare radius
15	orbit height left	32	malar length inferior	55	ectoconchion radius
16	orbit breadth left	33	malar length maximum	56	zygomaxillare radius
17	bijugal breadth	34	malar subtense		
18	nasal breadth	35	cheek heigth		

Table 4. Comparison of different options of the K-group method tr(W) applied to two sets of skull data. First to the Norse-Teita-Hainan-skulls, secondly to the skulls of three Asiatic and Oceanic populations

	n var	transf.	dist.	starting configuration	number of clusters	correct classifications
first data set (n=134)	8	RS	Eucl.	AL	3	72
	8	B	Dot Pr.	AL	5	84
	8	C	Eucl.	AL	4	99
	8	A	Eucl.	AL	3	105
	8	A	Eucl.	Ward	3	106
	46	RS	Eucl.	AL	3	103
	46	B	Dot Pr.	AL	3	88
	46	C	Eucl.	AL	6	122
	46	A	Eucl.	AL	7	108
	46	A	Eucl.	Ward	5	92
second data set (n=128)	8	RS	Eucl.	AL	5	54
	8	B	Dot Pr.	AL	3	71
	8	C	Eucl.	AL	4	63
	8	A	Eucl.	AL	4	75
	8	A	Eucl.	Ward	3	76
	46	RS	Eucl.	AL	4	72
	46	B	Dot Pr.	AL	4	82
	46	C	Eucl.	AL	9	106
	46	A	Eucl.	AL	7	97
	46	A	Eucl.	Ward	3	101

As (dis)similarity measures we used the Euclidean distance and the dot product. The Euclidean distance is defined by

$$d(x,y) = \sqrt{\sum_{i=1}^{n} (x_i - y_i)^2}$$

where n = number of variables
x,y = objects
x_i, y_i = scores of the objects on the variables

The Euclidean distance d(x,y) can also be written as

$$d(x,y) = \sqrt{\sum_{i=1}^{n} x_i^2 + \sum_{i=1}^{n} y_i^2 - 2 \sum_{i=1}^{n} x_i y_i}.$$

The quantities $\sqrt{\sum_{i=1}^{n} x_i^2}$ and $\sqrt{\sum_{i=1}^{n} y_i^2}$ are the Euclidean lengths of the vectors of scores of objects x and y, respectively. For many sets of variables, this length can be interpreted more or less as the "size" of the object (Mosimann, 1970). In such cases the dot product, defined by

$p(x-y) = \frac{1}{n} \sum_{i=1}^{n} x_i \cdot y_i$, is a similarity measure which is less sensitive to size differences than d(x,y), and possibly more sensitive for detecting "shape".

Three different transformations were used in the K-group method. Transformation A is division of the scores by the pooled within standard deviation σ. The true groups were used for this purpose. (In practical applications the true group structure is not known, of course, and another estimate of σ must be used. We expect this to lead only to little differences.) Transformation B is division of the scores by the Euclidean length $\sqrt{\sum_{i=1}^{n} x_i^2}$. The aim of this transformation is reduction of size influences. Transformation C is first standardizing the scores to z-scores by scores to z-scores

$z_{ij} = \frac{x_{ij} - \bar{x}_j}{\sigma_j}$, (where i indicates the individual and j the variable, and $\bar{x}_j$ and σ_j are the total sample mean and standard deviation, respectively, for variable j), and then subtracting the individual z-score averages, leading to C-scores $C_{ij} = z_{ij} - \bar{z}_i$. See Howells (1979) for this transformation.

From table 4 we concluded that for the tr(W) method taking the AL or Ward's method solution as initial configuration and using the Euclidean distance with variables transformed by A and C gave the best results.

2.2. Rest group, application of mode analysis (table 5)

The choice and results of the different TDC-combinations we have used are summarized in table 5.

As similarity measures we used the Euclidean distance and the Pearson product moment correlation. For the first data set we have also used the dispersion measure (the definition of this measure corresponds with the numerator of the Pearson product moment correlation), but this resulted two times (for both data sets) in chaining. The Pearson product moment correlation is defined by

$$r(x,y) = \frac{\sum_{i=1}^{n} (x_i-\bar{x})(y_i-\bar{y})}{\sqrt{\sum_{i=1}^{n} (x_i-\bar{x})^2 \sum_{i=1}^{n} (y_i-\bar{y})^2}}$$

Table 5. Comparison of mode analysis options applied to two sets of skull data. First to the Norse-Teita-Hainan skulls, secondly to the skulls of three Asiatic and Oceanic populations

	n var	Transf.	Dist.	K*	enc. ratio	number** of clusters	correct classifi- cations
first data set (n=134)	46	RS	Eucl.	$2\frac{1}{2}$	.19	5	73
	46	RS	Eucl.	1	.34	6-5	100
	46	RS	Eucl.	$\frac{1}{2}$	.16	8-3	90
	46	RS	Correl.	$2\frac{1}{2}$	.17	4	65
	46	RS	Correl.	1	.22	6-3	97
	46	C	Eucl.	1	.15	7-4	83
	46	C	Eucl.	$\frac{1}{2}$	.16	8-4	93
	46	C	Correl.	1	.27	9-3	104
	46	A	Eucl.	1	.18	5-3	84
	46	A	Eucl.	$\frac{1}{2}$	.16	8-3	90
	46	A	Correl.	1	.28	8-3	100
	46	A	Correl.	$\frac{1}{2}$	.63	12-3	88
second data set (n=128)	46	RS	Eucl.	1	.13	7-5	68
	46	RS	Eucl.	$\frac{1}{2}$	.18	10-4	87
	46	RS	Correl.	1	.43	12-4	85
	46	C	Eucl.	1	.25	8-6	51
	46	C	Eucl.	$\frac{1}{2}$	.27	9-6	60
	46	C	Correl.	1	.70	8-3	99
	46	A	Eucl.	1	.27	10-3	85
	46	A	Eucl.	$\frac{1}{2}$	.35	10-4	92
	46	A	Correl.	1	.35	8-3	93
	46	A	Correl.	$\frac{1}{2}$	.72	10-4	96

* K = density level

** As can be seen from this table, in most cases the original number of clusters is reduced to a smaller number. This was done by eliminating those clusters which included only a few individuals, in order to obtain a better idea about the group constitution of the whole data set.

Mode analysis was applied to raw scores (RS) and to transformed scores. Two kind of transformations were used: transformation A, division by the pooled within σ, and transformation C, C-scores.

The most important parameter to be chosen is the density parameter K, described in Wishart (1978). We concluded that, comparing the different density levels, the best results are generally obtained with $K=\frac{1}{2}$, especially for the second data set, which does not include the doubtful individuals.

2.3. Conclusions

Summarizing the foregoing results we may conclude that for the tr(W) method taking the AL or Ward's method solution as initial configuration and using the Euclidean distance gave the best results. For mode analysis, best results were obtained with density level $K=\frac{1}{2}$. In general, the variables transformed by A (division by the pooled within σ) and C (C-scores) gave the best results.

On the basis of these conclusions, we selected the following eight TDC-combinations as giving good results.

	Transformation	Distance measure	Cluster method
I	division by σ_W^*	Euclidean distance	K-group, viz. tr(W), average linkage solution as initial configuration
II	division by σ_W	Euclidean distance	K-group, viz. tr(W), Ward's method solution as initial configuration
III	division by σ_W	Euclidean distance	mode analysis ($K=\frac{1}{2}$)
IV	division by σ_W	Pearson product moment correlation	mode analysis ($K=\frac{1}{2}$)
V	C-scores**	Euclidean distance	K-group, viz. tr(W), average linkage solution as initial configuration
VI	C-scores	Euclidean distance	K-group, viz. tr(W), Ward's method solution as initial configuration
VII	C-scores	Euclidean distance	mode analysis ($K=\frac{1}{2}$)
VIII	C-scores	Pearson product moment correlation	mode analysis ($K=\frac{1}{2}$)

* σ_W = pooled within standard deviation

** C-scores (Howells, 1979); $C_{ij} = z_{ij} - \bar{z}_i$, where i = individual i and j = variable j

ACKNOWLEDGEMENTS

Special thanks are due to T. Snijders, Ph.D., for his constructive and detailed critique on an earlier draft of the manuscript, to R. Corluy, Ph.D., for some helpful suggestions and to Lidewij Leyten and Jogé Benjamins for their kind and accurate secretarial assistance.

Appendix

SOME HISTORICAL APPLICATIONS OF CLUSTER ANALYSIS IN PHYSICAL ANTHROPOLOGY

Author's name Material and measurements	Clustering methods and results
BOYCE, A.J. (1964) - Hominoids (Pongidae, Australopithecinae and Hominidae) - 71 craniometric variables	- Transformation into z-scores. - 5 distance (or similarity) coefficients were used i.e. correlation coefficient, mean squared distance, mean character difference, "size" distance and "shape" distance. - Pair-group method of cluster analysis (= AL*) (Sokal & Michener, 1958) - Results: 5 different pictures of the relationships. Most of the differences appear to be the result of the differences in relative emphasis they give to the size and shape components.
EDWARDS, A.W.F. & CAVALLI-SFORZA, L.L. (1965) - Taxonomy in general	- Mahalanobis distance. - Analysis of variance criterion for division. The measure of cluster density = variance. - They conclude that cluster analysis is the best method we know for treating data of evolutionary studies or of studies on the spatial distribution of organisms.
HOWELLS, W.W. (1970, 1973) - 151 local populations of the Pacific - 7 craniometric variables	- Transformation by principal component analysis. - Euclidean distance. - Hierarchical cluster method, based on group averages (= AL). - He finds subdivisions in Melanesians, Micronesians, Polynesians and Australians.

* AL = average linkage

BLACKITH, R.E. & REYMENT, R.A. (1971)
- More general article about the comparison of cluster methods, among others in physical anthropology

- The pair-group method (= AL) seems to be very good.
- Choosing agglomerative techniques and not divisive ones is logical in the sense that it commits one to begin the classificatory process by taking together single OTU's and such an attitude may be appropriate for, say, an archaeologist faced with an assortment of artifacts. In relation to the problems of a taxonomist it seems less desirable.

RAO, C.R. (1971)
- Castes and tribes belonging to 3 different states in India

- Transformation of variables into shape factors.
- Mahalanobis distance.
- Result: almost complete separation of the 3 groups.

GOWER, J.C. (1972)
- About anthropometry and measures of taxonomic distance

- Standardization, among others division by the variate standard errors.
- Czeckanowsky DD, Coefficient of Racial Likeness CRL and Mahalanobis distance.
- Two broad classes of analytical methods are appropriate for distances and similarities:
  1. traditional latent root and vector statistical methods such as component analysis.
  2. analytical methods that provide a hierarchical display of the sample individuals: monothetic and polythetic, divisive and agglomerative cluster methods.

HIERNAUX, J. (1972)
- 5 ethnic groups of Tchad (10 anthropometric variables)
- 15 populations of the Kivu-Rwanda-Burundi area (9 anthropometric variables)

- Mahalanobis distance.
- Principal coordinate analysis.
- Conclusion: the technique used made representation of biological diversity possible.

OXNARD, C. (1973) - Investigation of the structure of the shoulder of primates - 1000 specimens, 17 variables	- Full matrix of D^2's. - Minimum spanning tree was used, with the information contained within the single cluster analysis as ancillary. The principal value of the minimum spanning tree is that all the information required for the single linkage cluster analysis is contained within it.
CAVALLI-SFORZA, L.L. (1974) - 10 populations, 58 blood genes - 10 populations, 26 external measurements and observations	- Using the 58 genes and applying cluster analysis 3 main groups emerge: Africans, Europeans and a group containing Orientals, Oceanians and native Americans. - Using the 26 external measurements and applying cluster analysis, 2 groups emerge: one group, including Europeans, Orientals and native Americans, a second group including Oceanians and Africans.
PIETRUSEWSKY, M. (1974) - 4 prehistoric populations and 15 more recent populations - 13 cranial variables	- Mahalanobis distance. - Average pair-group clustering technique (= AL). - Results: -males were distinguished from females. -no close affinities exist between Neolithic and modern populations from Southeast Asia.
ZEGURA, S.L. (1975) - 609 Eskimoid crania (353 male and 256 female) - 74 craniometric variables	- Mahalanobis distance. - Unweighted pair-group clustering (AL) producing a dendrogram, where both axes have a meaning: one axis contains information about within group relationships, the other about between group relationships.
BOWLES, G.T. (1977) - 519 populations (males) - 9 anthropometric variables	- Variables were transformed into uncorrelated variables by principal component analysis. - Mahalanobis distance. - The "Nature's Group Program" was used, the most-alike series form pairs and most-alike pairs form clusters and groups. - The approach supports most previous regional studies.

PIETRUSEWSKY, M. (1977) - 17 populations from the Pacific - 14 craniometric variables	- Mahalanobis distance. - Average pair-group clustering technique (= AL). - Two important clusters were found: one including the Melanesian populations, the other including Polynesian and Micronesian populations.
NEI, M. (1978) - 12 human populations - gene frequency data: 11 protein loci 11 blood group loci	- Cluster analysis was done, using 1. Fich & Margoliash's method (= compute the average squared deviations of estimated distances from observed distances for all possible trees and choose the one that minimizes the average squared deviation) for determining the topology. 2. Nei's method for determining the branch lengths. - A certain pattern of divergence of Caucasians, Negrids and Japanese was found.
HAUSMAN, A.J. (1980, 1982) - Khoisian crania from a variety of geographic areas and time periods throughout southern Africa (at least 9000 years of South-African prehistory) - 37 cranial measurements	- Mahalanobis distance among group centroids - Unweighted pair-group AL method. The cophenetic correlation coefficient (r) measures the agreement between the original similarity matrix and the similarity values implied by the dendrogram - Conclusions: - coastal and inland people are different, - cluster analysis demonstrates some clusters that coincide with cultural and geographic boundaries, but there is considerable overlap.
HOWELLS, W.W. (1979) - 14 populations, especially from South East Asia, (males) - Craniometric variables	- Transformation: C-scores: correction for scale and individual differences, - Mahalanobis distance values (from scores on 20 PC's) - Single linkage. Agglomerative cluster method - Conclusion: Ainu and Japanese are distinct, also Jomon peoples are distinct.

OMOTO, K. (1981) - Negritos (Aetas) of the Philippines and 8 other populations - Allele frequency data for 17 polymorphic loci	- Genetic distance (Cavalli-Sforza & Edwards-method) - Conclusion: Negritos are more related to the Asian-Pacific populations than to the African group

References

Anderberg, M.R.: 1973, Cluster analysis for applications. Academic Press, New York.

Becker, P.W.: 1978, Recognition of Patterns. Springer Verlag, Wien/New York.

Binder, D.A.: 1978, Bayesian Cluster Analysis. Biometrica, 63, 1, 31-38.

Blackith, R.E. and Reyment, R.A.: 1971, Multivariate morphometrics. Academic Press, London, New York.

Blashfield, R.K.: 1976, Mixture Model Tests of Cluster Analysis: Accuracy of Four Agglomerative Hierarchical Methods. Psychol. Bull., 83, 3, 377-388.

Blashfield, R.K., & Aldenderfer, M.S.: 1978, The literature on cluster analysis. Multivariate Behavioral Research, 13, 271-295.

Bock, H.H.: 1975, Automatische Klassifikation. Vandenhoek & Ruprecht, Göttingen (W-Germany).

Bolshev, L.N.: 1969, Cluster Analysis. Bulletin de l'Institute International de Statistique, 43, 409-425.

Bowles, G.T.: 1977, The people of Asia. Scribner, New York.

Boyce, A.J.: 1964, The value of some methods of numerical taxonomy with reference to Hominoid classification. Systematics Association Publication 6: 47-65.

Burr, E.J.: 1968, Cluster sorting with mixed character types. I. Standardization of character values. Aust. Computer J., 1, 97-99.

Cattell, R.B.: 1966, Handbook of multivariate experimental psychology. Rand McNally, Chicago.

Cavalli-Sforza, L.L.: 1974, The genetics of human populations. Scientific American 231: 81-89.

Cohen, J.A.: 1969, A coefficient of agreement for nominal scales. Educ. and Psychol. Measurement, 20, 37-46.

Cole, A.J. (Ed.): 1969, Numerical Taxonomy. Academic Press, London/New York.

Compstat 78, Proceedings in Computational Statistics, 3rd Symposium held in Leiden 1978. Ed. by L.C.A. Corsten, J. Hermans, Wien Physica-Verlag, 1978.

Cormack, R.M.: 1971, A Review of Classification. J. Royal Statist. Soc. Series A, 134, 321-367.

Corruccini, R.S.: 1973, Size and Shape in Similarity Coefficients Based on Metric Characters. Am. J. Phys. Anthrop., 38, 743-754.

Corruccini, R.S.: 1975, Multivariate analysis in biological anthropology: some considerations. J.of Hum.Evol., 4, 1-19.

Crovello, T.J.: 1968, A numerical taxonomic study of the genus Salix, section Sitchenses. Univ. of Calif. Publications in Botany, 1-61.

Cunningham, K.M. & Ogilvie, J.C.: 1972, Evaluation of hierarchical grouping techniques: a preliminary study. The Computer Journal, 15, 209-213.

Day, N.E.: 1969, Estimating the components of a mixture of normal distributions. Biometrica, 56, 463-474.
Degens, P.O. & Federkiel, H.: 1978, A Monte Carlo study on agglomerative large sample clustering. In COMPSTAT 78.
Dubes, R. & Jain, A.K.: 1976, Clustering Techniques: The users dilemma. Pattern Recognition, 8, 247-260.
Dubes, R. & Jain, A.K. Validity studies in clustering methodologies. Pattern Recognition, 11, 235-254.
Edelbrock, C.: 1979, Mixture Model Tests of Hierarchical Clustering Algorithms, The problem of classifying everybody. Multivar.Beh. Res., 14, 367-384.
Edwards, A.W.F. & Cavalli-Sforza, L.L.: 1965, A method for cluster analysis. Biometrics, 21, 362-375.
El Gazzar, A. et.al.: 1968, The taxonomy of Salvia: a test of two radically different numerical methods. J.Linn.Soc. (Bot.), 60, 383,237-250.
Everitt, B.S. et.al.: 1971, An attempt at validation of traditional psychiatric syndromes by cluster analysis. Brit.J.Psychiat., 119, 399-412.
Everitt, B.S.: 1974, Cluster analysis. Heinemann Educational Books Ltd. London.
Everitt, B.S.: 1975, Multivariate analysis: The need for data, and other problems. Brit. J. Psychiat., 126, 237-240.
Everitt, B.S.: 1980, Cluster analysis. Quality and Quantity, 1, 75-100.
Farris, J.S.: 1969, On the cophenetic correlation coefficient. Systematic Zoology, 18, 279-285.
Fisher, L. & van Ness, J.W.: 1971, Admissible clustering procedures. Biometrika, 58, 91-104.
Frank, D.: 1978, Inferences concerning cluster structure. In COMPSTAT 78.
Friedman, H.P. & Rubin, J.: 1967, On some invariant criteria for grouping data. J.Am.Stat.Ass., 62, 1159-1178.
Garside, R.F. & Roth, M.: 1978, Multivariate statistical methods and problems of classification in psychiatry. Brit. J. Psychiat., 133, 53-67.
Gnanadesikan, R.: 1977, Methods for statistical data analysis of multivariate observations. Wiley, New York.
Gower, J.C.: 1966, Some distance properties of latent root and vector methods used in multivariate analysis. Biometrika, 53, 325-338.
Gower, J.C.: 1967, A comparisom of some methods of cluster analysis. Biometrics, 23, 623-637.
Gower, J.C.: 1971, A general coefficient of similarity and some of its properties, Biometrics, 27, 857-871.
Gower, J.C.: 1972, Measures of taxonomic distance and their analysis. In: The assessment of population affinities in man. Weiner J.S. and Huizinga, J. (eds.) Clarendon Press, Oxford: 1-24.

Green, P.E. et al.: 1976, Cluster analysis in test market selection. Management Science, 13, 8, 387-400.

Guttman, R. et al.: 1967, Cross-Ethnic variation in dental, sensory and perceptual traits: A nonmetric multibivariate derivation of distances for ethnic groups and traits. Am.J. Phys.Anthrop., 27, 259-276.

Hall, A.V.: 1969, Avoiding informational distortion in automatic grouping programs. Syst.Zool., 18, 318-329.

Harary, F. & Ross, I.C.: 1957, A procedure for clique detection using the group matrix. Sociometry, 20, 205-215.

Hartigan, J.: 1975, Clustering algorithms. Wiley, New York.

Hartigan, J.A.: 1978, Asymptotic distributions for clustering criteria. The Annals of Statistics, 6, 117-131.

Hausman, A.J.: 1982, The biocultural evolution of Khoisian population of Southern Africa. Am.J.Phys.Anthrop. 58, 316-330.

Hiernaux, J.: 1972, The analysis of multivariate biological distances between human populations: principles, and applications to sub-Saharan Africa. In: The assessment of population affinities in man. Weiner, J.S. and Huizinga, J. (eds.) Clarendon Press, Oxford: 96-114.

Hirschfield, W.J. et al.: 1973, A method of deriving subgroups of a population: A study of craniofacial taxonomy. Am.J.Phys. Anthrop., 39, 279-290.

Hodson, F.R. et al.: 1966, Some experiments in the numerical analysis of archaeological data. Biometrika, 53, 3 and 4, 311-324.

Howells, W.W.: 1970, Anthropometric grouping analysis of Pacific peoples. Arch. Phys. Anthropol. Oceania 5, 192-217.

Howells, W.W.: 1973, Cranial variation in man. Papers of the Peabody Museum of Archaeology and Ethnology 67.

Howells, W.W.: 1973, The Pacific islanders. Scribner's Sons, New York.

Howells, W.W.: 1978, Quantitative Systematik menschlicher Populationen. Ein Ueberblick. Homo 29, 217-228.

Howells, W.W.: 1979, Physical anthropology of the prehistoric Japanese. Paper presented at Beardsley Symposium, Ann Arbor.

Jain, A.K. & Dubes, R.: 1978, Feature definition in pattern recognition with small sample size. Pattern Recognition, 10, 85-97.

Jardine, N. & Sibson, R.: 1971, Mathematical Taxonomy. Wiley, London.

Jensen, R.E.: 1969, A dynamic programming algorithm for cluster analysis. Op.Res., 1034-1056.

Johnson, S.E.: 1967, Hierarchical clustering schemes. Psychometrika, 32, 241-254.

Jolicoeur, P. & Mosimann, J.E.: 1960, Size and shape variation in the painted turtle. A principal component analysis. Growth, 24, 339-354.

Kanal, L.N.: 1972, Interactive pattern analysis and classification systems: A survey and commentary. Proceedings of the IEEE, 60, 1200-1215.

Korhonen, P.: 1978, Experiments with cluster analysis criteria based on the within groups scatter matrix. In COMPSTAT 78.

Kruskal, J.B.: 1964, Nonmetric multidimensional scaling: a numerical method. Psychometrika, 29, 115-129.

Kuyper, F.K. & Fisher, L.A.: 1975, A Monte Carlo study of the accuracy of a hierarchical grouping procedure. Biometrics, 31, 1777-1783.

Lambert, J.M. & Williams, W.T.: 1962, Multivariate methods in plant ecology. IV. Nodal analysis. J.of Ecology (50), 775-802.

Lambert, J.M. & Williams, W.T.: 1966, Multivariate methods in plant ecology. VI. Comparison of information-analysis and association-analysis. J.of Ecology (54), 635-664.

Lance, G.N. & Williams, W.T.: 1967, A general theory of classificatory sorting strategies: I. Hierarchical systems. Computer Journal, 9, 373-380.

Lazersfeld, P.F.: 1950, The logical and mathematical foundation of latent structure analysis. In: S.A. Stouffer et.al. (eds.): Measurement and prediction. Princeton University Press, Princeton, N.J.

Lee, K.L.: 1979, Multivariate tests for clusters. J.Am.Stat.Ass., 74, 708-714.

Ling, R.F.: 1972, Theory and construction of K-clusters. Computer Journal, 15, 326-332.

Ling, R.F.: 1973, A probability theory of cluster analysis. J.Am. Stat.Ass., 68, 341, 159-164.

Mahalanobis, P.C.: 1936, On the generalized distance in statistics. Proceedings of the National Institute of Science, India, 12, 49-55.

Marriott, F.H.C.: 1971, Practical problems in a method of cluster analysis. Biometrics, 27, 501-514.

McNaughton-Smith, P. et al.: 1964, Dissimilarity analysis. Nature, 202, 1034-1035.

McQueen, J.: 1967, Some methods for classification and analysis of multivariate observations. Proc. 5th Berkeley Symp., 1, 281-297.

McRae, D.J.: 1971, MICKA, a Fortran IV iterative K-means cluster analysis program. Beh.Sci., 16, 423-424.

McRae, D.J.: 1973, Clustering multivariate observations. Unpublished doctoral dissertation, University of North Carolina.

Mezzich , J.: 1978, Evaluating clustering methods for psychiatric diagnosis. Biol. Psychiat., 13, 265-281.

Milligan, G.W.: 1980, An examination of the effect of six types of error perturbation on fifteen clustering algorithms. Psychometrika, 45, 3, 325-342.

Milligan, G.W.: 1981, A review of Monte Carlo tests of cluster analysis. Multiv.Beh.Res., 16, 379-407.

Milligan, G.W. & Isaac, P.D.: 1980, The validation of four ultrametric clustering algorithms. Pattern Recognition, 12, 41-50.

Mojena, R.: 1977, Hierarchical grouping methods and stopping rules: an evaluation. Computer Journal, 20, 359-363.

Mosimann, J.E.: 1970, Size allometry: size and shape variables with characterizations of the lognormal and generalized gamma distributions. J.Am.Stat.Association, 65: 930-945.
Needham, R.M.: 1965, Automatic classification: models and problems. In: Mathematics and Computer Science in Medicine and Biology, 111-114, London: H.M.S.D.
Nei, M.: 1979, The theory of genetic distance and evolution of human races. Jap.J.Human Genet. 23: 341-369.
Nunnally, J.C.: 1967, Psychometric Theory. McGraw-Hill, New York.
Omoto, K.: 1981, The genetic origins of the Philippine Negritos. Current Anthrop. 22: 421-422.
Oxnard, C.: 1973, Form and pattern in human evolution. Some mathematical, physical, and engineering approaches. The Univ. of Chicago Press, Chicago & London.
Paykel, E.S.: 1971, Classification of depressed patients: a cluster analysis derived grouping. Brit.J.Psychiat., 118, 275-288.
Paykel, E.S.: 1972, Depressive Typologies and Response to Amitryptyline. Brit.J.Psychiat., 120, 147-156.
Pietrusewsky, M.: 1974, Neolitic populations of Southeast Asia studied by multivariate craniometric analysis. Homo 25, 207-250.
Pietrusewsky, M.: 1977, Etude des rélations entre les populations du Pacifique par les méthodes d'analyse multivariée appliquées aux variations crâniennes. L'Anthropologie, 21, 67-97.
Pilowsky, I. et al.: 1969, The classification of depression by numerical taxonomy. Br.J.Psychiat., 115, 937-945.
Pritchard, N.M. & Anderson, A.J.B.: 1971, Observations in the use of cluster analysis in botany with an ecological example. J.Ecol., 59, 727-747.
Rand, W.M.: 1971, Objective criteria for the evaluation of clustering methods. J.Am.Stat.Ass., 66, 846-850.
Rao, C.R.: 1952, Advanced statistical methods in biometrical research. Wiley, New York.
Rao, C.R.: 1971, Taxonomy in anthropology. In: Mathematics in the archaeological and historical sciences. Proceedings of the Anglo-Romanian Conference, Mamaia, 1970. Eds. Hodson, F.R., Kendall, D.G. and Ttu, P. Edinburgh, 19-29.
Schnell, G.D.: 1970, A phenetic study of the suborder Lari (Aves).I. Methods and results of principal components analyses. Systematic Zool., 19, 35-57.
Scott, A.J. & Symons, M.J.: 1971, Clustering methods based on likelihood ratio criteria. Biometrika, 27, 387-397.
Sheals, J.G.: 1964, The application of computer techniques to Acarine taxonomy: a preliminary examination with species of the Hypoaspis-Androlaelaps complex (Acarina). Proc.Linn.Soc. Lond. 176 (1), 11-21.
Shepard, R.N.: 1962, The analysis of proximities: Multidimensional Scaling with an unknown distance function. II. Psychometrika, 27 (3), 219-246.
Sneath, P.H.A. & Sokal, R.R.: 1973, Numerical Taxonomy. Freeman & Co., San Francisco.

Sokal, R.R. & Michener, C.D.: 1958, A statistical method for evaluating systematic relationships. Computer J. 14, 162-165.

Sokal, R.R. & Sneath, P.H.A.: 1963, Principals of numerical taxonomy. Freeman, San Francisco.

Späth, H.: 1980, Cluster analysis algorithms. Wiley, New York.

Spielman, R.S.: 1973, Do the natives all look alike? Am.Naturalist 107, 694-708.

Strauss, J.S. et.al.: 1973, The use of clustering techniques for the classification of psychiatric patients. Brit.J.Psychiat., 122, 531-540.

Stringer, P.: 1967, Cluster analysis of non-verbal judgment of facial expressions. Brit.J.Math.Stat.Psychol., 20, 71-79.

Symons, M.J.: 1981, Clustering criteria and multivariate normal mixtures. Biometrics, 37, 35-43.

True, D.L. & Matson, R.G.: 1970, Cluster analysis and multi-dimensional scaling of archaeological sites in northern Chile. Science, 169, 1201-1203.

Tryon, R.C.: 1939, Cluster analysis. Edwards Brothers. Ann Arbor, Michigan.

Uytterschaut, H.T. & Wilmink, F.W.: 1982, The efficiency of principal component analysis as a variable ordination and selection procedure prior to the sexdiagnosis of human skulls. Z. Morph. Anthr., 73, 113-123.

Van Meurs, A. et al.: 1978, Cluster-analyse. Geografisch instituut R.U. Utrecht, (Netherlands).

Van Ness, J.W.: 1973, Admissible clustering procedures. Biometrika, 60, 422-424.

Van Vark, G.N. & van der Sman, P.G.M.: 1982, New discrimination and classification techniques in anthropological practice. Z.Morph.Anthrop., 73, 21-36.

Wallace, C.S. & Boulton, D.M.: 1968, An information measure for classification. Computer Journal, 11, 185-194.

Watson, F.L.S. et al.: 1966, Angiosperm taxonomy: a comparative study of some novel numerical techniques. J.Linn.Soc. (Bot.), 59, 380, 491-501.

Wilkens, L.T. & McNaughton-Smith, P.: 1964, New prediction and classification methods in criminology. J.of Research in Crime and Delinquency, 1, 19-32.

Williams, W.T. & Lambert, J.M.: 1959, Multivariate methods in plant ecology. I. Association analysis in plant communities. J.Ecol., 47, 83-101.

Williams, W.T. et al.: 1979, Controversies concerning the criteria for taxonomic strategies. Computer J., 14, 162-165.

Wishart, D.: 1969, Mode analysis. In Cole (ed.): Numerical Taxonomy, 282-308.

Wishart, D.: 1969, FORTRAN II Programs for 8 methods of cluster analysis (CLUSTAN I). Comp. Contribution, 39, State Geological Survey University of Kansas.

Wishart, D.: 1971, A generalized approach to cluster analysis. Part of Ph.D.Thesis, University of St.Andrews.

Wishart, D.: 1978, Clustan user manual. Inter University Research Councils Service, Report no. 47.
Wolfe, J.H.: 1970, Pattern clustering by multivariate mixture analysis. Multiv.Beh.Res., 5, 329-350.
Wolfe, J.H.: 1971, A Monte Carlo study of the sampling distribution of the likelihood ratio for mixtures of multinormal distributions. Naval Personnel and Training Research Laboratories. Technical Bulletin, STB 72-2, (San Diego, California 92151).
Zegura, S.L.: 1975, Taxonomic congruence in Eskimoid populations. Am.J.Phys.Anthrop. 43, 271-284.

SOME ASPECTS OF ALLOCATION AND DISCRIMINATION

N.A. Campbell

Division of Mathematics and Statistics, CSIRO,
Wembley 6014, Western Australia

INTRODUCTION

The application of multivariate statistical methods to problems in physical anthropology involves both the allocation of new individuals and the description of group differences. It is not uncommon to find techniques which are appropriate for the description of group differences, such as canonical variate analysis, being used to allocate new individuals (see, e.g., Rightmire, 1979).

This paper discusses the allocation of individuals. In particular, the role of appropriate probabilities for determining group membership and the typicality of an individual are considered. Data which relate to the allocation of the Border Cave adult cranium (Rightmire, 1975, 1979) are used to illustrate the ideas.

2. ALLOCATION OF AN INDIVIDUAL - KNOWN PARAMETERS

The basic ideas may be introduced by considering an idealized representation involving two bivariate Gaussian populations (see Figure 1). Assume that the means and covariances are known, and that the covariances are the same for the two populations. Assume also that a new individual is equally likely a priori to belong to either population.

2.1. Minimum Mahalanobis Distance and Forced Allocation

Given that a new individual belongs to one of the reference populations, one possible allocation procedure is to assign the individual to the population to which it is closer. Mahalanobis distance is used, since the variables are in general correlated, and this distance includes both the variances and correlations in

G. N. van Vark and W. W. Howells (eds.), Multivariate Statistical Methods in Physical Anthropology, 177–192.

the calculations. With this approach, a definite allocation decision is made; such a procedure is referred to as a forced allocation procedure.

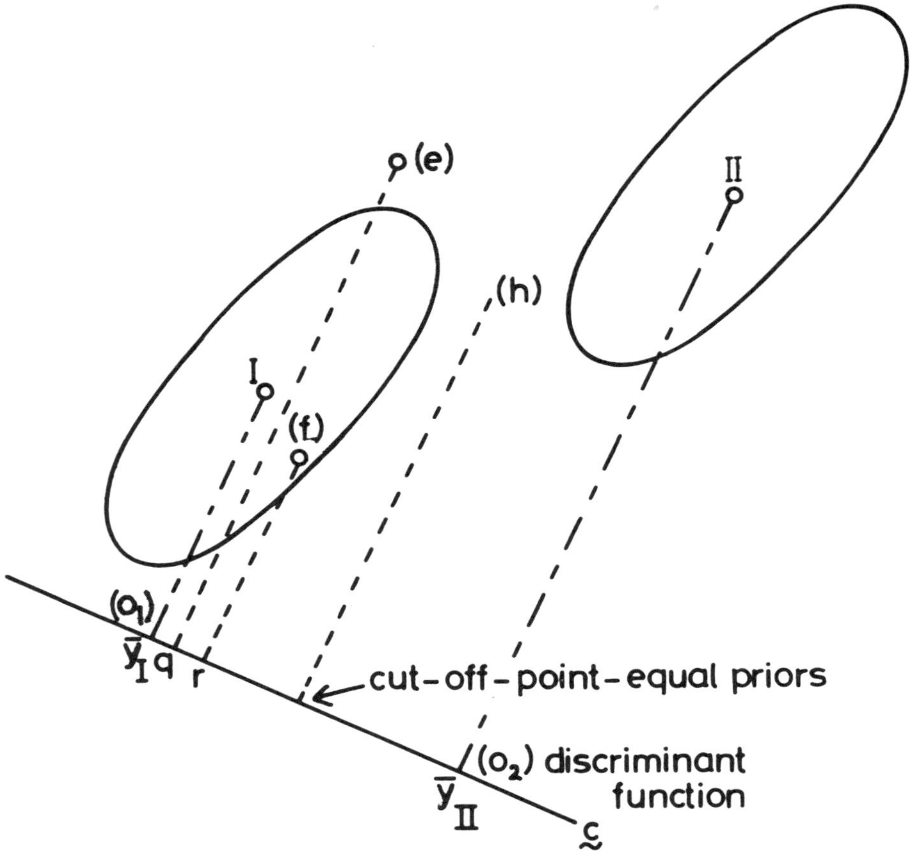

Figure 1. Representation of the discriminant function for two populations and two variables. The population means I and II and associated 95% probability contours are shown. The vector $\underset{\sim}{c}$ is the discriminant vector. The points $\bar{y}_I$ and $\bar{y}_{II}$ represent the discriminant means for the two populations. The points (e), (f) and (h) represent three new individuals to be allocated. The points (q) and (r) are the discriminant scores for the individuals (e) and (f). The point (O_1) is the discriminant mean $\bar{y}_I$.

In Figure 1, new individuals (e) and (f) are both closer to population I; for forced allocation, both would be allocated to this population. Individual (h) is marginally closer to population I than it is to population II; a forced allocation procedure would assign it to population I. However a more reasonable decision would be to place it in a region of doubt (see page 178).

2.2. The Linear Discriminant Function

The allocation procedure may also be considered in terms of the linear discriminant function -- the linear combination of the original variables which maximizes the distance between the population means, relative to the within-populations variation. With this approach, the population means are projected onto the discriminant function to give the mean score for each population. The score for a new individual is determined in the same way, and this score is then compared with the mean score for each population. For forced allocation, the individual is assigned to the population to which it is nearer. When the *a priori* probabilities differ, the rule for assignment is modified accordingly. For known populations, the rule using the linear discriminant function minimizes the total probability of misallocation.

Figure 1 shows the linear discriminant, together with the cut-off boundary under the assumption of equal prior probabilities. Individuals (e) and (f) have discriminant scores which are closer to the mean score for population I than they are to the mean score for population II, while the score for individual (h) falls just within the cut-off boundary for population I.

2.3. The Linear Discriminant Function - Canonical Representation

An alternative representation of the information in Figure 1 may illustrate these points more clearly. Figure 2 shows the same configuration of populations and new individuals, except that the original variables have been transformed to new variables which are uncorrelated and which have unit variance within each population (see, e.g., Campbell and Atchley, 1981, Section 3). The 95% concentration ellipses become 95% concentration circles, while the discriminant function may be drawn to pass through the mean of each population. In this representation, often referred to as a canonical representation, the distance between the means of the two populations is equal to the population Mahalanobis distance. The advantage of this representation is that the common Euclidean or Pythagorean distance is now the appropriate distance; the Euclidean distance in the canonical representation is identical to Mahalanobis distance in the more familiar representation.

In the canonical representation, individuals (e) and (f) again are closer to the mean of population I than they are to the mean of population II. Individual (h) falls roughly half-way between, and again its allocation would be considered doubtful.

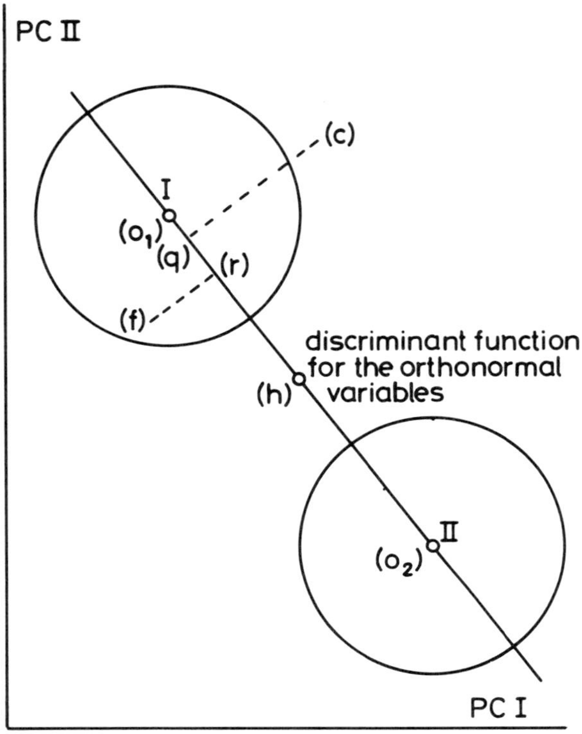

Figure 2. Alternative representation of the discriminant function. The axes PCI and PCII represent orthonormal linear combinations of the original variables. The 95% probability ellipses become 95% probability circles in the space of the orthonormal variables. The population means I and II for the discriminant function for the orthonormal variables are equal to the discriminant means $\bar{y}_I$ and $\bar{y}_{II}$ in Figure 1. Pythagorean distance can be used to determine the distances from the new individuals to the populations means.

2.4. Probabilistic Assessments

To make more meaningful statements about the affinities of the individuals for the reference populations, some probabilistic assessments must be made. One aspect to be considered is the relative affinities of a new individual for each of the reference populations. The appropriate probabilities are referred to as posterior probabilities of membership of each of the reference populations. Another aspect to be considered is the affinity of the individual for each population in turn, without reference to the other populations. The appropriate probabilities are referred to as typicality probabilities.

2.5. Posterior Probabilities of Population Membership

The posterior probabilities of membership of each of the reference populations are related to the values of the probability density function for the given data vector for each of the populations. Specifically, let $f(\underset{\sim}{x}_m;\underset{\sim}{P}_k)$ denote the value of the probability density function for the vx1 vector $\underset{\sim}{x}_m$ for the kth reference population (k=1,...,g) with known parameters $\underset{\sim}{P}_k$, and let p_k denote the prior probability that the individual belongs to the kth population. Then the posterior probability of membership of the kth population is given by

$$pr(k;\, \underset{\sim}{x}_m) = p_k\, f(\underset{\sim}{x}_m;\, \underset{\sim}{P}_k)/\sum_{j=1}^{g} p_j\, f(\underset{\sim}{x}_m;\underset{\sim}{P}_j). \tag{1}$$

For the multivariate Gaussian densities assumed in this Section, $\underset{\sim}{P}_k$ consists of the mean vector $\underset{\sim}{\mu}_k$ and the common covariance matrix $\underset{\sim}{\Sigma}$. Write the individual squared Mahalanobis distance for the vector $\underset{\sim}{x}_m$ for the kth population as

$$\Delta^2_{k,m} = (\underset{\sim}{x}_m - \underset{\sim}{\mu}_k)^t\, \underset{\sim}{\Sigma}^{-1}\, (\underset{\sim}{x}_m - \underset{\sim}{\mu}_k). \tag{2}$$

Then the value of the probability density function for the vector $\underset{\sim}{x}_m$ for the kth population may be written as

$$f(\underset{\sim}{x}_m;\, \underset{\sim}{\mu}_k,\, \underset{\sim}{\Sigma}) = (2\pi)^{-v/2}|\underset{\sim}{\Sigma}|^{-\frac{1}{2}}\, \exp(-\tfrac{1}{2}\Delta^2_{k,m}). \tag{3}$$

2.6. Posterior Probabilities and the Linear Discriminant Function

In Figure 2, the individual Mahalanobis distance for individual (e) from population I is the difference (o_1)-(f). The squared Mahalanobis distance for individual (e) may be partitioned into a component due to the distance (o_1)-(q) in the direction of the discriminant vector, and a component due to the distance (q)-(e) in the direction orthogonal to the discriminant vector. That is, $\Delta^2_{1,e} = \{(o_1)-(e)\}^2 = \{(o_1)-(q)\}^2 + \{(q)-(e)\}^2$. For population II, $\Delta^2_{2,e} = \{(o_2)-(e)\}^2 = \{(o_2)-(q)\}^2 + \{(q)-(e)\}^2$. Note that the distance (q)-(e) is common for both populations. For the multivariate Gaussian densities assumed in this Section, the term $\exp(-\frac{1}{2}\Delta^2_{k,e})$ may be factorized as $\exp[-\frac{1}{2}\{(o_k)-(q)\}^2]\, \exp[-\frac{1}{2}\{(q)-(e)\}^2]$, k=1,2. It follows from the form of equation (1) that only the first component of the squared distance is relevant for the calculation of the posterior probabilities $pr(k;\underset{\sim}{x}_m)$. The component $(o_k)-(q)$ is simply the deviation of the discriminant score for the new individual from the discriminant mean for the kth population.

Hence for two multivariate Gaussian densities with equal covariance matrices and known parameters, posterior probabilities of population membership based on the individual squared Mahalanobis distances for each of the populations are equivalent to those based on the population discriminant scores. For the latter, the squared distances used are the squares of the deviations of the score for a new individual from the mean scores for the populations.

For more than two populations with common covariance matrix and known parameters, the posterior probabilities based on the individual squared Mahalanobis distances are equivalent to those based on the canonical variate scores. For the latter, the squared distances used are the sums of squares of the deviations of the canonical variate score for a new individual from the mean scores for the populations, with the summation over all $h = \min(g-1,v)$ canonical vectors.

2.7. Typicality Probabilities

The typicality probability for the kth population may be formulated in terms of the individual Mahalanobis distance, as the probability of observing a distance greater than the Mahalanobis distance for the given data vector relative to the kth population. More formally, the typicality probability for the kth population is the probability of observing data vectors with values of the probability density function greater than the value of the probability density function for the given data vector for the individual to be allocated.

For the multivariate Gaussian densities assumed in this Section, the typicality probability is calculated by referring the individual squared Mahalanobis distance for the kth population to the chi-squared distribution on v degrees of freedom.

In Figures 1 and 2, individual (e) lies outside the 95% concentration contour for population I; the associated typicality probability is less than 0.05. Individual (f) lies well within the 95% concentration contour and would be considered quite typical.

2.8. Typicality Probabilities and the Linear Discriminant Function

The typicality probability for the kth population depends on the total Mahalanobis distance and cannot be calculated from the discriminant function alone. In Figure 2, the Mahalanobis distance for individual (e) is given by $(o_1)-(e)$. This distance has a component $(o_1)-(q)$ in the direction of the discriminant vector, and a component $(q)-(e)$ in the direction orthogonal to the discriminant vector (see Section 2.6). Individual (e) has a large value for the individual Mahalanobis distance and hence a small typicality probability; its distance component in the direction of the discriminant vector is small, while that in the orthogonal direction is large. Hence inferences regarding the typicality of an individual based only on the discriminant scores may be misleading.

To calculate the typicality probabilities from the results of a (multiple) discriminant analysis, v discriminant (canonical) vectors must be calculated. When $g-1 < v$, discriminant (canonical) vectors corresponding to the $v-g+1$ null roots must be determined.

In Figure 2, if a vector orthogonal to the discriminant vector is also used in the calculations, then the individual Mahalanobis distance again results.

2.9. Typicality Probabilities and Posterior Probabilities

The typicality probabilities and posterior probabilities of population membership reflect two different aspects of the allocation procedure. Under the assumptions made in this Section, the definitions of both probabilities involve the associated probability density functions $f(\underset{\sim}{x}_m; \underset{\sim}{\mu}_k, \underset{\sim}{\Sigma})$, and these in turn depend on the individual squared Mahalanobis distances $\Delta^2_{k,m}$.

The posterior probabilities give the probabilities of the new individual belonging to each of the reference populations, relative to the other reference populations considered. The typicality probabilities provide a measure of the similarity of the new individual to each of the reference populations, without considering the other reference populations. The typicality probabilities consider the closeness of the individual to (the means of) each of the reference populations, relative to other individuals in the population.

The representations in Figures 1 and 2 show individual (e) to be closer to population I in the direction of the discriminant function than is individual (f). And yet individual (e) lies outside the notional 95% concentration contour for population I. Individual (f) is closer to population I than is individual (e); this is obvious in Figure 2, where individual (f) has the smaller Euclidean distance. Of course, individual (f) is also closer to population II than is individual (e).

In Figures 1 and 2, the posterior probability of individual (e) belonging to population I is greater than the posterior probability of individual (f) belonging to population I. However individual (e) is further from the mean of population I than is individual (f). In fact, individual (e) would be considered atypical for both populations, even though it is closer to population I than it is to population II.

2.10. Unequal Covariance Matrices

The discussion to date has assumed that the covariance matrix is the same for all populations. When this assumption cannot be made, the formulae in (2) and (3) still hold, except that the common $\underset{\sim}{\Sigma}$ is replaced by the individual $\underset{\sim}{\Sigma}_k$. The results in (1) and in Section 2.7 are as before.

However, the equivalence between the calculation of posterior probabilities from the $\Delta^2_{k,m}$ and from the linear discriminant function (see Section 2.6) no longer holds.

3. ALLOCATION OF A NEW INDIVIDUAL - PARAMETERS TO BE ESTIMATED

It is unusual in practical applications for the population parameters to be known. The common situation is that individuals are available from a number of reference groups, and an allocation scheme is constructed on the basis of these individuals. It is usually assumed that the underlying distributions are multivariate Gaussian, though this assumption is rarely examined; quantile-quantile (Q-Q) plots of individual squared Mahalanobis distances (see, e.g., Gnanadesikan, 1977, p. 172; Campbell, 1980) are used to examine this assumption in the example in Section 4.

3.1. The Estimative Approach - Ignoring Sampling Variation

The commonly adopted approach is to assume that each population follows a multivariate Gaussian distribution, with the population parameters being replaced by their corresponding estimators. That is, $\underset{\sim}{\Sigma}$ in (2) and (3) is replaced by the sample covariance matrix $\underset{\sim}{V}_p$, while the vectors of means $\underset{\sim}{\mu}_k$ are replaced by the sample means $\underset{\sim}{\bar{x}}_k$ calculated from n_k observations. Both the typicality probabilities and the probabilities of group membership involve the sample squared Mahalanobis distance from an individual $\underset{\sim}{x}_m$ to each of the group means, given by

$$D^2_{k,m} = (\underset{\sim}{x}_m - \underset{\sim}{\bar{x}}_k)^t \underset{\sim}{V}_p^{-1} (\underset{\sim}{x}_m - \underset{\sim}{\bar{x}}_k). \qquad (4)$$

The notation $\underset{\sim}{V}_p$ is adopted for the sample covariance matrix; it may denote either the pooled within-groups sample covariance matrix on $n_f = \sum_{k=1}^{g} (n_k-1)$ degrees of freedom or the individual covariance matrix $\underset{\sim}{V}_k$ on n_k-1 degrees of freedom. The posterior probability of membership of the kth group is then given by (1), with the $f(\underset{\sim}{x}_m;\underset{\sim}{P}_k)$ replaced by $f_G(\underset{\sim}{x}_m; \underset{\sim}{\bar{x}}_k,\underset{\sim}{V}_p)$, the sample analogue of (3), with $\Delta^2_{k,m}$ replaced by $D^2_{k,m}$. Using the same approach of replacing population parameters by their estimators, typicality probabilities are calculated by referring the sample $D^2_{k,m}$ to the chi-squared distribution on v degrees of freedom. Aitchison, Habbema and Kay (1977) refer to the approach in which population parameters are replaced by their corresponding estimators as the estimative approach.

3.2. The Predictive Approach - Considering Sampling Variation

Sample means and covariances are involved in the calculations, but this is ignored in the distributions considered above. If $\underset{\sim}{x}_m$ represents a new individual from the kth group, then standard distribution theory (see, e.g., Kshirsagar, 1972, p. 126) gives the result that

$$v^{-1}(n_f-v+1)n_f^{-1}(n_k+1)^{-1}n_k\, D^2_{k,m} \text{ is distributed as } F(v,n_f-v+1). \quad (5)$$

The probability associated with the observed value of F in (5) gives the typicality probability.

The calculations for the posterior probabilities of group membership are based on multivariate Student densities. The probability density function $f_S(\underset{\sim}{x}_m;\ \bar{\underset{\sim}{x}}_k,\ \underset{\sim}{V}_p)$ for the kth group is given by

$$\pi^{-v/2}\Gamma\{(n_f+1)/2\}\Gamma\{(n_f-v+1)/2\}^{-1}|n_k^{-1}(n_k+1)n_f\,\underset{\sim}{V}|^{-\frac{1}{2}}\{1+n_k\,(n_k+1)^{-1}\,n_f^{-1}D^2_{k,m}\}^{-(n_f+1)/2}$$

The posterior probability of membership of the kth group is then given by (1), with the $f(\underset{\sim}{x}_m;\ \underset{\sim}{P}_k)$ replaced by $f_S(\underset{\sim}{x}_m;\ \bar{\underset{\sim}{x}}_k,\underset{\sim}{V}_p)$.

There is strong evidence, both theoretical (Aitchison, Habbema and Kay, 1977, Section 3; Murray, 1977; Anderson, 1958, Section 6.5.5) and applied (Aitchison, Habbema and Kay, 1977, Section 4; Aitchison and Dunsmore, 1975) to suggest that the predictive approach is to be preferred to the estimative approach.

3.3. Posterior Probabilities and the Linear Discriminant Function

For the estimative approach, the equivalence between the calculation of the posterior probabilities from the overall squared Mahalanobis distances and from the discriminant (canonical variate) scores again results.

The calculation for the predictive approach involves a power of $D^2_{k,m}$ rather than an exponential function of $D^2_{k,m}$ and this does not factorize as in Section 2.6. Hence for this preferred approach, both the typicality and posterior probabilities must be calculated directly from the individual Mahalanobis distances.

4. EXAMINING DISTRIBUTIONAL ASSUMPTIONS FOR EACH OF THE REFERENCE GROUPS

To examine whether the data show reasonable agreement with a multivariate Gaussian distribution, a quantile-quantile (Q-Q) probability plot of the individual squared Mahalanobis distances for each observation in a group may be used (see, e.g., Gnanadesikan, 1977, p. 172). The appropriate quantiles are those of a Beta distribution with parameters $v/2$ and $(n_f-v)/2$ (Small, 1978). Healy (1968) and Campbell (1980) plot the cube root of the ordered squared distances against the quantiles of an univariate Gaussian distribution.

For multivariate data, the presence of an atypical observation(s) will tend to decrease the Mahalanobis distance for the atypical observation and distort the rest of the plot (Campbell, 1980). Atypical observations may be accommodated in the analysis by adopting robust multivariate procedures. These procedures give full unit weight to reasonable observations, with atypical observations being given reduced weight. For multivariate data, the weights depend on the individual Mahalanobis distances, with a robust average value and a robust estimate of the covariance matrix being used in the calculations.

The robust estimators adopted here are called w-estimators, following Hampel and Tukey. Their form is such that observations with unduly large Mahalanobis distances are given reduced weight in the calculations. Theoretical aspects are discussed in Maronna (1976), Hampel (1977) and Huber (1977a,b; 1981), while Campbell (1980) discusses more applied aspects. The equations given in that paper are adopted here.

5. APPLICATION OF RESULTS IN PREVIOUS SECTIONS

The data analysed here relate to the allocation of the Border Cave adult cranium. Rightmire (1979) compares this cranium with eight reference populations of modern South Africans: Bushman males and females; Hottentot males; Zulu males and females; Southern Sotho males and females; and Venda males. Rightmire (1975, 1979) uses canonical variate analysis to allocate the Border Cave skull; distances calculated from some or all of the canonical variates are compared with the percentiles of the chi-squared distribution.

In this Section, individual Mahalanobis distances, typicality and group membership probabilities are presented, based on the results outlined in previous Sections.

5.1. Distributional Assumptions

A Gaussian Q-Q plot of Mahalanobis distances for Zulu males (n=40) is shown in Figure 3. The distances are calculated by resubstitution using the usual means and covariances, with the cube roots of the squared Mahalanobis distances being plotted against Gaussian

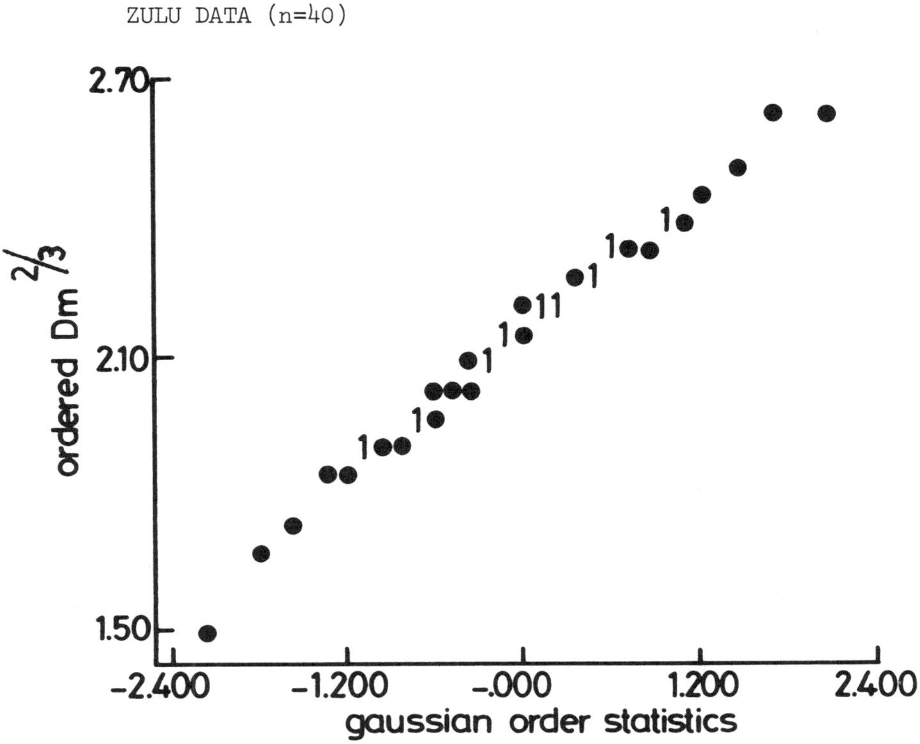

Figure 3. Gaussian Q-Q plot of usual Mahalanobis distances for Zulu male data (n=40). The number 1 indicates 1 overprinting.

quantiles. The linearity of the plot suggests that the multivariate Gaussian assumption is reasonable for this group. When robust calculations are used, all observations are given unit weight.

Application of the robust procedures discussed by Campbell (1980) to these data do not show any atypical observations. To illustrate the influence an atypical value may have on the procedure, the value for variable 5 for observation 8 for the Zulu data was reduced by 5 units (the standard deviation for v5 is 2.44); the value is still within the range of values for the other observations. There is no clear indication of an atypical value in the Gaussian Q-Q plot of usual Mahalanobis distances in Figure 4(a).

When robust estimates of the means and covariances are used in the calculations, a Q-Q plot clearly indicates one atypical observation (Figure 4(b)); the associated weight for observation 8 is 0.14. The estimated means and standard deviations are little

Modified Zulu data

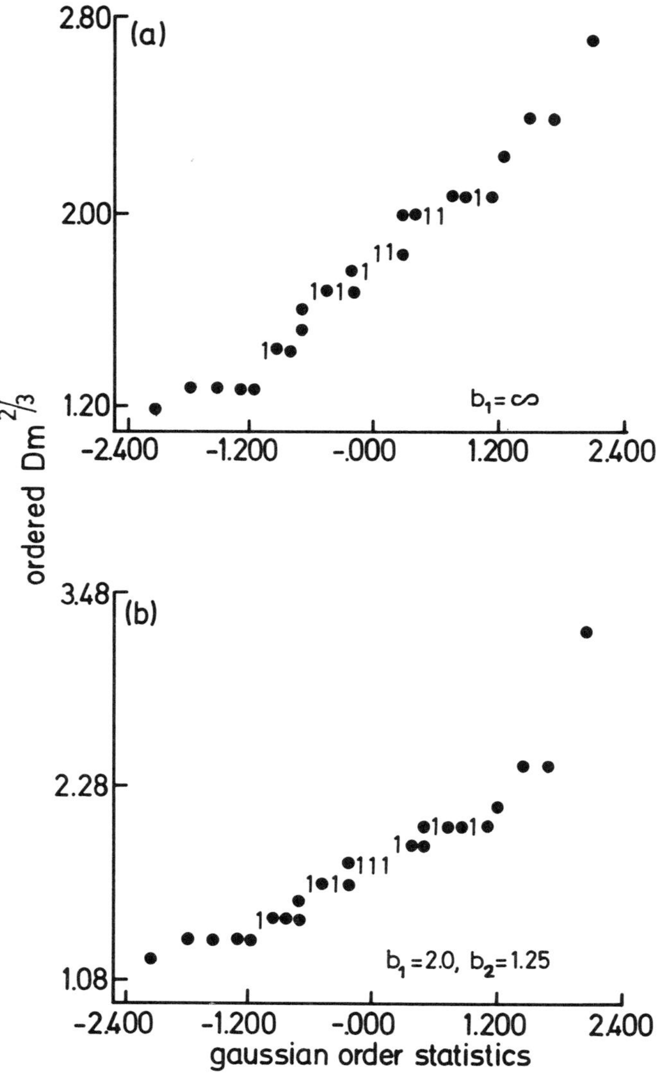

Figure 4. Gaussian Q-Q plot of Mahalanobis distances for modified Zulu male data (for observation 8, variable 5, 22 is changed to 17): (a) using distances calculated from usual estimates of means and covariances; (b) using distances calculated from robust estimates of means and covariances.

affected, while the estimated correlation r(4,5) increases from 0.890 to 0.938; the correlation r(4,5) for the original Zulu data is 0.949.

5.2. Allocation of Border Cave Specimen - Equal Covariance Matrices

The individual squared Mahalanobis distances for the comparison of the Border Cave skull with each of the reference groups are given in Table 1; the associated typicality and posterior probabilities of group membership are also given.

The posterior probabilities of group membership suggest that the Border Cave skull is male and is more closely related to the Khoisan peoples (Bushman and Hottentot) than to Negro populations (Zulu, Sotho, Venda). However the small typicality probabilities would indicate that some caution is necessary with interpretation. Referring the D^2 values to the chi-squared distribution leads to even more conservative conclusions.

5.3. Allocation of Border Cave Specimen - Unequal Covariance Matrices

A graphical comparison of the covariance matrices (Campbell, 1981), indicates that the correlations are similar for all groups, while the variances for the females are consistently smaller than those for males. The Bushman groups have smaller variances than those for other groups of the corresponding sex. Hence an assumption that the covariance matrices differ from group to group seems more reasonable than that made in the previous Section.

The squared Mahalanobis distances and associated probabilities based on the individual covariance matrices are given in Table 2. The posterior probabilities of group membership indicate that, when compared with the other reference groups, the Border Cave specimen is most likely to be a Bushman male. The typicality probabilities support the conclusion that Border Cave is best referred to as Bushman-like (see Rightmire, 1981). Note that if the typicality calculations were based on the chi-squared distribution, then the Border Cave skull would be atypical for all the modern South African groups considered here.

Table 1. Squared Mahalanobis distances (D^2) and associated probabilities relating the Border Cave individual to eight groups of modern South African crania - assuming EQUAL covariance matrices

	Bushman		Hottentot	Zulu		Sotho		Venda
	Males n=22	Females n=17	Males n=18	Males n=40	Females n=32	Males n=35	Females n=30	Males n=33
D^2 calculated from 9 variables for male and female samples	18.8	36.3	17.8	26.7	27.7	23.3	32.9	28.5
Posterior probabilities of group membership, assuming equal prior probabilities, using multivariate Student densities	0.335	0.000	0.577	0.010	0.007	0.045	0.001	0.005
Typicality probabilities, using the F distribution	0.050	0.000	0.069	0.004	0.003	0.012	0.001	0.002
"Typicality" probabilities, using the chi-squared distribution	0.027	0.000	0.038	0.002	0.001	0.006	0.000	0.000

Table 2. Squared Mahalanobis distances (D^2) and associated probabilities relating the Border Cave individual to eight groups of modern South African crania - assuming UNEQUAL covariance matrices

	Bushman		Hottentot	Zulu		Sotho		Venda
	Males n=22	Females n=17	Males n=18	Males n=40	Females n=32	Males n=35	Females n=30	Males n=33
D^2 calculated from 9 variables for male and female samples	20.9	40.06	38.9	43.3	37.8	22.9	52.2	31.6
Posterior probabilities of group membership, assuming equal prior probabilities, using multivariate Student densities	0.910	0.041	0.015	0.000	0.007	0.023	0.001	0.003
Typicality probabilities, using the F distribution	0.292	0.150	0.132	0.003	0.016	0.098	0.004	0.032
"Typicality" probabilities, using the chi-squared distribution	0.013	0.000	0.000	0.000	0.000	0.006	0.000	0.000

References

Aitchison, J. and Dunsmore, I.R.: 1975, Statistical Prediction Analysis. Cambridge, Cambridge University Press.

Aitchison, J., Habbema, J.D.F. and Kay, J.W.:1977, A critical comparison of two methods of statistical discrimination. Appl. Statist., 26, 15-25.

Anderson, T.W.: 1958, An Introduction to Multivariate Statistical Analysis. New York, Wiley.

Campbell, N.A.: 1980, Robust procedures in multivariate analysis. I. Robust covariance estimation. Appl. Statist., 29, 231-237.

Campbell, N.A.: 1981, Graphical comparison of covariance matrices. Aust. J. Statist., 23, 21-37.

Campbell, N.A. and Atchley, W.R.: 1981, The geometry of canonical variate analysis. Syst. Zool., 30, 268-280.

Gnanadesikan, R.: 1977, Methods for Statistical Data Analysis of Multivariate Observations. New York, Wiley.

Hampel, F.R.: 1977, Modern trends in the theory of robustness. Math. Operationsforsch. Statist., Ser. Statistics, 9, 425-442.

Healy, M.J.R.: 1968, Multivariate normal plotting. Appl. Statist., 17, 157-161.

Huber, P.J.: 1977a, Robust covariances. In: Statistical Decision Theory and Related Topics II (Shanti S. Gupta and David S. Moore, eds.), pp. 165-191. New York, Academic Press.

Huber, P.J.: 1977b, Robust Statistical Procedures. Philadelphia Siam.

Huber, P.J.: 1981, Robust Statistics. New York, Wiley.

Kshirsagar, A.M.: 1972, Multivariate Analysis. New York, Marcel Dekker.

Maronna, R.A.: 1976, Robust M-estimators of multivariate location and scatter. Ann. Statist., 1, 51-67.

Murray, G.D.: 1977, A note on the estimation of probability density functions. Biometrika, 64, 150-152.

Rightmire, G.P.: 1975, New studies of post-Pleistocene human skeletal remains from the Rift Valley, Kenya. Amer. J. Phys. Anthrop., 42, 351-369.

Rightmire, G.P.: 1979, Implications of Border Cave skeletal remains for later Pleistocene human evolution. Curr. Anthrop., 20, 23-35.

Rightmire, G.P.: 1981, More on the study of the Border Cave remains. Curr. Anthrop., 22, 199-200.

Small, N.J.H.: 1978, Plotting squared radii. Biometrika, 65, 657-658.

INTERPRETATION AND TESTING IN MULTIVARIATE STATISTICAL APPROACHES TO PHYSICAL ANTHROPOLOGY: THE EXAMPLE OF SEXUAL DIMORPHISM IN THE PRIMATES

Charles E. Oxnard

The Graduate School, and the Departments of Biological Sciences and Anatomy, University of Southern California, USA.

ABSTRACT: During consideration of general problems of testing and interpretation of multivariate statistical methods in physical anthropology, the question of bias due to pooling of sex subgroups has been an issue. Though, as expected, recent tests have confirmed the validity of such pooling, they have also revealed unexpected information about structural sexual dimorphism itself. Thus, canonical variate analysis of primate overall bodily proportions shows that sexual differences are most marked in proportions representing breadths. In conjunction with high-dimensional displays, they further demonstrate that a number of different structural sexual dimorphisms exist rather widely in the entire Order. This finding has implications for our understanding of the causal mechanisms of structural sexual dimorphism in living species, for our attempts to assess sexual differences in fossils, and, by extension, for our understanding of change in non-structural sexual dimorphisms in humans.

INTRODUCTION: TESTING AND INTERPRETATION

The evaluation and interpretation of multivariate statistics in studies of anatomical structure is a vexing question in physical anthropology today. Their usage has been characterized as lying anywhere from one end of a spectrum implying that they can give major insights in anthropological problems over and above those of visual assessment (e.g. Oxnard, 1973, 1978), to the other end of the spectrum suggesting that they provide only obfuscation and confusion under the guise of mathematical sophistication (e.g. Day,

G. N. van Vark and W. W. Howells (eds.), Multivariate Statistical Methods in Physical Anthropology, 193–222.

1977). That these opposite opinions can be held at the same time is partly (a) because the difference between results and interpretations stemming from results has not always been clearly understood, partly (b) because it is not common for published presentations to include full details of all the testing necessary before reliance can be placed upon results, and partly (c) because it is not always recognised that many different types of multivariate statistical analysis exist and that they can be applied in different ways for a variety of problems.

Two decades of studies from our laboratories now allow review and assessment of some of the practical problems in anthropological evaluation and interpretation of the results of canonical variate analysis (summarized in Oxnard, 1983a). Thus, interpretations resulting from study of dimensions of the primate shoulder (Oxnard, 1963; Ashton and Oxnard, 1963, 1964a, b; Ashton, Healy, Oxnard and Spence, 1965) have now been evaluated by a longitudinal set of investigations. Some of these investigate new materials on the shoulder using the same approach (e.g. canonical variate analysis applied to new shoulder data in primates and other mammals, Oxnard, 1967, Oxnard, 1968, Ashton Flinn, Oxnard and Spence, 1971). Others study the original materials using approaches independent of multivariate statistics (e.g. neighborhood limited classification, experimental stress analysis, Oxnard and Neely, 1969, Oxnard, 1973). As our current investigations proceed equivalent longitudinal studies for testing and evaluation are becoming available for each of the other anatomical regions in which we are interested (e.g. the pelvis, Zuckerman, Ashton, Flinn, Oxnard and Spence, 1973, see: Ashton, Flinn, Moore, Oxnard and Spence, 1981).

Our studies have also been evaluated by a parallel set of investigations (summarized in Oxnard, 1983a). These assess the degree to which study of neighboring anatomical regions (e.g. the arm, the forearm, the upper limb as a whole, Ashton, Flinn and Oxnard, 1975; Ashton, Flinn, Oxnard and Spence, 1976; Oxnard, 1983b) when investigated by measurement and canonical variate analyses, provide results compatible with the interpretations offered by the prior canonical variate studies of the shoulder. And similar parallel investigations are being attempted for other anatomical regions (e.g. for the pelvis: Oxnard, 1983b).

Yet again, our interpretations have been assessed by new studies (summarized in Oxnard, 1983a) which relate to the clustering of variables (anatomical elements) rather than to the arrangement of groups (genera of primates - more fully described in Oxnard, 1983b). Interpretations should make sense in terms of anatomies as well as animals.

Finally, some investigators believe that particular multivariate statistical approaches may give conflicting results when applied to given anatomical regions by different investigators (e.g. Day, 1977). Though it will certainly be the case that individual investigators will define different measures of a particular anatomical region, given that the measures in each study cover approximately the same anatomical parts there should be marked similarities between the results of different laboratories. This, too, has been re-examined (Oxnard, 1983a) because the accidents of coincident activities of several workers recently have resulted in approximate replication of investigations. The result is that very similar findings are indeed achieved by different investigators (e.g. on the arm and forearm, Ashton, Flinn, Oxnard and Spence, 1976 and Feldesman, 1976; on part of the hominoid pelvis, e.g. Zuckerman, Ashton, Flinn, Oxnard and Spence, 1973 and McHenry and Corruccini, 1975; on the talus in higher primates, e.g. Day and Wood, 1968 and Oxnard, 1972 together with Lisowski, Albrecht and Oxnard, 1974, 1976; and on the anthropoid shoulder, e.g. Ashton Flinn, Oxnard and Spence, 1971 and Corruccini and Ciochon, 1976, this last as outlined in Oxnard, 1977).

The result of all these investigations has been not only to confirm and greatly extend the interpretations resulting from the initial studies but also to increase our general confidence in our use of canonical variate analysis.

In the process of performing all of these studies, much internal testing has also been carried out. Such tests usually receive only cursory description in physical anthropology due to limitations imposed by publication. They have certainly not been fully recognised by all investigators. But they are important parts of our investigations; they were briefly described in Oxnard (1973) and have been drawn together recently in Oxnard (1983a). In this category are tests required because of the existence of differently aged specimens, of subspecific or geographic variations, of pathological lesions, and of course, of the two sexes in each species. In some cases, e.g. pathology, most problems can be eliminated by restricting samples to those specimens not showing the confounding factor. In other cases e.g. sex and age, specific tests are required because it is not easy to eliminate these phenomena from data (tests summarised in Table 1).

Study of one set of these tests, those for sex differences, has lead to an unexpected result. It was determined that

TABLE 1: TESTING OF MULTIVARIATE MORPHOMETRIC ANALYSES

Simple tests:

- Errors of replication.
- Inter-observer errors.
- Inter-instrument errors.
- Differences between sex, age, subspecific, pathological, and other confounding groupings.
- Normality of data, and if not normal, appropriate transformation.
- Equality of variance, and if not equal, appropriate transformation.
- Univariate, bivariate and multivariate searches for outliers.

More complex tests:

- Principal components anaylsis on individual groups to study normality and homogeneity of dispersion.
- Principal components analysis on individual groups to identify hidden subgroups of specimens.
- Principal components analysis to summarise information from group means alone.

Tests internal to ultimate analysis:

- Significance of latent roots.
- Variance co-variance tests.
- Homogeneity of dispersion matrices.
- Significance and percent of information within individual axes.
- Tests of significance of positions of individual groups.

Tests of ultimate analysis:

- Groupings of animals:
  - Based on all genera separately.
  - Based upon pooled groups from classifications, e.g. all Old World monkeys, all apes.
  - Based upon function (often locomotion), e.g all quadrupeds, all leapers.
  - Based upon combinations, e.g. all leaping prosimians.
  - Based upon other factors, e.g. all forms apparently showing neoteny, or all forms from the same geographic area.

TABLE 1: CONTINUED

Groupings of anatomies:
- Individual anatomical regions, e.g. pelvis.
- Individual functional regions, e.g. lower humerus, upper radius and ulna = elbow joint.
- Anatomical form of variable, e.g. all transverse variables, all longitudinal variables.
- Metrical form of variable, e.g. all angles, all indices, all measures.

Sample size and numbers of variables:
- Large samples versus large plus small samples.
- Analyses with and without data from single specimens.
- Analyses with and without particular peculiar variables (e.g. tail length especially variable).

Tests relating to overall bodily size:

- Tests comparing analyses of measures with other variables.
- Tests of extent to which canonical variates represent size.
- Test of extent to which individual factors represent size.
- Tests of manipulating size (e.g. through regression adjustment).

Methods of display:

- Plots of one, two or three three canonical axes.
- Dendrograms of generalised distance connections.
- Minimum spanning trees of generalised distance connections.
- Models, usually three-dimensional, of generalised distance connections.
- High dimensional displays of patterns of canonical or factor axes.

Independent corroboration

- Use of several appropriate multivariate methods.
- Use of methods with independent axiomatic bases, e.g. neighborhood limited classification.
- Combinations of methods, e.g. neighborhood limited classification and discriminant function analysis.

sexual differences, though fairly often statistically significant, were in general small and that males and females of each given group might therefore be pooled in the main studies without overly biasing them. But further study, resulting in part from the discussion at the original meeting of the Intercongress of the International Union of Anthropological and Ethnological Sciences in Amsterdam in 1981, has shown that additional sexual differences exist. Although they do not especially affect interpretations of prior canonical variate analyses of primate structure as carried out in our laboratories, they could well have implications for our general understanding of the phenomenon of sexual dimorphism.

SEXUAL DIMORPHISM IN PRIMATES

Characters peculiar to each sex but not directly part of the sexual apparatus are found in the majority of animals. This sexual dimorphism has been widely studied, especially in humans, and encompasses many different phenomena, not only structural but functional, not only biological but behavioral, not only social but psychological, cultural and so on (e.g. Montagu, 1974, Friedman, Richert and Van de Wiele, 1974).

One very obvious subset of such characters involves the size and shape of the body. Among the primates (e.g. Schultz, 1969, Napier and Napier, 1967) males may be more than twice the size of females, for example, in baboons, proboscis monkeys, orang utans and gorillas. Proportions differ between males and females; for instance, male orang utans have a chest girth over 200 percent of trunk length, but females only about 170 percent.

This structural dimorphism is well established - differences in size go with differences in proportions, and both are related to robusticity. The idea, a single phenomenon of sexual dimorphism with differential expression, seems to be especially confirmed by the studies of Schultz who examined proportional measurements of various major segments of the body of many representative genera of the Order (Schultz, 1969). It is also the specific finding of the only major study of several genera that utilizes more than simple statistical methods (the detailed study by Wood, 1975, using correlation, regression and Penrose's size and shape factors for studying measurements of five species: humans, chimpanzees, gorillas, baboons, and colobs). Wood reports that the "idea that considerable differences in shape exist between males and females must be rejected"; he suggests that such differences in shape as appear to exist are simply the result of differential size.

In earlier studies (reviewed in Schultz, 1969) sexual dimorphism was interpreted as having greater expression in terrestrial primates than in arboreal species, and in savannah-living forms as compared with forest-living species. It is easy to note particular genera to which these ideas do not apply. More recently, the degree of sexual dimorphism has been related to sexual competition within differently organized social groups (e.g. Gautier-Hion, 1975; Clutton-Brock, Harvey and Rudder, 1977; Leutenegger and Kelly, 1977; Harvey, Kavanagh and Clutton-Brock, 1978). Thus, those primates that live in complex polygynous groups, in which there may be overt competition within the sexes for access to the opposite sex, show greater degrees of sexual dimorphism. In contrast, among those primates that live in nuclear families, such competition may be less, and lesser degrees of sexual dimorphism exist. In each of these cases, troop defence, whether against conspecifics or predators, may be additionally implicated; in species in which reproductive groups consist of many males and females, males are often more involved in troop defence; in species with the nuclear family, defence seems to be more equally carried out by both males and females.

Other explanations for the phenomenon of sexual dimorphism can be readily suggested. For instance, those species that apparently display neoteny - retention of immature-looking forms into sexually mature stages of life - are often less conspicuous in their sexual dimorphism: many prosimians, certain New World monkeys, and lesser apes and humans among the hominoids. Sexual dimorphism may be related, for example, to the additional energetic, physiological, social, or psychological "costs" of the female reproductive process.

Some of the ideas cited above have been applied to human ancestors and other fossil primates (e.g. in assessing savannah as opposed to forest habitats for particular fossils as by Pilbeam, 1972, or positing nuclear family as opposed to polygynous social organizations for particular fossils as in Fleagle, Kay and Simons, 1980 and Gingerich, 1981).

The data on the overall proportions of many primates, collected by the late A. H. Schultz and examined by him univariately, allow the notion of sexual dimorphism as a single spectrum with differential expression to be tested in a more subtle way because multivariate statistical approaches have become available since his time. Accordingly, those data, kindly made available to me during Professor Schultz' lifetime, are here analyzed using the methods of canonical variate analysis (e.g. Ashton, Healy and Lipton, 1967; Oxnard, 1973), and the results are displayed using Andrews' high-dimensional technique (Andrews, 1972, 1973; Oxnard, 1975; Wilson, 1981 and this volume).

THE MATERIALS, AND THE DATA

Schultz' data are from 472 primate specimens representing 34 genera of the primates (Schultz, 1929, 1956; Ashton, Flinn and Oxnard, 1975; Oxnard, 1983b). The sexes of specimens are known from field records. Not every genus measured by Schultz can be used because of deficiencies in the sizes and representations of the samples. Such considerations reduce the number of usable genera to 18 and specimens to 455 (table 2).

The data were calculated from measurements of adult individuals in which the full permanent dentition was present. The techniques used to take the measuremelts ald to assess mensurational accuracy are described by Schultz (1929, 1956) and Oxnard (1983b). Schultz compounded his original measurements into 23 ratios describing the overall shapes of the various bodily regions (table 3). Unfortunately, the original measures are no longer available, and the original measurements cannot be regenerated.

INITIAL UNIVARIATE STUDIES

The univariate data were logarithmically transformed and this removed an association that was found to exist between the standard deviations of the dimensions and their means. A number of preliminary statistical examinations were carried out (univariate distribution plots, bivariate scatterplots, multivariate scatterplots, statistical tests of distributions in those genera with large enough samples to allow this, investigations of large groups separately from more limited groups, and so on, as outlined in Oxnard, 1978 and 1983a). These preliminary studies confirmed that features liable to perturb subsequent statistical examinations were not present in the data.

All 23 ratios were examined univariately. Ten showed very little sexual differentiation at all. Seven ratios involving the lengths of the different bodily regions indicated several individual statistically significant differences between the sexes but all were very small. Six ratios containing bodily breadths showed many differences between the sexes that were both statistically significant and big.

As expected from Schultz (1969), Napier and Napier, (1967), and Wood (1975) the major differences between the sexes proved to be in shoulder breadth and hip breadth, relative shoulder

TABLE 2: NUMBERS OF SPECIMENS IN EACH SEX SUBGROUP OF EACH GENUS

Genera	Females	Males
Nycticebus	4	5
Galago	3	7
Tarsius	9	2
Aotus	6	6
*Alouatta**	2	2
Cebus	11	14
Saimiri	23	26
Ateles	47	27
Leontocebus	10	14
Macaca	17	10
*Cercocebus**	2	1
Presbytis	7	7
Nasalis	15	10
Hylobates	37	41
Pongo	8	5
Pan	17	9
Gorilla	6	5
Homo	20	20
Total 18	Total 244	Total 211

*The numbers of specimens in these two genera are very small. The genera were included in the study because the differences between their sexes are very large.

TABLE 3: DIMENSIONS USED IN MORPHOMETRICS OF OVERALL BODILY PROPORTIONS

Schultz' data on upper limbs

Chest circumference relative to trunk height
Shoulder breadth relative to trunk height
Upper limb length relative to trunk height
Intermembral index
Brachial index
Hand length relative to trunk height
Hand breadth relative to hand length
Thumb length relative to hand length

Schultz' data on lower limbs

Hip breadth relative to trunk height
Lower limb length relative to trunk height
Intermembral index
Crural index
Length of foot relative to limb length
Length of foot relative to trunk height
Foot breadth relative to foot length

Schultz' data on head, neck and trunk

Chest circumference relative to trunk height
Shoulder breadth relative to trunk height
Hip breadth relative to trunk height
Chest index: chest breadth as percent of saggital chest diameter
Head size: average head diameter relative to trunk height
Face height relative to trunk height
Upper face height relative to average head diameter
Cephalic index: head breadth as percent of head height
Interocular index: inner eye breadth as percent of face breadth
Relative Ear size: ear height x ear breadth as percent of head length x total head height

Subanalyses performed without duplication of variables

breadth being often markedly greater in males than in females, and relative hip breadth being often greater in females than in males. The chest index was generally higher in males than in females, but the degree of difference was less than for the afore-mentioned ratios. Relative foot breadth was less informative still, being only slightly significantly different in nine genera. And the remaining two ratios of head breadth (cephalic index and relative facial breadth) were least informative of all of this group, being different in only four and five genera respectively. If additional information is present in these data, then it is in the intercorrelations among them, and it cannot thus be revealed by univariate statistical analyses.

MULTIVARIATE STATISTICAL STUDIES: CANONICAL VARIATE ANALYSIS

Canonical variate analyses were used to examine these dimensions in three groups: (a) all 23 ratios characterizing all areas of the body, (b) seven including only length measures of the body, and (c) six comprizing only breadth measures.

The study of all 23 dimensions, although based on a reduced number of genera and including sex-subgroups, provided a pattern of information about the distributions of the genera entirely similar to that found in the earlier study using all genera and with the sex-subgroups pooled (Ashton, Flinn and Oxnard, 1975). The separations of the genera were essentially similar to the overall systematic arrangement of the primates; the separations between the sexes were negligibly small (Oxnard, 1981, 1983b).

A plot of the first against the second canonical variate of the study of length dimensions also demonstrated very little difference between the sexes (figure 1). Inspection of the remaining canonical variates and the entire generalized distance matrix confirmed this.

But plots of the first three canonical variates of the study of breadth dimensions separated the sexes markedly from one another (figure 2 and Oxnard, 1983b). Figure 2 also suggests that the study of breadth dimensions effects a number of different patterns of separation between the sexes. Thus about half of the primate genera show sexual separations that are parallel with one another; that is, whatever is the position of the males of a given group, the females differ in each direction of canonical space so that lines joining males and females of each genus are approximately parallel in each plot. A number of genera show separations between the sexes that do not fit this pattern; in some of these the lines of separation

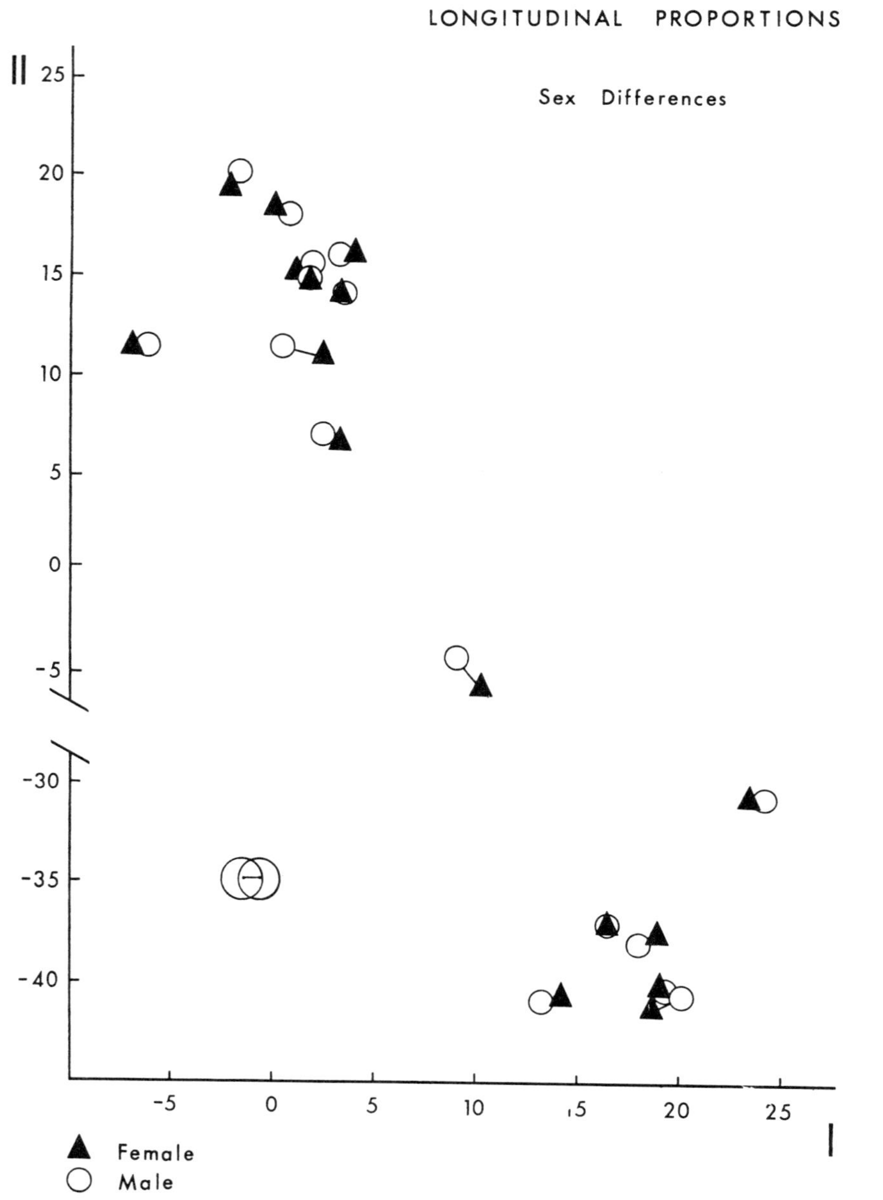

Figure 1: Plot of canonical variates one and two for the study of length dimensions of primates. Sexes of individual genera are joined by straight lines. Markers indicate standard deviation unit scales.

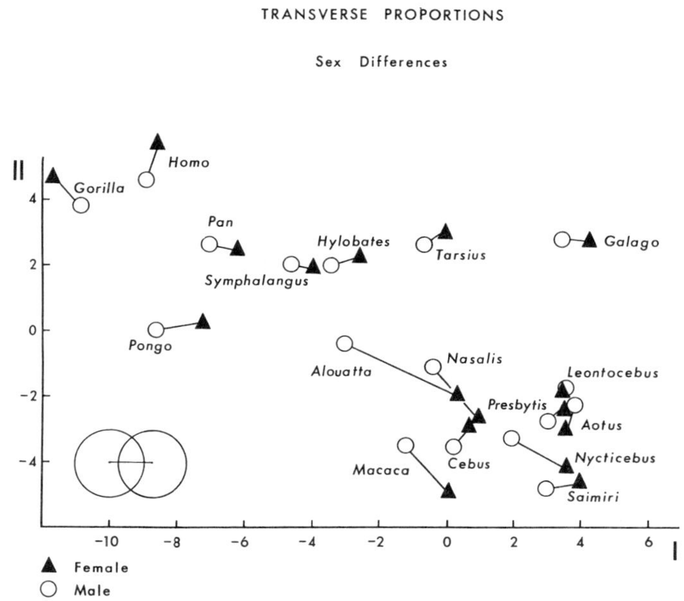

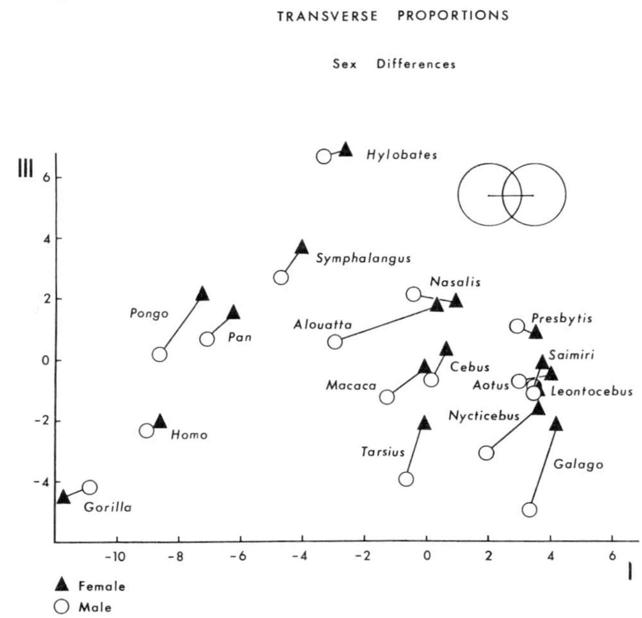

Figure 2: Plot of canonical variates one, two and three for the study of breadth dimensions of primates. Sexes of individual genera are joined by straight lines. The plot is displayed in two parts because of the confusion that would otherwise result from overlapping groups. Markers indicate standard deviation.

between the sexes are not parallel with the majority (e.g. douroucoulis, proboscis monkeys); in some they are actually at right angles to the main pattern (e.g. capuchins, humans); in one it is even reversed (e.g. gorillas).

Though the genera themselves are primarily separated by the early canonical variates in this study of breadth dimensions, additional separations of the sex sub-groups exist in higher canonical variates. It is for this reason that a high-dimensional display (first described by Andrews, 1972, 1973, summarized in a non-technical way by Oxnard, 1975, and now further described in this volume by Wilson), was used to examine the entire sub-structure of the canonical spaces (7 and 6 variates, respectively, in each study). Andrews′ method involves embedding the first "k" canonical variates values for the mean of a cluster of specimens within the first "k" of the infinite series of coefficients of a sine-cosine function. The remaining coefficients are then truncated. The shape of the resulting plot, from -pi to +pi, reflects the multi-dimensional position, within the "k" canonical variates, of the mean of that cluster. Although, therefore, the actual positions of clusters in multi-dimensional space cannot be seen directly, it is possible to note when some clusters are similar in their multi-dimensional positions to others, when some clusters lie between others, when some clusters are located entirely peripherally, when some clusters are located as the result of common patterns of canonical variate values different to others, and so on.

In order to understand Andrews′ display as used here, let us look at a series of high-dimensional plots derived from theoretical data. Figure 3 shows three curves which differ by one, two and three standard deviation units, respectively, in a first canonical variate, a second, a third, and all three canonical variates taken together. These plots provide us with a feel both for one, two and three standard deviations worth of separation in single axes, and for similar separations that are in several axes together.

With these theoretical plots in mind, it is evident that the canonical variates derived from the study of length dimensions (figure 4) provide high-dimensional plots that demonstrate rather small differences between the sexes.

In total contrast is the picture (figures 5, 6 and 7) seen in the high-dimensional examination of all canonical axes in the study of breadth dimensions. The separations between the sexes are big and fall into three categories.

The first of these (figure 5) demonstrates that, whatever

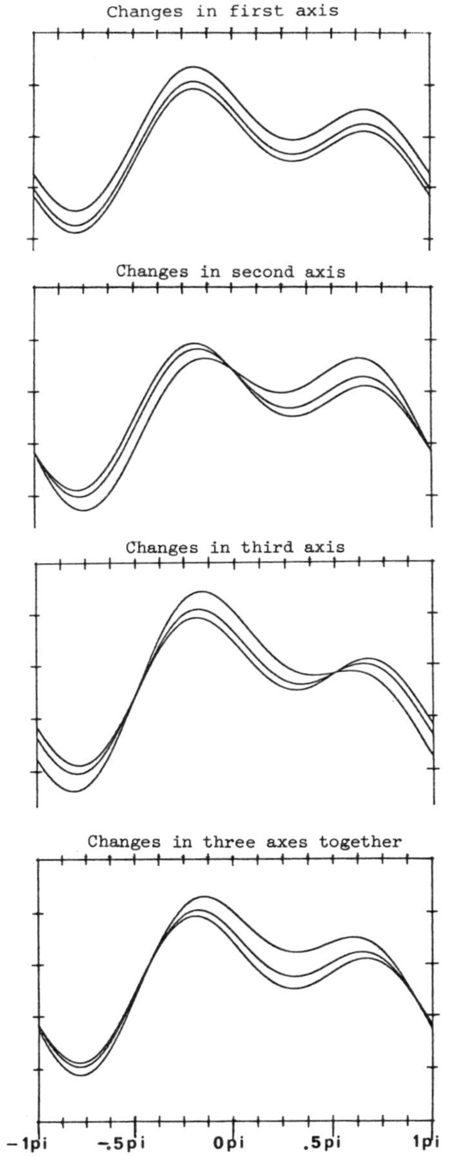

Figure 3: Andrews high-dimensional plots for theoretical data. The central plot is one standard deviation unit different from the nearer plot and two from the farther. In the first three frames, these differences are in each single canonical axis; in the fourth frame they are shared between all three canonical variates. This figure, and all of the remainder, are drawn to the same scale.

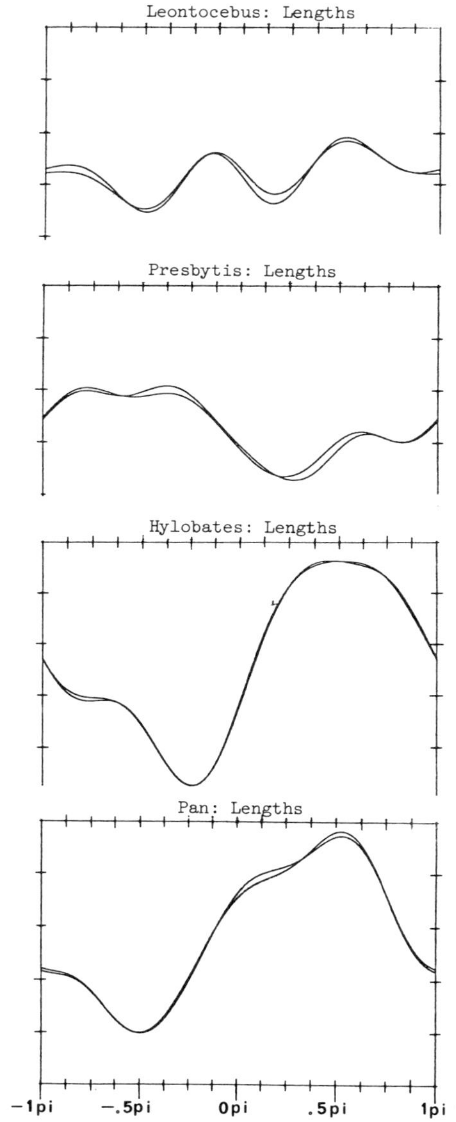

Figure 4: Andrews' high-dimensional plots for sex pairs for four representative genera from the analysis of bodily lengths. Comparison with figure 3 indicates that the differences between the sex pairs are extremely small, not only in a genus such as Leontocebus which displays only very minor sexual dimorphism, but also in a genus like Pan which is quite markedly dimorphic.

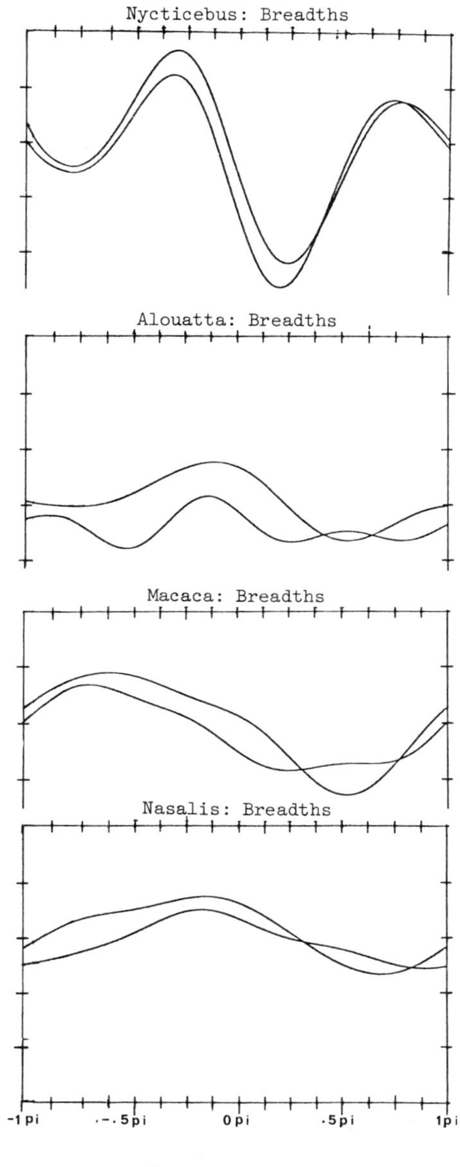

Figure 5: Andrews' high-dimensional plots from the analysis of breadth dimensions for sex pairs of the first group of genera. The comparison with figures 3 and 4 indicate that the differences between the sex pairs are large and display a common pattern.

the specific curvatures of the plots for any given genus, the separations between the sexes display similarities by being parallel for about three-quarters of the plots, and then crossing twice in the last 0.5 pi units near the right hand end of the plots. This pattern is shared by the sex sub-groups for the genera: slow lorises, howler monkeys, macaques, proboscis monkeys and mangabeys; these represent each major taxonomic group of the primates except the hominoids.

A second pattern of difference between the sexes is reflected in a series of plots that cross four times, two on the left hand side and two on the right (figure 6). This pattern is shared by several genera from each major group of anthropoids: douroucoulis, squirrel monkeys, capuchins, spider monkeys, langurs and gibbons.

Finally, there remain several genera which do not share a common pattern of difference between the sexes. Each of these is quite different from the two patterns just enumerated, and, in addition, each is also different from each of the others. They include bush-babies and tarsiers, but, in particular, they include humans and our closest living relatives, gorillas and chimpanzees, and, much more distantly, orang utans (figure 7).

A RETURN TO THE ORIGINAL VARIABLES

What are the arrangements of the original data that provide these several patterns of sexual dimorphism?

As a first step, we can determine the pattern of differences in the canonical variates that give rise to the groups of similar high-dimensional plots. The plots of figures 5 and 6, for instance, result from similar patterns of differences between the sexes among the values in the canonical axes (table 4) for each of the two groups of genera. For example, the difference between males and females is negative in the second axis for all genera in the first group, and positive in that same axis for all members of the second group. Similar patterns exist for other axes.

And each of the genera that display their own unique patterns (figure 7) display different patterns in each variate (table 4). Again, for example, the differences between males and females are different for each genus in the patterns displayed in axes two, three, four and six. Each of these axes shows a mixture of positive, negative and zero differences between the sexes.

Although all of this is easily seen with the hindsight of

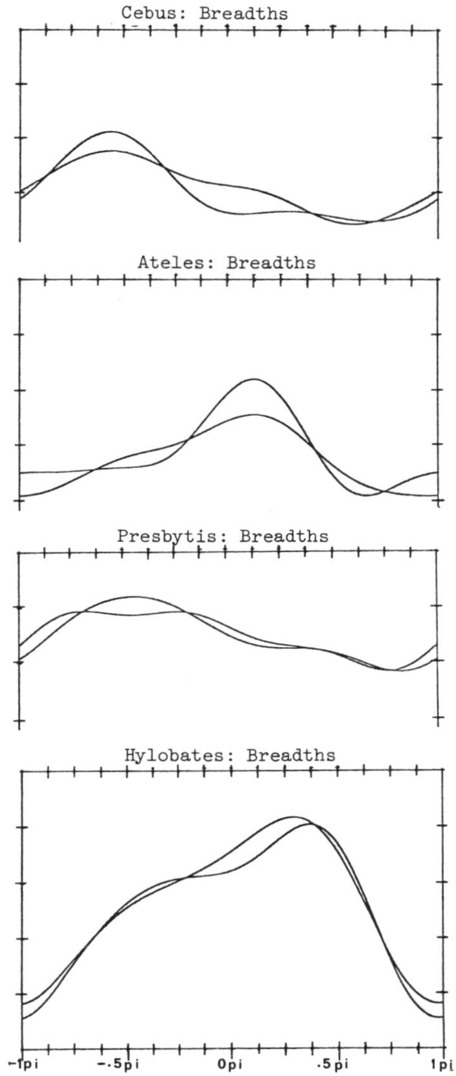

Figure 6: Andrews' high-dimensional plots from the analysis of breadth dimensions for sex pairs of the second group of genera. The comparison with figures 3 and 4 indicate that the differences between the sex pairs are large and display a common pattern, though one different from figure 5.

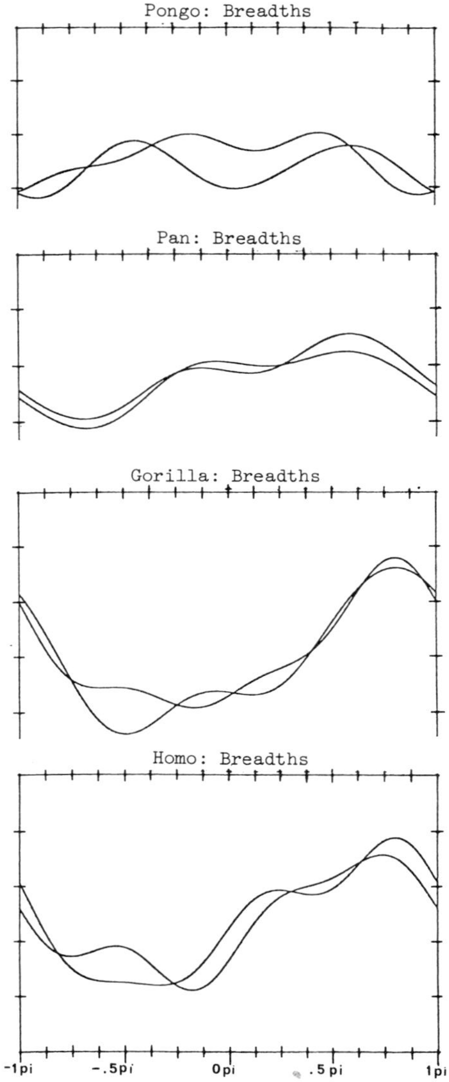

Figure 7: Andrews' high-dimensional plots from the analysis of breadth dimensions for sex pairs of hominoid genera. The comparison with figures 3 and 4 indicate that the differences between the sex pairs are large and that each genus displays its own pattern unique among the primates.

TABLE 4: DIRECTION OF DIFFERENCE IN CANONICAL VARIATES VALUES FROM MALE TO FEMALE FOR SELECTED GENERA

	Canonical variates values in each axis					
Genera	1	2	3	4	5	6
Nycticebus	+	-	+	=	+	+
Alouatta	=	-	+	=	+	+
Macaca	+	-	+	=	+	+
Cercocebus	+	=	+	-	+	-
Nasalis	+	-	-	-	+	=
Saimiri	+	+	+	+	+	+
Cebus	+	+	+	+	+	=
Ateles	+	+	+	+	+	+
Presbytis	+	+	-	+	+	-
Hylobates	+	+	=	+	+	-
Pongo	+	+	+	+	+	-
Pan	+	=	+	-	+	=
Gorilla	-	+	-	+	+	-
Homo	+	+	=	-	+	=

A more positive value for the mean of a female group as compared with a male group is indicated by + in the table.

the high-dimensional displays, it was not clearly evident before they were computed.

Second, the study can reach even further back and discover what are the patterns of original variables contributing to each set of separations between the sexes. The major contributing variables (based only upon the existence, or otherwise of especially large loading factors) are shown in table 5 which indicates that the separations in each variate depend principally upon only two or three individual variables. Again, the two groups of sex subgroups (figures 5 and 6) are characterized by common differences in the contributions of the variables. For example, in each of the first two groups of genera, relative shoulder breadth and chest index are among the chief determining variables in axes one and five. But the differences between these two groups of genera relate to the ways that these two variables interact; thus in axes one and five both contribute negatively in the first group of genera, but one contributes negatively and the other positively in the second group of genera. Similar contrasts exist for other contributing variables in other axes. These types of interactions could never be detected from univariate examination.

In the same way, the unique patterns existing between the sexes in the hominoids (figure 7) are due to individual differences in the contributions of the variables from one genus to the next. Thus the differences between the sexes in each pair of comparisons (humans v gorillas, humans v chimpanzees, gorilla v chimpanzees) are in different sets of axes, and have different combinations and signs of original variables (table 6).

GENERAL IMPLICATIONS OF THESE FINDINGS

These findings (more fully described in Oxnard, 1983c) have implications for our understanding of sexual dimorphism. No longer can we adhere to the view that this is a single spectrum with differential expression (Schultz, 1969) nor to the view that it is related primarily to size differences between the sexes (Wood, 1975). It is, for instance, of considerable interest that the differences between the sexes revealed in these studies are as big in bush-babies and tarsiers (animals with no obvious external dimorphism in size) as they are in orangutans and howler monkeys (animals with the biggest obvious external dimorphism in size among the primates). Equally, other single major explanatory ideas such as troop defence or social organization would seem to be, by themselves, inappropriate causal phenomena. Even a multifactorial

TABLE 5: PRINCIPAL CONTRIBUTIONS OF VARIABLES TO EACH CANONICAL AXIS

FIRST GROUP OF GENERA

Major contributions of variables

Axes	shoulder	chest	hip	foot	face	head breadth
1	-	-			-	
2		-	+		-	
3			+	-		
4						
5	-	-				
6						-

SECOND GROUP OF GENERA

Axes	shoulder	chest	hip	foot	face	head breadth
1	-	+			+	
2		+	-		+	
3			-	-		
4			-			
5	-	+				
6						-

(+ = larger in female)

TABLE 6: PRINCIPAL DIFFERENCES BETWEEN CONTRIBUTIONS OF VARIABLES TO CANONICAL AXES IN THREE UNIQUE GENERA

	Major contributions of variables					
Homo v *Gorilla*	shoulder	chest	hip	foot	face	head breadth
Axis 1	-	-			+	
Axis 3			+	-		
Homo v *Pan*						
Axis 2		+	-		+	
Axis 5	-	+				
Axis 6						+
Gorilla v *Pan*						
Axis 1	+	+			-	
Axis 2		-	+		-	
Axis 3			-	+		
Axis 4			-			

(+ = larger in first named genus)

explanation is inadequate if we assume that the same multifactorial pattern exists among all genera.

It is more likely that many different factors have had effects upon the evolution of sexual dimorphism and, interacting in different ways, have produced various sexual dimorphisms among the genera. Thus, though there is no single consistent arrangement of social organization, ecological niche, feeding pattern, reproductive efficiency, or even developmental transformation that fits the results here, each of these taken together in different weightings for each genus may be truly implicated. It is also possible that other, hitherto unconsidered, factors may be involved.

The findings lead on to the idea that some structural sexual differences (additional to those presumably present as a result of the initial sexual dimorphism shared with mammalian progenitors of primates) must have evolved more than once within each group. The two major patterns, because of the range of particular genera sharing them, must have evolved in parallel a number of times. And the various unique patterns which particularly describe many of the hominoids indicate that, whatever was the original sexual dimorphism prior to their separate evolution, there must have been a number of divergences in sexual dimorphisms since.

The findings also mean that when we attempt to view differences between the sexes in pre-human, pre-hominid, or even pre-hominoid fossils, we can no longer assume (as do a number of workers: Fleagle, Kay and Simons, 1980; Johanson and Edey, 1980; Gingerich, 1981) that a single spectrum of sexual dimorphism with differential expression covers all genera. Now we must be prepared to ask: is the sexual dimorphism of a particular fossil population like that of humans, or of gorillas, or of chimpanzees, or even of orang utans (a species presumably very far removed from human ancestry), or of one of the two general patterns that exist more widely among the primates? For whatever sexual dimorphism exists in the fossil group it cannot be similar to all of these at once. Indeed, we might even ask: is the sexual dimorphism of the fossil group something that resembles that of no extant form at all?

Finally, the findings mean that we may have to re-evaluate aspects of non-structural sexual dimorphisms in humans. As long as humán structural sexual differences were thought to be part of an overall primate pattern that might be as old as 70 million years (or even more) then the genetic basis of that dimorphism could be supposed to be equally old. This would provide an ancient, and perhaps very rigid, set of biological constraints within which other sexual differences (for instance social, cultural, psychological, intellectual) might have to be set.

Once, however, it is apparent that structural differences have under-gone separate genetic differentiation even since the time of a human-ape progenitor (certainly enormously less than the 70 million years of the entire primates, or even the 120 million years of the original mammals - perhaps only a very few million years for the gorilla-chimpanzee-human triad), then we must recognize that the genetic constraints acting on those structural differences are quite new and flexible. To an even greater degree therefore, genetic constraints acting on the new non-structural sexual dimorphism unique to humans may be even less rigid. A much bigger part may be epigenetic and environmental. The production of change in non-structural human sexual differences may be much less hampered by underlying biological rigidities, and much more facilitated by non-biological phenomena such as individual decision-making, family, social and cultural modification, education and legislation, than some of us may have thought.

Before we can be certain about any oi these suggestions, a bigger and better study than is provided for by Schultz′ data (large though that data set is) is needed, with larger samples for each sex, greater representation of primate genera, more detailed morphometric definitions of each form and further, more direct statistical analyses than the few that are all that we have carried out here. But, though such a study, if performed with adequate samples and at the species, even sub-species level would be enormous (requiring, possibly, a team of investigators) at least doing it now looms far more interesting than merely certifying the existence of a single spectrum of sexual difference common, in greater or lesser degree, to all primates.

SUMMARY

During the process of testing and evaluation of prior multivariate statistical studies (by canonical variate analysis) of 23 relative measurements (taken by Professor A. H. Schultz) of various bodily parts of 34 primate genera, the question of bias due to pooling of sex subgroups has been investigated.

This has resulted in studies that confirm that univariate statistical differences between the sexes in bodily proportions of 18 primate genera are negligible for ten dimensions, small for seven relative lengths, but significant and big for six relative breadths.

Canonical variate analysis of all 23 ratios, and more specifically of the seven length ratios confirms the findings from their univariate examination: that little differentiation between the sexes exists in these data.

Canonical variate analysis of the six relative breadths, however, reveals two major patterns of difference between the sexes common to two sets of genera, together with 7 patterns of difference unique to specific genera (including each hominoid genus).

The degree of these multivariate differences is unrelated to size differences between the sexes; thus the largest differences are found in orang utans and bush babies, two genera at opposite ends of the scale of size difference between the sexes.

The patterns of these structural sexual dimorphisms are due less to variations in the original variables, and more to interactions among them. This is presumably why they have not been observed in extensive prior univariate examinations.

The existence of a number of different structural sexual dimorphisms has implications for our understanding of the causal mechanisms of sexual dimorphisms in living species, for our attempts to assess sexual differences in fossils, and, by extension, for our understanding of change in non-structural sexual dimorphisms in humans.

This result has arisen through extensive testing of data for other purposes. It especially demonstrates that information can come from a multivariate statistical approach even in the absence of obvious univariate pointers.

ACKNOWLEDGEMENTS

This paper has depended upon the kindness of the late Professor A. H. Schultz in making available his original data on the overall bodily proportions of primates.

This research stems from the scientific stimulus long provided to me by Lord Zuckerman, OM, KCB, FRS, President of the Zoological Society and Honorary Professor, University of East Anglia, and Professor E. H. Ashton, University of Birmingham, U.K. The paper includes a discussion of data, results and ideas stemming from collaborative investigations carried out with Professor Ashton, Dr. R. M. Flinn and Mr. T. F. Spence of the University of Birmingham, and with Professors

F. P. Lisowski of the University of Hong Kong and G. H. Albrecht of the University of Southern California. Thanks are also due to Professors W. W. Howells of Harvard and G.N. van Vark of Amsterdam for helpful discussion, and for the invitation to give the original presentation at the Intercongress of the International Union of Anthropological and Ethnological Sciences in Amsterdam in April, 1981.

Professor F. P. Lisowski, Professor Marc Feldesman of the University of Oregon, and Dr. Nina Jablonski of the University of Hong Kong provided many critical comments upon drafts of the manuscript. Dr. Norm Campbell, a participant in the Intercongress and in this volume, is especially to be thanked for a great deal of work on a prior draft of this paper.

In addition, the work depends upon computational, drafting and secretarial tasks undertaken by Hugh C. N. Oxnard and David C. G. Oxnard.

The studies were supported by NSF grants GS 30508, DEB 81939, and DEB 24366, and by the University of Southern California.

REFERENCES

Andrews, D. F.: 1972, Biometrics, 28, pp. 125-136.
Andrews, D. F.: 1973, in "Discriminant analysis and applications". Ed. T. Cacoullos, pp. 37-59. New York: Academic Press.
Ashton, E. H., Flinn, R. M., Moore, W. J., Oxnard, C. E. and Spence, T. F.: 1981, Trans. Zool. Soc. Lond. 36, pp. 1-98.
Ashton, E. H., Flinn, R. M. and Oxnard, C. E.: 1975, J. Zool. Lond. 175, pp. 73-105.
Ashton, E. H., Flinn, R. M., Oxnard, C. E. and Spence, T. F.: 1971, J. Zool. Lond. 163, pp. 319-350.
Ashton, E. H., Flinn, R. M., Oxnard, C. E. and Spence, T. F.: 1976, J. Zool. Lond. 179, pp. 515-556.
Ashton, E. H., Healy, M. J. R. and Lipton S.: 1957, Proc. Roy. Soc. B. 146, pp. 552-572.
Ashton, E. H., Healy, M. J. R., Oxnard, C. E. and Spence, T. F.: 1965, J. Zool. Lond. 147, pp. 406-429.
Ashton, E. H. and Oxnard, C. E.: 1963, Trans. Zool. Soc. Lond. 29, pp. 553-650.
Ashton, E. H. and Oxnard, C. E.: 1964a, Proc. Zool. Soc. Lond. 142. pp. 1-28.
Ashton, E. H. and Oxnard, C. E.: 1964b, Proc. Zool. Soc. Lond. 142, pp. 49-66.
Clutton-Brock, T. H., Harvey, P. H. and Rudder, B.: 1977, Nature, 269, pp. 797-800.
Corruccini, R. S. and Ciochon, R. L.: 1976, Amer. J. Phys.

Anthrop. 45, pp. 19-38.
Day, M. H.: 1977, "Guide to fossil man". London: Cassell.
Day, M. H. and Wood, B. A.: 1968, Man, 3, pp. 440-455.
Feldesman, M.: 1976, Univ. Oregon. Anthrop. Pap. 10, pp. 1-154.
Fleagle, J. G., Kay, R. F. and Simons, E. L.: 1980, Nature, 287, pp. 328-330.
Friedman, R. C., Richart, R. M. and Vande Wiele, R. L.: 1974, "Sex differences in behavior". London, Wiley.
Gautier-Hion, A.: 1975, Mammal. 39, pp. 365-374.
Gingerich, P. D.: 1981, Amer. J. Phys. Anthrop. 56, pp. 217-234.
Harvey, P. M., Kavanagh, M. and Clutton-Brock, T. H.: 1978, J. Zool. 186, pp. 475-485.
Johanson, D. and Edey. M.: 1980, "Lucy: the beginnings of humankind". New York, Simon and Schuster.
Leutenneger, W., Kelly, J. T.: 1977, Primates, 18, pp. 177-186.
Lisowski, F. P., Albrecht, G. H. and Oxnard, C. E.: 1974, Amer. J. Phys. Anthrop. 41, pp. 191-215.
Lisowksi, F. P., Albrecht, G. H. and Oxnard, C. E.: 1976, Amer. J. Phys. Anthrop. 45, pp. 5-18.
McHenry, H. M. and Corruccini, R. S.: 1975, Amer. J. Phys. Anthrop. 43, pp. 263-270.
Montagu, A.: 1974, "The Natural Superiority of Women". New York, Colliers.
Napier J. R., and Napier, P. R.: 1967, "A Handbook of the Living Primates". London, Academic Press.
Oxnard, C. E.: 1963, Symp. Zool. Soc. Lond. 10, pp. 165-182.
Oxnard, C. E.: 1967, Amer. J. Phys. Anthrop. 26, pp. 219-240.
Oxnard, C. E.: 1968, J. Morph. 126, pp. 249-290.
Oxnard, C. E.: 1972, Amer. J. Phys. Anthrop. 37, pp. 3-12.
Oxnard, C. E.: 1973, "Form and pattern in human evolution: some mathematical, physical and engineering approaches". Chicago: Univ. Chicago Press.
Oxnard, C. E.: 1975, "Uniqueness and diversity in human evolution: morphometric studies of australopithecines". Chicago: Univ. Chicago Press.
Oxnard, C. E.: 1977, Amer. J. Phys. Anthrop. 46, pp. 367-374.
Oxnard, C. E.: 1978, Ann. Rev. Ecol. Syst. 9, pp. 219-241.
Oxnard, C. E.: 1981, Homo, 32, pp. 149-176.
Oxnard, C. E.: 1983a, Zeitschr. f. Morphol. u. Anthrop. 73, pp. 237-278.
Oxnard, C. E.: 1983b, "The Order of Man: a biomathematical anatomy of the primates". Hong Kong, Hong Kong Univ. Press.
Oxnard, C. E.: 1983c, Amer. J. Primatol. 4, pp. 1-22.
Oxnard, C. E. and Neely, P. M.: 1969, J. Morph. 129, pp. 117-148.
Pilbeam, D. R.: 1972, "The Ascent of Man". New York, Macmillan.
Schultz, A. H.: 1929, Contr. Embryol. pp. 213-257.
Schultz, A. H.: 1956, Primatol. 1, pp. 887-964.
Schultz, A. H.: 1969, "The Life of Primates". New York,

Universe Books.
Wilson, S.: 1981, "Towards an understanding of data in physical anthropology". Amsterdam, Proc. Intercongr. Internat. U. Anthrop. Ethnol. Sci.
Wood, B.: 1975, "An analysis of sexual dimorphism in primates". Ph.D. Diss. Univ. London.
Zuckerman, S., Ashton, E. H., Flinn, R. M., Oxnard, C. E. and Spence, T. F.: 1973, Symp. Zool. Soc. Lond. 33, pp. 71-165.

A REPORT ON THE HERITABILITY OF SOME CRANIAL MEASUREMENTS AND NON-METRIC TRAITS

Torstein Sjøvold

Osteological Research Laboratory, S-171 71 Solna, Sweden

The human skull offers a rigid unit for the study of past populations and for personal identification. Consisting of at least 21 individual bones, united according to anatomical principles peculiar to man, the skull develops during childhood and youth to reach a size and expression which in general, but not completely in detail, may be regarded as final. Because of genetical differences between the major races of the world (Caucasian, Mongoloid, Negroid), principal differences in the appearance between typical skulls from these groups may easily be pointed out in terms of morphology. These differences are the result of different kinds of skull growth in the different groups, and may be expressed in terms of facial flatness, cranial indices, or degrees of prognatism. By tradition in craniometry, and more or less supported by direct or indirect evidence such as historical records, differences in mean skull measurements between subpopulations from neighbouring geographical regions are similarly regarded as expressions of population differences related to differences in the genetic composition of the populations.

The differences between mean measurements are, furthermore, the result of how the different bones of the skull grow in the different populations. In this manner an individual skull may be said to express the general skull measurements characteristic of the population to which it belongs, modified by individual deviations from the mean measurements of the population. These deviations are limited by the possibilities for individual development of the different bones which, in turn, may be due to genetic and environmental factors.

In addition to dimensions regarding size or shape of the bones of the skull, each bone may exhibit variations along its borders or on the surface of the bone, which are known as non-metric traits.

G. N. van Vark and W. W. Howells (eds.), Multivariate Statistical Methods in Physical Anthropology, 223–246.

This kind of variation may be explained as phenotypical or individual variation due to differences in ossification, formation of ossification centres, or in the passage of nerves or vessels through or between the bones. Normally, such characters can only be recorded as either present or absent, though when present, for some traits a graded expression may be observed. At the present time there are indications that much of this kind of variation in the human skull is genetically determined, though formally this has been demonstrated in only a few cases in man (Torgensen 1951a, 1951b; Suzuki & Sakai 1960).

Craniometry has by tradition been mostly concerned with measurements defined by early anthropologists such as Martin (1928). These measurements are defined between formally described landmarks on the skull. Because of their vast numbers, it may be argued that many measurements have been defined simply because of convenience. There may simply exist suitable landmarks on the skull between which a measurement may be defined. In practice, a suitable subset of measurements is selected for an analysis (Rösing 1974), though it has not yet been proved that the single measurements or any subset of measurements is of any genetic value, that is, if the measurements are to any degree hereditary features, as far as man is concerned. Genetic studies of measurements of the head of the living have provided evidence that even skull dimensions are hereditary (e.g., Bernhard et.al. 1978), and similar support is provided by studies of the heritability of skull measurements in the house mouse (Leamy 1974). With regard to the heritability of non-metric traits, studies on inbred and wild populations of mice have demonstrated genetic factors in connection with the manifestation of the traits (Grüneberg 1963, Berry 1964). More recently, heritability estimates for non-metrical skull traits have been published for random-bred house mice (Self & Leamy 1978), and for the Cayo Santiago Rhesus macaques (Cheverud & Buikstra 1981), demonstrating that the traits are to some extent under genetic control.

The difficulty of investigating the heritability of cranial measurements and non-metric traits in man is that suitable skull series of individuals with known family relationships are almost nonexistent. With regard to teeth, which are easily studied in the living, the heritability of both measurements and non-metrical traits have been studied (Mizoguchi 1977a, 1977b). The present report is the first study of the heritability of measurements and non-metric traits of the human skull, based on a collection of skulls from individuals with known identity and family relationships.

MATERIAL

The skull collection utilized for this study derives from the village Hallstatt in Austria, situated in the Eastern Alps, approximately 70 km SE of Salzburg. In this village, a tradition has prevailed during at least the last 200 years, that skulls and long bones, from graves in the churchyard opened in order to give room for new burials, are exhumed, cleaned, and with respect to the skull, even bleached and decorated with the name of the individual as well as with artistically painted flowers and leaves, and subsequently placed in a charnel house facing the churchyard. Previously, this tradition was widespread in Austria and Bavaria in Germany, but after the turn of the last century and even after World War I, most collections of decorated and identified skulls were reburied. At present, the Hallstatt collection is the only collection left, and only single skulls remain from other collections. On the other hand, it appears that the Hallstatt collection is the largest series that has ever existed, comprising some 700 decorated skulls, of which about 600 are still displayed in the charnel house in Hallstatt, where the tradition of decorating skulls is still carried out, the rest of the skulls having been taken to different museums in Austria. Furthermore, the collection seems to be fairly complete, judging from studies of photographs taken of the collection in 1922, though some single skulls have been illegally removed. With respect to accumulation, the bulk of the skulls derives from the period 1825-1900, with but some 50 skulls having been decorated during the period 1775-1825, and approximately the same number during this century. When the painting of skulls was flourishing during the 19th century, no social stratification could be observed among the painted skulls, though members of the upper social groups dominated during the first decades when the skull painting was carried out.

Having identified a skull according to the name and even kind of decoration painted on a skull, the pedigree of the individual was reconstructed by means of the church records on births, deaths and marriages. If possible each family line was followed five generations back in time. Unfortunately, however, the identification of a number of skulls could not be ascertained, either because the name on the skull had disappeared, or only the initials of the name had been put on the skull, or, in some cases, the skull was decorated, but not named. In some cases when initials had been utilized, on the other hand, identification was possible because of singular combinations of initials, or on the basis of the decoration of the skull, which changed in a characteristic way during time.

Among the skulls, 346 turned out to fall into pedigrees for which the skulls of at least two relatives had been identified. By relative, in this connection, is meant any kind of family relationship, such as sisters and brothers, parents, grandparents, or grand-grandparents, and even brothers and sisters of the parental generations. A total of 91 pedigrees were erected, in which the skulls of between 2 and 10 family members had been identified, with a mean number of identified skulls in each of 4.29, the median number being 4.75. These pedigrees, together with some other 100 pedigrees in which no relative had been identified in the collection, showed the surprising fact that inbreeding was virtually nonexistant in Hallstatt, contrary to the common belief. In only one of the pedigrees the same couple appeared in both the mother's and the father's lines, who happened to have one pair of common grand-grandparents. On the other hand, Hallstatt has to be regarded as a semi-isolate until late 1800, and studies of the church records reveal that common relatives exist in both parental lines for most of the individuals, though many generations back in time. Most marriages were between inhabitants from Hallstatt, who generally numbered 1700-1800 individuals, and if not, a wife was most often taken from a village in the same valley downstream. The pedigrees and further details on the population are given by Sjøvold (1983).

Because of the two centuries during which the collection has been accumulating, different patterns of identified skulls from several generations may be present in the pedigrees. In this way a variety of different kinds of family relationships are available for study. In what follows, however, the study of cranial heritability was restricted to family relationships of first degree.

METHODS

The heritability (denoted h^2) of a character is defined as the ratio of additive to phenotypic genetical variance (Falconer 1960). For estimation of the heritability of a measurement based on one parent and offspring, it may be shown that this is given by the relation

$$h^2 = \frac{2cov_{OP}}{var_P} = 2b_{OP}$$

where cov_{OP} denotes the covariance between offspring and one parent, and var_P the variance of the parents. The term b_{OP} therefore is simply the regression coefficient of offspring on parent. When there is no sexual dimorphism, such as for offspring and parent of the same sex, the regression coefficient equals the correlation between the two kinds of relatives. However, as sampling fluctuations had to be considered, the heritability estimates were calculated in terms of the regression coefficient proper. As in some families there were more than one offspring of the parents, the mean values of the offspring were utilized. The regression coefficients calculated were subsequently tested for significance according to the F distribution with 1 and N-3 degrees of freedom (Williams 1959), N being the number of pairs of relatives utilized.

As the possibility of assortative mating among the parents could not be excluded, the correlation between the measurements were calculated for the complete pairs of parents and tested for significance according to the t-test with N-2 degrees of freedom. Because the effects of assortative mating could be expected to produce either positive or negative correlations, the two-sided test was utilised.

The measurements studied were selected primarily in order to test the amount of genetic information in a set of measurements put forward by Howells (1966, 1973), having in part been introduced by him. A number of these measurements are measured by means of coordinate calipers. In the following list of measurements, reference is generally made to Howells (1973), where the complete definitions are given, although some of the measurements are, in turn, quoted or modified from other sources. These are, on the other hand, referred to in connection with his definitions.
The different radii defined all refer to Howells, being defined as distances from the transmeatal axis. This axis, in turn, is the line connecting the centres of the greatest circles to be inscribed in the right and left auditory meatus. To be quite frank, the measurements 21-24 were measured from the line connecting right and left porion, the radii calculated by triangulation, because a modified Aichel caliper as described by Howells (1973) was not available at the time of measuring. Later checks did, however, show very high correlations between the two kinds of measurements, so that the heritability estimates may be applied to both. The remaining radii (no. 25-29) were all measured by means of a modified large coordinate caliper with conical end points.
The measurements taken are as follows:

1. Mastoid height, MDH, Howells 1973.
2. Bimaxillary breadth, ZMB, Howells 1973.
3. Bimaxillary subtense, SSS, Howells 1973.
4. Bimaxillary subtense fraction, ZMF. Defined as an analog to the subtense fractions FRF, PAF and OCF of Howells 1973. Measured along measurement no. 2 (ZMB) from the left zygomaxillare to the point where measurement no. 3 (SSS) falls. Expresses asymmetry of the maxillae, and is directly read off the coordinate calipher when taking the other two measurements.
5. Bifrontal breadth, FMB, Howells 1973.
6. Nasio-frontal subtense, NAS, Howells 1973.
7. Bifrontal subtense fraction, FMF. Defined similarly to no. 4 (ZMF), measured from the left frontomalare anterior.
8. Malar length, inferior, IML, Howells 1973.
9. Malar length, maximum, XML, Howells 1973.
10. Malar subtense, MLS, Howells 1973.
11. Cheek height, WMH, Howells 1973.
12. Nasion-bregma chord (Frontal chord), FRC, Howells 1973.
13. Nasion-bregma subtense (Frontal subtense), FRS, Howells 1973.
14. Nasion-subtense fraction, FRF, Howells 1973.
15. Bregma-lambda chord (Parietal chord), PAC, Howells 1973.
16. Bregma-lambda subtense, PAS, Howells 1973.
17. Bregma-subtense fraction, PAF, Howells 1973.
18. Lambda-opisthion chord (Occipital chord), OCC, Howells 1973.
19. Lambda-opisthion subtense (Occipital subtense), OCS, Howells 1973.
20. Lambda-subtense fraction, OCF, Howells 1973.

21. Bregma radius, BRR. Defined as the perpendicular from the transmeatal axis to bregma. The measurement was introduced in order to have a fixed point to which to measure, as vertex used in the following measurement may be situated anywhere on the parietals, bregma and lambda included.
22. Vertex radius, VRR. Howells 1973.
23. Nasion radius, NAR, Howells 1973.
24. Subspinale radius, SSR, Howells 1973.
25. Dacryon radius, DKR, Howells 1973.
26. Zygoorbitale radius, ZOR, Howells 1973.
27. Frontomalare radius, FMR, Howells 1973.
28. Ectoconchion radius, EKR, Howells 1973.
29. Zygomaxillare radius, ZMR, Howells 1973.

Some measurements related to the maxilla, particularly to alveolar measuring points, such as prosthion and molar alveolus, unfortunately had to be disregarded. The reason for this was that the skull series showed evidence of general severe tooth loss and resorption of the alveoli, so that these measurements were only rarely possible to obtain. In addition, measurements related to the simotic subtense had to be omitted because of the lack of a simometer.

The set of non-metric traits studied in the present report are all either absent or present in an individual, though traits exist which may be further subdivided when present. Most traits are bilateral, and may occur in one side of the individual, or in both. In the present connection, however, the concern is whether or not the traits are hereditary features and therefore, for bilateral traits, a trait is regarded as present when observed in at least one side.

Non-metric traits are regarded as being generated by the action of a large number of genetic and environmental factors all with small and additive effects, and a trait is manifested permanently if the cumulative effect of the different factors at a critical age of the individual exceeds a certain value, or "threshold" (Grüneberg 1963). It is furthermore assumed that all factors are additive, that there is no dominance, and that there is no genetic-environmental variance. Provided these conditions are satisfied, it is permissible to utilize all different pairs of first degree relatives, given, however, that the trait frequencies of the different groups of relatives are homogeneous. When there are several family members, it does not matter if the same individual contributes to several pairs of relatives (Falconer 1965), though when it is possible to plan the experimental design, repeated use of the same individual should be avoided, as two pairs with one common member are not independent. In this case, however, it was regarded as most important to detect as many family members as possible within a given number of skulls.

Estimates of non-metric trait heritabilities were calculated by means of the method of Pearson (1900), utilizing the tetrachoric tables by Everitt (1914). Because of the great number of additive effects which are thought to generate a trait, the sum of these effects is approximately normally distributed according to the Central Limit Theorem, and pairs of relatives will be binormally distributed with respect to these sums, which are denoted the liability to the trait. The binormal distribution surface of liabilities of relatives is divided into four parts by means of the threshold values for each of the relatives. This is shown in figure 1, where the figure *a* denotes the number of pairs where both relatives lack the trait, *b* and *c* are the numbers of discordant pairs, and *d* is the number of concordant pairs possessing the trait.

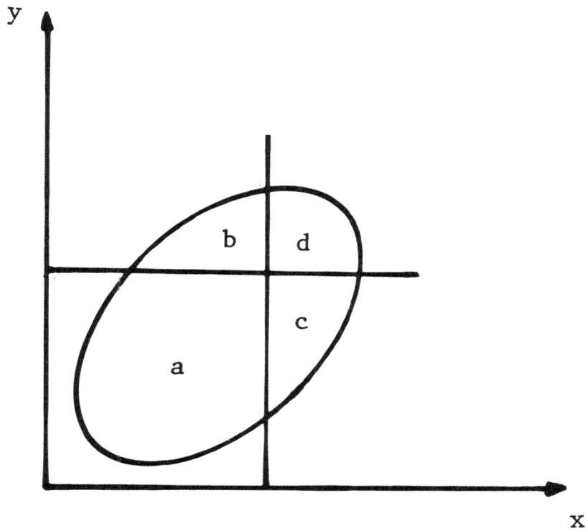

Figure 1. The principle of the tetrachoric correlation demonstrated by the elliptic binormal distribution being divided into four quadrants by thresholds imposed on the marginal distributions of liabilities, indicated by x and y, so that individuals with liabilities above the threshold value possess the trait, and those below do not. In the binormal case, therefore, the figure *d* corresponds to concordant pairs of individuals (relatives) possessing the trait, the figure a corresponds to concordant pairs lacking the trait, and *b* and *c* indicate discordant pairs. The term h appearing in the text means the distance from the mean of the distribution to the threshold, measured in standard deviation units.

The method of Pearson calculates the correlation in the underlying distribution based on the numbers *a*, *b*, *c* and *d*, and is known, because of the tabulation by Everitt, as the tetrachoric correlation. What is actually calculated is the correlation of liabilities to a trait, and not of the trait itself.

The method of tetrachoric correlation is regarded as the most direct method for calculating correlation in liability (Edwards 1960, 1969). Initially, however, an approximation to correlation in liability by Falconer (1965, 1967) was essayed, but this resulted most often in negative estimates of heritabilities due to a paradox inherent in the method. One reason was that many traits were fairly common and the heritability not very high, and Falconer's method is not suited to such traits (Cavalli-Sforza & Bodmer 1971).

With regard to the binormal distribution of liabilities shown in figure 1, it turns out that the estimate of the tetrachoric correlation is minimized if *b* and *c* are equated, each given the value of half the sum of discordant pairs. This is permissible because different kinds of first degree relatives are being combined, and because otherwise the correlation may be influenced by how the discordant pairs are calculated, and not by the more important relation between the number of discordant to concordant pairs. This procedure also simplifies the formulas for estimating the tetrachoric correlation, which may now be given in the form

$$\frac{d}{N} = \frac{(b+d)^2}{N^2} + \sum_{i=1}^{6} \tau_i^2 \rho^i$$

where N is the total number of pairs, τ_i $(i=1,\ldots,6)$ are coefficients related to the fraction $(b+d)/N$, tabulated by Everitt (1914), and ρ is the unknown correlation, the equation being solved by iteration (Newton's solution). The variance of this estimate is, after simplification, given by the formula

$$\text{var } \rho = \frac{2}{N^3 \chi_o^2} \left(\frac{ab + bd}{4} + \Psi^2(2ad + ab + bd) - \Psi(ab - bd) \right)$$

where

$$\chi_o = \frac{1}{2\pi\sqrt{1-\rho^2}} \exp\left(-\frac{h^2}{1+\rho}\right)$$

$$\Psi = \frac{1}{\sqrt{2\pi}} \int_0^{\beta} \exp\left(-\frac{z^2}{2}\right) dz$$

$$\beta = h\sqrt{(1-\rho)/(1+\rho)}$$

where h is the distance of the threshold from the mean, in terms of standard deviations, tabulated by Everitt (1914) in connection with the corresponding τ_i values. Since the variances of trait frequencies of relatives are equal when trait frequencies are equal (because of the binomial distribution), the estimates h^2 of the heritability of liability to a trait is calculated according to the formula

$$h^2 = 2\rho$$

because first degree relatives are concerned (Cavalli-Sforza & Bodmer 1971).

The traits analysed in the present report were those originally given by Berry & Berry (1967), as these 30 traits have been widely used in later studies, or have at least formed an important group of traits used. The traits are named in full in table 3-4, and defined in the above mentioned paper by Berry & Berry. In the present report there has been no sub-division of the traits according to the age of the individuals, because one purpose of this study was to inquire into the possible genetic value of the traits when analysed on the same basis as they have been utilized by other authors for population studies.

Only one small deviation from the original list of traits has been made, as trait no. 4 concerns the absence, and not the presence, of the parietal foramen. By retaining the other definitions of the traits, it turns out that in some few cases the trait sought is the commoner of the two alternatives possible. All individuals studied have attained ages when traits should be permanently manifested, if regarded as present. As parents and children are involved, it should be remembered that the material spans several generations, and in many cases the age of children greatly surpasses that of the parents. Prior to estimation of heritability of liability of non-metric traits, usual chi-square tests for homogeneity of trait frequencies were applied. Since standard regression was applied in the metric case, no need was found to test mean values between parents and children for equality.

HERITABILITY OF SKULL MEASUREMENTS

The means and the standard deviations of the measurements according to the different combinations of parent-offspring are given in table 1. In this table, for each measurement, the number of pairs utilized for the calculation varies according to the combination of relatives, because of the fact that only 26 complete pairs of skulls of parents could be identified in the material. The remainder of the pairs of relatives derives from pedigrees in which the skull of one parent only had been identified. As for the sons and the daughters, on the other hand, both sons and daughters were represented in a number of families but, naturally, in some families skulls from only sons and in other from daughters only existed. The practical effect of this is that the two samples of fathers and mothers with their sons and daughters, respectively, are somewhat overlapping. They may therefore be considered as obtained by sampling with replacement from the skull series. This also explains the variation in the means and standard deviations.

Variation in the number of pairs of relatives also occurs within each sample. The main reason for this is that damage to

Table 1. Mean measurements with standard deviations (in mm) for the differents family relationships studied. The number N of pairs for each comparison is given in brackets. For abbreviations, see text

Measure-ments	Fathers	Sons	N	Fathers	Daughters	N	Mothers	Sons	N	Mothers	Daughters	N
1 MDH	27.7±2.0	27.5±2.0	(49)	27.5±2.7	24.0±2.6	(38)	24.2±2.4	27.4±2.2	(51)	25.2±2.9	27.7±2.3	(25)
2 ZMB	89.8±3.9	90.5±5.2	(50)	90.2±4.0	84.5±4.5	(28)	87.3±4.3	91.6±5.4	(35)	86.9±4.6	86.7±5.0	(21)
3 SSS	26.1±3.1	25.3±3.1	(48)	24.7±3.0	23.2±3.4	(28)	24.1±2.7	25.6±3.1	(34)	23.1±3.0	23.7±2.1	(21)
4 ZMF	44.9±2.5	45.0±2.4	(51)	45.6±2.5	42.5±2.5	(28)	43.6±1.9	45.3±2.9	(36)	43.6±2.9	43.6±2.2	(21)
5 FMB	104.3±2.4	104.4±3.5	(64)	99.0±4.2	100.8±3.7	(43)	100.8±3.7	104.0±3.5	(61)	100.5±3.4	99.4±3.6	(33)
6 NAS	19.9±2.7	19.7±2.4	(65)	19.7±2.5	19.6±2.7	(43)	20.1±2.2	19.5±2.6	(61)	19.1±1.9	18.7±2.2	(33)
7 FMF	52.0±1.6	52.0±2.0	(66)	51.8±2.5	49.2±2.1	(43)	49.8±2.0	51.6±2.1	(61)	50.0±1.8	49.6±1.9	(33)
8 IML	34.0±3.3	34.0±2.9	(50)	33.9±4.3	30.7±3.2	(40)	31.8±3.4	33.4±2.7	(51)	32.5±2.5	30.7±2.8	(26)
9 XML	51.5±3.6	53.0±2.5	(49)	52.6±3.8	47.6±4.3	(38)	48.8±4.1	53.0±3.3	(51)	48.1±3.7	47.9±3.9	(24)
10 MLS	9.6±1.1	9.6±1.2	(49)	9.7±1.5	8.2±1.7	(38)	8.8±1.4	9.5±1.5	(51)	8.5±1.2	8.8±1.7	(24)
11 WMH	20.1±3.3	20.8±2.2	(49)	20.0±3.8	18.9±2.4	(39)	18.9±2.4	20.8±2.5	(50)	19.4±2.1	(18.2±2.9	(27)
12 FRC	114.0±6.2	112.2±4.8	(54)	111.9±5.8	106.3±5.8	(42)	109.0±4.4	112.3±5.3	(56)	107.8±3.3	106.5±4.5	(31)
13 FRS	28.8±2.7	27.3±2.3	(54)	27.8±2.8	26.6±2.5	(42)	26.8±2.6	27.1±2.7	(56)	26.4±2.3	26.6±2.6	(31)
14 FRF	52.7±3.7	52.8±3.1	(54)	51.8±4.9	47.1±3.6	(42)	50.3±4.0	53.6±4.0	(56)	51.3±4.1	47.9±3.9	(31)
15 PAC	111.6±6.3	111.2±5.7	(54)	110.4±6.8	106.8±5.8	(42)	106.0±6.9	110.5±5.5	(56)	106.6±7.0	106.0±6.1	(31)
16 PAS	23.2±2.6	23.5±2.8	(54)	23.6±3.1	22.3±2.7	(42)	22.8±2.8	23.5±2.8	(56)	23.1±2.5	22.8±2.5	(32)
17 PAF	61.3±4.4	60.5±4.6	(54)	61.5±4.9	58.8±4.9	(42)	56.6±5.8	60.0±4.5	(56)	58.1±6.5	56.1±4.5	(32)
18 OCC	94.1±6.5	94.5±4.6	(52)	94.0±6.3	92.7±4.8	(41)	92.0±5.0	94.8±4.1	(56)	94.6±7.0	92.6±4.0	(28)
19 OCS	29.1±3.6	28.8±2.8	(53)	28.3±3.8	27.9±2.4	(40)	28.0±4.2	27.6±2.9	(56)	28.2±3.4	27.4±2.3	(28)
20 OCF	49.8±5.7	51.3±6.1	(52)	50.8±7.7	47.4±5.5	(40)	48.7±7.1	51.0±6.4	(56)	46.9±6.0	46.5±7.3	(28)
21 BRR	116.6±5.1	115.0±4.2	(54)	114.2±5.9	108.6±5.2	(43)	110.0±4.0	115.0±4.3	(56)	109.8±5.4	108.3±4.2	(31)
22 VRR	117.7±5.7	117.0±4.2	(54)	116.4±5.8	110.6±5.5	(43)	111.5±4.3	117.0±4.2	(56)	111.9±6.4	110.4±4.8	(31)
23 NAR	93.3±5.2	92.4±3.9	(54)	92.0±5.6	87.8±4.8	(43)	90.0±4.2	91.9±4.1	(56)	88.7±4.2	87.6±4.1	(31)
24 SSR	94.0±6.2	93.1±3.6	(35)	92.2±4.7	88.0±5.8	(23)	87.8±4.7	91.8±3.7	(38)	88.5±3.7	88.9±4.6	(17)
25 DKR	82.5±5.0	81.4±3.3	(59)	82.1±4.4	78.4±4.1	(41)	78.5±3.9	81.1±3.5	(50)	77.6±3.4	77.7±3.9	(24)
26 ZOR	78.5±4.6	78.6±2.6	(56)	79.4±4.4	75.7±4.3	(39)	75.9±3.9	78.6±3.5	(43)	76.0±4.2	75.8±3.9	(26)
27 FMR	77.0±4.0	76.1±2.9	(59)	77.2±3.9	73.1±3.3	(40)	72.4±3.9	75.4±3.4	(50)	73.5±3.1	73.0±3.6	(30)
28 EKR	70.6±4.0	69.8±2.5	(55)	70.4±3.8	66.5±3.4	(39)	67.0±3.0	69.2±3.2	(43)	67.0±3.9	67.1±3.2	(24)
29 ZOR	68.8±4.6	68.6±2.9	(55)	68.9±5.0	65.5±4.2	(39)	65.2±4.2	68.2±3.6	(43)	66.8±4.1	65.6±3.8	(26)

skulls would most often prevent some measurements to be taken. The most regrettable cases were those when a measurement on a parent having several offspring was not obtainable.

Some comments should, however, be given with respect to the reduction of pairs utilized, e.g., for measurements no. 2, 3, 4 and no. 24. These are all measurements related to the subspinale. When the subspinale region was damaged or broken, which happened fairly frequently, measurements related to this landmark had to be omitted.

The estimates of the heritabilities, calculated according to the formula $h^2 = 2\rho$, are given in table 2. In the rightmost column, estimates of phenotypic correlations between parents are given. Among these correlations, 14 are positive and 15 negative. Two of the correlations appear to be rather high. These correlations concern measurement no. 2 (zygomaxillary breadth, $r = 0.4156$), and no. 16 (parietal chord subtense, $r = 0.3200$). Neither of the correlations, however, turn out to be significant, not even at the 75% level. One reason for the lack of significance may perhaps be the rather small number of skulls from both parents, being but 13 in the first case and 25 in the second. On the other hand, one or perhaps two significant test results may be expected to occur by chance among 29 tests, if there is in fact no correlations between the parents, even if, as in this case, the tests are not independent because of the correlations between the measurements. The correlations between the measurements, furthermore, explain some systematic tendencies with respect to the sign of the estimates of the correlations between parents, e.g., the negative signs of no. 26-29.

The pattern of the heritabilities differs among the combinations of relatives. In part this may again be explained by varying numbers of pairs of relatives, being highest with respect to the combinations in which sons are involved. Even in these cases, however, the number of pairs of relatives is generally less than 60. Because of the limited number of observations, the standard deviations of the estimates are also generally fairly large, and it is meaningless to try to refer to exact values for the heritabilities. The more important information in this connection therefore concerns the question which measurements may be regarded as significantly hereditary features.

By selecting the 90% level for testing the heritabilities, it turns out that the pattern of significant heritabilities differs from the one combination of relatives to the other. With respect to the father-son relationship, the measurements related to the bimaxillary breadth, subtense and subtense fraction (no. 2, 3, 4), the lengths of the malar bone and the cheek height (no. 8, 9 and 11), the parietal chord (no. 15), the occipital chord and subtense (no. 18, 19), and the four measurements of the distances from the biauricular line (no. 21, 22, 23, 24) are significant. For the combination of fathers and daughters the length of the mastoid process (no. 1), the nasio-frontal subtense (no. 6), all three measurements related to the malar bone (no. 8, 9, 10), and the

Table 2. Estimates of heritability with standard deviations of skull measurements calculated from the four different combinations of parent-offspring relationships with numbers of pairs N given in brackets. In the rightmost columns, estimates of the corresponding correlations between parents are given. For abbreviations, see text

	Fathers				Mothers				Correlation between parents	
	Sons		Daughters	N	Sons		Daughters	N		
1 MDH	0.1875±0.2796	(49)	0.9064±0.2794[1]	(38)	0.3192±0.2558[5]	(51)	0.4660±0.3250[5]	(25)	-0.0595	(22)
2 ZMB	0.9308±0.3608[2]	(50)	0.5852±0.4204[5]	(28)	0.4192±0.4304	(35)	0.3850±0.4938	(21)	0.4156	(13)
3 SSS	0.9334±0.2552[1]	(49)	0.5830±0.4208[5]	(28)	0.7586±0.3834[4]	(34)	0.1852±0.3096	(21)	0.1658	(13)
4 ZMF	0.6870±0.2522[1]	(51)	0.4380±0.3744[5]	(28)	0.1636±0.5110	(36)	-0.0352±0.4548	(21)	0.0775	(13)
5 FMB	0.2900±0.3596	(64)	0.5098±0.3292[5]	(43)	0.5192±0.2394[3]	(61)	0.1370±0.3742	(33)	-0.0857	(13)
6 NAS	0.0184±0.2242	(65)	0.9414±0.3066[1]	(43)	0.0172±0.3070	(61)	0.0588±0.4068	(33)	0.1557	(29)
7 FMF	0.4838±0.3828[5]	(66)	-0.0160±0.2676	(43)	0.3450±0.2654[5]	(61)	-0.0138±0.3852	(33)	-0.0553	(29)
8 IML	1.0314±0.2074[1]	(50)	0.4512±0.2282[4]	(40)	0.0512±0.2268	(51)	0.3700±0.4494	(26)	0.1162	(21)
9 XML	0.5332±0.1854[1]	(49)	1.0264±0.3416[1]	(38)	-0.1336±0.2306	(51)	0.7672±0.4148[4]	(24)	0.1178	(21)
10 MLS	0.1964±0.2976	(49)	0.9844±0.3528[1]	(38)	0.1500±0.3142	(51)	0.0406±0.6200	(24)	-0.2250	(21)
11 WMH	0.5340±0.1850[1]	(49)	-0.1692±0.2030	(39)	0.4294±0.2948[5]	(50)	0.2954±0.5676	(27)	0.0220	(22)
12 FRC	0.0202±0.2140	(54)	0.3342±0.3078	(42)	0.4372±0.3228[5]	(56)	0.0944±0.5010	(31)	0.1232	(25)
13 FRS	0.1202±0.2438	(54)	-0.0112±0.2834	(42)	0.0982±0.2816	(56)	-0.0288±0.4304	(31)	-0.1411	(25)
14 FRF	-0.3520±0.2248	(54)	0.3534±0.2246[5]	(42)	-0.1878±0.2762	(56)	0.0844±0.3934	(31)	-0.1457	(25)
15 PAC	0.4704±0.2440[4]	(54)	0.2366±0.2688	(42)	-0.5878±0.1990[1]	(56)	0.1390±0.3228	(31)	0.0766	(25)
16 PAS	0.3788±0.2960[5]	(54)	0.0604±0.2750	(42)	0.3792±0.2660[5]	(56)	0.1736±0.3556	(32)	0.3200	(25)
17 PAF	0.2812±0.2872	(54)	0.4902±0.2732[4]	(42)	-0.0966±0.2110	(56)	0.4234±0.2444[4]	(32)	0.1158	(25)
18 OCC	0.5816±0.1804[1]	(52)	-0.0080±0.2414	(41)	0.2486±0.2234	(56)	0.2804±0.2192[5]	(28)	-0.1056	(25)
19 OCS	0.7180±0.1986[1]	(53)	0.2202±0.3546	(40)	0.5240±0.1792[1]	(56)	0.0320±0.2594	(28)	-0.1786	(25)
20 OCF	0.0036±0.3000	(52)	0.3048±0.2274[5]	(40)	0.3884±0.2392[5]	(56)	-0.2290±0.4772	(28)	-0.1661	(25)
21 BRR	0.5264±0.2142[2]	(54)	0.1316±0.2750	(43)	-0.0292±0.2924	(56)	-0.0678±0.2910	(31)	-0.0552	(25)
22 VRR	0.5058±0.1936[2]	(54)	-0.2440±0.2922	(43)	-0.2020±0.2622	(56)	-0.0744±0.2780	(31)	-0.1229	(25)
23 NAR	0.3386±0.1984[4]	(54)	0.1380±0.2656	(43)	-0.6098±0.2534[2]	(56)	0.6324±0.3494[4]	(31)	0.0081	(25)
24 SSR	0.3690±0.1870[4]	(35)	0.4464±0.5260	(23)	0.2076±0.2634	(38)	0.3294±0.6348	(17)	-0.1681	(11)
25 DKR	0.2126±0.1724[5]	(59)	-0.0554±0.2970	(41)	0.1816±0.2556	(50)	0.9712±0.4544[1]	(24)	0.0690	(26)
26 ZOR	0.1332±0.1526	(56)	0.3688±0.3130[5]	(39)	0.4192±0.2718[5]	(43)	0.9376±0.3258[1]	(26)	-0.1206	(24)
27 FMR	-0.0424±0.1898	(59)	-0.1532±0.2708	(40)	-0.0694±0.2558	(50)	0.3202±0.4392	(30)	-0.1806	(26)
28 EKR	0.0820±0.1698	(55)	0.2538±0.2970	(39)	-0.0460±0.3316	(43)	0.5424±0.3538[5]	(24)	-0.1709	(22)
29 ZOR	0.0346±0.1718	(55)	0.4028±0.2728[5]	(39)	0.5446±0.2478[4]	(43)	0.5984±0.3558[5]	(26)	-0.1601	(24)

1) $p < 0.01$, 2) $p < 0.025$, 3) $p < 0.05$, 4) $p < 0.10$, 5) $p < 0.25$

parietal chord subtense (no. 16) demonstrate significant heritabilities. The mother-son relation demonstrates significant heritabilities with respect to the bimaxillary subtense (no. 3), the bifrontal breadth (no. 5) and the occipital subtense (no. 19). All these heritabilities are positive. With respect to the mother-son combination, however, the parietal chord (no. 15) and the nasion distance (no. 23) display significantly negative heritabilities. These were the only heritabilities at all which were found to be significantly negative, even when compared to the extremely low significance level of 75%. The fourth combination of relatives, the mother-daughter combination turned out to yield significant heritabilities with respect to the maximum malar lenght (no. 9), the bregma-subtense fraction (no. 17), the occipital chord (no. 18), the nasion distance (no. 23) and the zygoorbitale radius (no. 26). In sum, most of the measurements turned out to yield significant heritability in one or more of the combinations of relatives, but none of the heritability estimates was significant in all combinations.

The fluctuating numbers of pairs of relatives available for the heritability analysis made direct comparison of the estimates of heritability difficult. For example, the nasion distance (no. 23) was significant at the 90% level with respect to father-son comparison based on 54 pairs ($h^2 = 0.3386$), whereas the bimaxillary breadth (no. 2) with respect to father-daughter was insignificant, being based on 28 pairs ($h^2 = 0.5852$). In order to reveal to what extent a tendency or trend of heritability of otherwise insignificant estimates could be discovered, the estimates were compared with the 75% level of the test statistic. The result of this procedure was that a number of additional heritability estimates were found to display such a trend. In general, previous findings of significant heritabilities were confirmed for other combinations of relatives. For three measurements which previously did not show significant heritability estimates, a heritability trend was demonstrated in two of the combinations of relatives, that is, with respect to no. 7, 16 and 20. For another three measurements, significance at the 75% level was demonstrated in but one of the combinations of relatives, these being the frontal chord (no. 12), the frontal subtense fraction (no. 14) and the ectoconchion radius (no. 28). The above mentioned heritability of 0.5852 was also significant at the 75% level.

To sum up the results of the heritability analysis, two measurements did not turn out to provide significant heritability estimates at all, these being the frontal subtense (no. 13) and the frontomalare radius (no. 27). Caution should perhaps be made with respect to the frontal chord (no. 12), the frontal subtense fraction (no. 14) and the ectoconchion radius (no. 28) as just a suggestion of significant heritabilities could be demonstrated for one combination of relatives only. With two other exceptions significant heritability was demonstrated at least at the 75% level in two or three of the combinations of relatives. The two exceptions are the nasio-frontal subtense (no. 6) and the malar subtense (no. 10), which are highly significant (99%) in the father-daughter combination. The combination of mother-daughter displayed the lowest number of significant heritabilities, however probably due to the fact that these estimates were calculated from the lowest number of pairs of relatives.

In general, the results of the heritability studies of skull measurements are consistent with the findings of Leamy (1974) based on cranial and post-cranial measurements on random-bred mice. In that study, heritability estimates were calculated between parents and offspring being slaughtered at three different ages: 1 month, 3 months and 5 months. Though the sample sizes were found to be generally adequate, the smallest number of pairs corresponding to the largest numbers in this study, the largest being about 180, the precision of the estimates was found to be low. With increasing sample size, most of the heritability estimates turned out to be significant, at least when disregarding the estimates based on the 1-months offspring and the parents. Though the measurements differed from those of the present study, as a completely different species was investigated, a similar tendency might be expected if the material for the study of the human skull had been larger.

HERITABILITY OF NON-METRIC TRAITS

The incidences of the 30 non-metric traits are shown in table 3 according to the occurrence among parents and children of both sexes. The incidences are based on somewhat different samples from those utilized for the metric study, because the calculation in this connection was carried out by means of pairs of relatives of the same degree of relationship, i.e., parents and offspring, and brothers and sisters. Because of this, pairs of skulls from brothers and sisters were utilized, even if the skull from none of the parents had been identified.

Some comments should be made with respect to the frequencies of some traits and of how they were scored in practice. The highest nuchal line (no. 1) is not well defined by Berry & Berry (1967), being "more easily felt than seen". This criterion is unsatisfactory. In order to make scoring more objective, the occiput was lighted by a small torch, so that at least the low crest was revealed by its shadow. In many cases, however, the line could be directly observed.

The metopic suture (no. 6) is in principle an easy trait to score. In the Hallstatt material, however, many of the individuals had died in their seventies, and cases where most of the suture was obliterated were occasionally observed. In other cases, there was a suspicion that the suture had been completely obliterated in old age, for instance when all the other sutures of the skull vault had completely disappeared. A "trace" of the metopic suture within the glabellar region just above the nasion was not regarded

Table 3. Incidences of non-metric traits according to fathers, sons, mothers and daughters

Trait	Fathers		Sons		Mothers		Daughters	
		%		%		%		%
1. Highest nuchal line present	37/61	60.6	68/130	52.5	28/59	47.5	33/80	41.3
2. Ossicle at the lambda	14/55	25.5	23/125	18.4	15/51	29.4	11/73	15.1
3. Lambdoid ossicle present	45/63	71.4	79/123	64.2	24/51	47.1	27/67	40.3
4. Parietal foramen absent	42/61	68.9	70/127	55.1	28/58	48.3	48/81	59.3
5. Bregmatic bone present	0/62	0.0	0/128	0.0	0/59	0.0	1/81	1.2
6. Metopism	6/63	9.5	9/128	7.0	3/59	5.1	9/82	11.0
7. Coronal ossicle present	0/53	0.0	1/120	0.8	0/54	0.0	1/77	1.3
8. Epipteric bone present	12/66	18.2	23/134	17.2	11/62	17.7	24/79	30.4
9. Fronto-temporal articulation	0/59	0.0	2/125	1.6	1/57	1.8	0/78	0.0
10. Parietal notch bone present	6/67	9.0	25/114	21.9	15/63	23.8	19/81	23.5
11. Ossicle at asterion	22/69	31.9	33/132	25.0	14/60	23.3	12/76	15.8
12. Auditory torus present	0/61	0.0	0/132	0.0	0/61	0.0	0/84	0.0
13. Foramen of Huschke present	8/61	13.1	13/127	10.2	18/59	30.5	11/80	13.8
14. Mastoid foramen exsutural	42/56	75.0	96/123	78.0	41/56	73.2	53/77	68.8
15. Mastoid foramen absent	6/61	9.8	15/132	11.4	10/58	17.2	14/80	17.5
16. Posterior condylar canal patent	55/64	85.9	113/131	86.3	50/61	82.0	72/81	88.9
17. Condylar facet double	14/62	22.6	18/118	15.3	7/52	13.5	7/76	11.8
18. Precondylar tubercle present	3/67	4.5	10/126	8.1	6/60	10.0	5/73	6.8
19. Anterior condylar canal double	29/66	43.9	62/134	42.3	30/63	47.6	36/79	45.6
20. Foramen ovale incomplete	3/62	4.8	7/137	5.1	5/60	8.3	6/88	6.8
21. Foramen spinosum open	9/61	14.8	19/130	14.6	12/59	20.3	18/82	22.0
22. Accessory lesser palatine foramen present	46/61	75.4	92/126	73.0	37/51	72.5	46/75	61.3
23. Palatine torus present	11/60	16.7	28/120	23.3	20/55	36.7	23/70	32.9
24. Maxillary torus present	3/32	9.4	9/72	5.9	1/17	5.9	1/37	2.7
25. Zygomatico-facial foramen absent	16/58	27.6	22/126	17.5	8/55	14.6	19/78	24.3
26. Supraorbital foramen complete	24/58	41.4	41/127	32.3	26/63	41.3	29/82	35.4
27. Frontal notch or foramen present	12/64	18.8	23/129	17.8	13/58	22.4	10/69	14.5
28. Anterior ethmoid foramen exsutural	31/63	49.2	46/125	44.6	27/69	39.1	41/84	48.8
29. Posterior ethmoid foramen absent	2/60	3.3	3/135	2.2	2/57	3.5	2/78	2.6
30. Accessory infraorbital foramen present	22/59	37.3	46/125	36.8	22/55	40.0	30/81	37.0

as a metopic suture proper. The "trace" has the form of a series of transverse, short sutures, which may or may not be connected in the median-sagittal plane. In some skulls this trait is quite conspicuous but, by means of a technique similar to that applied to the highest nuchal line, it was discovered that this trace of the metopic suture existed in every skull, being in some cases invisible by direct inspection. With regard to traits no. 13 and 14, both concerning the mastoid foramen, Sjøvold (1973) pointed out that the definitions of these traits were not independent. For completeness, however, both were included in the present study. The scoring of sutural ossicles comprised ossicles surrounded by a free suture, as opposed from sutural cut-offs. For no. 17, even bifaceted condyles were scored. With respect to the anterior (hypoglossal) canal (no. 19), incomplete division of the canal was included as a trait. The developmental nature of this trait has been demonstrated by Dodo (1980) who found both complete and incomplete division of the canal in fetuses, but in a slightly lower frequency than in adults. For similar reasons, with regard to the supraorbital foramen (no. 26), a notch being almost closed, or Ω-shaped, was also counted as a trait. The frontal notch or foramen (no. 27) was only scored if there was a well defined second notch or foramen present. Rather often, a protruding glabellar area contributed to a notch-like structure at the medial border of the supraorbital margin. Such a structure was found in almost every individual and was not regarded as the trait in question.

The palatine torus (no. 23) turned out to be more difficult to judge than expected. In many skulls there were more or less irregular elevations and thickening of bony tissue at the midline of the palate, which was not regarded as a real torus palatinus. In the material, most individuals were completely toothless, with more or less complete resorption of the alveolar arcade, and in some cases a bony elevation of the palate was interpreted as a reaction in the form of bone apposition because of stress in the chewing apparatus connected with tooth loss. The cases which had the character of a swelling were scored as the torus. At any rate, the incidences were far higher than those observed by the Berrys. Because of dental status of the material, the number of possible observations of the maxillary torus (no. 24) was considerably reduced. The division of the incidences in the four groups was made in order to test the homogeneity of incidences between parents and children, and between the sexes. Most of the traits turned out to be homogeneous in incidence. Three of the traits, the highest nuchal line (no. 1), lambdoid ossicles (no. 3), and the palatine torus (no. 23), did, however, show heterogeneity between the groups. This was confirmed when testing the incidences between the sexes. A fourth trait, the foramen of Huschke (no. 13), was also found to be heterogenous between the groups. This could be directly explained by a peculiarly high incidence of the group of mothers.

Sexual heterogeneity has previously been reported for these traits. Finnegan (1972), for instance, reported the same kind of sexual dimorphism for the highest nuchal line, the foramen of Huschke and the palatine torus as observed in this report. Similarly, Berry (1975) confirmed this with regard to the highest nuchal line and the palatine torus, as well as for lambdoid ossicles. Significantly higher incidence among females than males for the palatine torus was even observed by Corruccini (1974) with respect to Caucasian American males. With regard to the foramen of Huschke, Berry (1975) found this trait to be age dependent. Differences in the mean age of mothers and daughters in the present case may therefore explain some of the difference in incidences. With respect to the other three traits, for which sexual difference in incidence was found, it is reassuring that there is the same trend towards higher or lower incidence both with regard to parents and children of the same sex.

Some traits were excluded from the estimation of heritability as being very rare, such as the bregmatic bone (no. 5), the coronal ossicle (no. 7), and the auditory torus (no. 12), the last trait being not observed at all. Another rare trait, the fronto-temporal articulation (no. 9) did only occur in three cases, two of which, however, were brothers.

With exception of the traits 5, 7 and 12, estimation of the heritability of liability of other traits were carried out. With regard to traits for which sexual dimorphism had been observed, the sexes were treated separately. In general, this led to a considerable reduction of the number of pairs of relatives that was used for the estimation, which in most cases was just above, or just below 300. Because the sexes had to be treated separately, a reduction of the number of pairs of relatives by approximately 67 was observed, the remaining pairs being further subdivided according to sex.

In some traits no concordant pair of relatives possessing the trait was observed. This was interpreted as obvious lack of heritability and the heritability was regarded as zero. Formally, however, because the structure of the tetrachoric function for estimation of the correlation in liability given above, this would result in a negative value of the correlation in liability ρ, as the right-hand side of the equation would be equal to zero. In some other cases, though concordant pairs were observed, negative estimates of heritability were arrived at. However, as any pair of first degree relatives was used in the calculations, negative heritabilities are difficult to interpret in a way similar to the negative heritabilities of the parent-offspring heritabilities based on measurements. On the other hand, negative estimates of non-metric heritabilities certainly do not imply that a strong family likeness exists, and such values were therefore regarded as implying lack of heritability. The heritabilities of 20 of the non-metric traits studied are given in table 4, together with corresponding standard deviations. Among the remaining traits, no. 5, 7 and 12 had been excluded for being very rare or for not having been observed at all, as mentioned above. With respect to no. 20 (incomplete foramen ovale) and no. 29 (lack of posterior ethmoid foramen) no concordant pair of relatives possessing the trait was observed. In addition, both traits were fairly rare. Trait no. 1 was analysed for each sex, but neither attained a positive value of heritability. The same applies for females with respect to trait no. 3, both sexes with regard to no. 13, and with respect to males concerning trait no. 23. In some of the cases the great reduction of the number of pairs of relatives when studying each sex separately could perhaps explain lack of positive heritability. With respect to trait no. 3 (lambdoid ossicles), only 24 pairs of relatives could be studied, because in other individuals this trait was not possible to judge as present or absent due to obliteration of the lambdoid suture. With respect to no. 13, the foramen of Huschke, however, three concordant pairs of relatives only were observed, two among females and one across the sexes.

Table 4. Estimates of the heritability of 20 non-metric skull traits. The remaining 10 traits have been omitted as not having been observed, as being veryr rare, or as yielding negative estimates of heritabilities

Trait	$h^2 \pm$ SD
2. Ossicle at the lambda	0.238 ± 0.242
3. Lambdoid ossicle present, ♂♂	0.563 ± 0.439
4. Parietal foramen absent	0.313 ± 0.125
6. Metopism	0.344 ± 0.376
8. Epipteric bone present	0.008 ± 0.219
9. Fronto-temporal articulation	0.954 ± 0.513
10. Parietal notch bone present	0.152 ± 0.222
11. Ossicle at asterion	0.555 ± 0.196
14. Mastoid foramen exsutural	0.171 ± 0.215
16. Posterior condylar canal patent	0.096 ± 0.188
17. Condylar facet double	0.140 ± 0.269
19. Anterior condylar canal double	0.008 ± 0.168
21. Foramen spinosum open	0.282 ± 0.112
22. Accessory palatine foramen present	0.196 ± 0.205
23. Palatine torus present ♀♀	0.091 ± 0.738
24. Maxillary torus present	0.679 ± 0.524
25. Zygo-facial foramen absent	0.269 ± 0.231
26. Supraorbital foramen complete	0.378 ± 0.183
28. Anterior ethmoid foramen exsutural	0.182 ± 0.176
30. Accessory infraorbital foramen present	0.062 ± 0.188

The traits 15 and 18 yielded heritability estimates which were approximately zero ($h^2 = -0.05$ and $h^2 = -0.007$ respectively), whilst 5 concordant pairs of relatives in connection with no. 27 did not suffice to demonstrate any degree of positive heritability.

The heritability estimates in table 4 are generally fairly low, and the standard deviations are large. With the exception of one rare trait (no. 9), observed in two brothers, the heritability estimates indicate that no greater concentration of traits occur in some few families only. This confirms the finding of Berry (1975) who did not find familial concentrations of traits in skulls from 10 families from a crypt in St. Bride's church in London, with the foramen of Huschke as an exception. That trait, however, turned out to be age dependent. The lack of family clustering denies a simple gene type of control of trait expression (Berry 1975).

The estimates of heritability of the traits are generally lower than those found for the metric traits. The same situation was observed by Self & Leamy (1978) who found this to be in agreement with the fact that measurements are generally highly correlated in the phenotype whereas nonmetric traits display low, if any, correlation. Mutational events are therefore apt to change the heritability of several measurements simultaneously because of the correlations, whereas this is not as easy to determine with respect to almost uncorrelated traits. The estimates of heritabilities found are, on the other hand, generally lower than those obtained by Cheverud & Buikstra (1981), though calculated by means of a different method. Their estimates are also connected with large standard deviations, so that just 6 out of 14 traits for which heritability estimates were produced, yielded values significantly greater than zero on the 90% level. However, most of the heritability estimates of non-metric skeletal traits arrived at by other authors concern different traits than those of the present report, as well as other species.

As it follows from Pearson (1900) that the tetrachoric correlation coefficient is approximately distributed according to the normal distribution, the heritabilities obtained in table 4 may be tested for significance in a simple manner. Choosing the 90% level, 7 traits are found to be significant, that is, nos. 3, 4, 9, 11, 21, 24 and 26. Among these, two concern sutural ossicles (lambdoid and asterionic) and three concern foramina for nerves and vessels (parietal, spinosum, supraorbital). The two others concern the fronto-temporal articulation and the maxillary torus.

Because of the large standard deviations, the heritabilities found to be significant turned out to be between 0.28 and 0.95. According to Self and Leamy (1978), whose sample sizes were far larger than in the present case, most traits displayed significant heritabilities, though the mean level of heritability was approximately 0.20, consistently calculated by means of three different methods. That the heritabilities generally were found to be low was interpreted as associated with major fitness traits (Falconer 1960). For this reason, indications of heritability in the present material was sought by regarding values of heritability being significant at the 75% level. The additional traits found to be significant at this level were nos. 2, 6, 10, 14, 22, 25 and 28. Among these, two relate to ossicles (lambda and parietal notch bone), and four to foramina (location of mastoid, palatine, zygo-facial, anterior ethmoid). The last trait is metopism. This group of traits displays heritabilities that overlap the first one, ranging from 0.15 to 0.34.

Six traits showed heritabilities that were not significant at all, ranging from 0.01 to 0.15. One of these is the epipteric bone (no. 8). In this report, however, epipteric bones were regarded as present whatever the location of the bone (squamosal, epipteric proper, sphenoid), and further analysis of the data may clarify the question of heritability. Another trait, the posterior condylar canal (no. 16), was found to be age-dependent in Caucasian males and females by Corruccini (1974). This finding was reconfirmed by Berry (1975), though not statistically significant.

It is surprising that the division of the anterior condylar canal was not found to be significant. This trait was found to yield significant heritability estimates both in random-bred mice (Self & Leamy 1978) and in macaques (Cheverud & Buikstra 1981). Division or tendencies of division of the hypoglossal canal was, on the other hand, rather frequent, with an overall frequency in the material of 45.9%, and this may tend to conceal heriditary patterns of the trait.

The reason that the palatine torus was not found to be significant may be the small number of pairs for the heritability estimate, amounting to 34 pairs only. Among these, however, three concordant mother-daughter pairs were observed; also in one family, all three sisters possessed the torus. In that family, unfortunately, the skull of the mother was not identified. It is therefore likely that significant heritability was not found for this trait because of sample size since, as seen from the formulas for the tetrachoric correlation and its variance, the square and the cube of the sample size are important denominators in these expressions. A similar argument may be made for the small sample of 24 pairs of family members with respect to the female heritability of the ossicle at the lambda, four of which were mother-daughter pairs, but where the small sample size in connection with the formula for the tetrachoric correlation even prevented a positive estimate of heritability to be calculated.

Non-metric traits are certainly not all heriditary characters simply because some of them are. For population studies, traits having been shown to carry genetic information should preferrably be utilized. Some rare traits may even be incorporated by referring to similar, more common traits being hereditary, for instance bregma and coronal ossicles. Traits for which the heritability is not at all significant should be excluded, unless there are other circumstances indicating that some genetic information may exist. It is certainly fortunate if the heritability of a trait is reconfirmed by different authors, but if reconfirmation fails, this may not be very serious, as heritability estimates are population specific (Cheverud & Buikstra 1981). Consequently, in some populations heritability estimates may be easier to obtain than in others, and, as seen, sample size may be quite important.

Some traits should be excluded from population studies, at least until some degree of genetic basis has been shown. In the first place, this concerns the highest nuchal line, which failed to show positive heritability. Corruccini (1974) omitted this trait, referring to Merkel (1871) who showed that the trait is dependent on the form of the superior nuchal line, though possessing muscular reality. Berry (1975), furthermore, found the trait unsatisfactory with regard to objectivity of scoring, and recommended that it is abandoned. Auditory tori were not observed at all, and are left inconclusive. The foramen of Huschke may be age-dependent (Berry 1975) and only a few concordant pairs of relatives were observed for this trait.

The mastoid foramen should not be scored as present or absent, but as exsutural or not exsutural (with absence as "not exsutural"). Double condylar facet and precondylar tubercles may relate to mechanical causes in connection with the atlanto-occipital joint, and not to genetic causes, whereas the imcomplete foramen ovale seems to be a simple deficiency of ossification.

At the moment epipteric bones, the frontal notch (foramen), doubling of the anterior condylar canal and infraorbital foramen should also be excluded. However, with respect to these traits, forthcoming studies of the Hallstatt cranial collection, utilizing different kinds of relatives as well as laterality of bilateral traits, may elucidate whether or not some genetic information still has remained undetected.

SUMMARY AND CONCLUSIONS

The present report presents evidence about the heritability of 29 measurements related to the shape of the human skull as measured by coordinate calipers, and of 30 non-metric, cranial traits. The measurements are mainly defined by Howells (1966, 1973) and the non-metric traits have been introduced by Berry & Berry (1967). The result of the two kinds of study, which were based on first degree relatives, is that most of the measurements, but not all, are significantly hereditary characteristics. Though significant heritabilities were not found for more than one or two of the parent-offspring relationships studied, this may be due to small sample sizes, and it was noted that significant heritabilities were confirmed by trends towards heritability of the measurement with respect to other relationships studied.

Above all, there seems to be a fair amount of heriditary factors related to the size of the malar bone, perhaps due to the muscular attachment of the masseter and the crossection of the temporal muscle. Other important measurements concern the zygomaxillary breadth, subtense and subtense fraction, the occipital chord and subtense, and the radii from the transmeatal axis with the exception of the frontomalar and ectoconchion radii. There were no significant correlations between the parents for the measurements.

The non-metric traits were studied with the individual as the unit. The main results of the analysis are that ossicles in the lambdoid and parieto-occipital sutures, metopism, fronto-temporal articulation, palatine and maxillary tori are hereditary characteristics. The same applies to a group of foramina comprising presence or absence of the parietal, the accessory lesser palatine, the zygomatico-facial and the supraorbital foramen, location of the mastoid and anterior ethmoid foramen in relation to sutures, and closed or open foramen spinosum. Epipteric bones, the frontal notch or foramen, as well as division of the anterior condylar canal await further analysis with respect to the subdivision of trait expression.

Measurements are not all heriditary characters because of being possible to measure, and non-metric traits are not all heriditary because they are non-metric. Previous studies of the relative merits of measurements and non-metric traits have primarily been based on assumptions and not on formal knowledge of the genetic basis of the characters studied (Rightmire 1972, 1976; Corruccini 1976). By utilizing information about the heritability of both kinds of characters, the apparent controversary between users of cranial measurements and non-metric traits can be settled.

ACKNOWLEDGEMENTS

The author is indepted to the following persons for access to the skull collection and permission to study the decorated skulls from Hallstatt: In Hallstatt, the Reverends Johann Weidinger and Gustav Stögner of the Catholic Church, the late Mr. Karl Höplinger and Mr. Franz Zahler, curators of Hallstatt Museumsverein; in Innsbruck, Professor Werner Platzer and Dr. Karl Mager, Anatomisches Institut der Universität; in Vienna, Dr. Paul Spindler and Dr. Herbert Kritscher, Naturhistorisches Museum, and Director Dr. Klaus Beitl and Dr. Franz Grieshofer, Österreichisches Museum für Volkskunde. The study was supported by grants from the Swedish Natural Science Research Council, by a grant from the Knut and Alice Wallenberg Foundation, and by a Doctoral grant from the University of Stockholm.

References

Bernhard, W., G. Bräuer, V.P. Chopra & A. Hancke: 1977, Quantitativ genetische Analyse morphologischer Merkmale der Weichteilregionen des menschlichen Kopfes. HOMO 30, pp. 26-43.

Berry, A.C.: 1975, Factors affecting the incidence of non-metrical skeletal variants. J.Anat. 120, pp. 519-535.

Berry, A.C. & R.J. Berry: 1967, Epigenetic variation in the human cranium. J.Anat. 101, pp. 361-379.

Berry, R.J.: 1964, The evolution of an island population of the house mouse. Evolution 18, pp. 468-483.

Cavalli-Sforza, L.L. & W.F. Bodmer: 1971, The Genetics of Human Populations. Freeman, San Francisco.

Cheverud, J.M. & J.E. Buikstra: 1981, Quantitative genetics of skeletal non-metrics in the Rhesus macaques on Cayo Santiago. I. Single trait heritabilities. Am.J.Phys.Anthrop. 54, pp. 43-49.

Corruccini, R.S.: 1974, An examination of the meaning of cranial discrete traits for human skeletal biological studies. Am. J.Phys.Anthrop. 40, pp. 425-446.

Corruccini, R.S.: 1976, The interaction between non-metric and metrical cranial variation. Am.J.Phys.Anthrop. 44, pp. 285-294.

Dodo, Y.; 1980, Appearance of bony bridging of the hypoglossal canal during fetal period. J.Anthrop.Soc. Nippon 88, pp. 229-238.
Edwards, J.H.: 1960, The simulation of Mendelism. Acta Genet. Stat. Medica 10, pp. 63-70.
Edwards, J.H.: 1969, Familial predisposition in man. Brit.Med.Bull. 25, pp. 58-64.
Everitt, P.F.: 1914, Tables of the tetrachoric functions for fourfold correlation tables. Biometrika 7, pp. 437-451.
Falconer, D.S.: 1960, An Introduction to Quantitative Genetics. Ronald, New York.
Falconer, D.S.: 1965, The inheritance of liability to certain diseases estimated from the incidence among relatives. Ann.Hum. Genet. 29, pp. 51-71.
Falconer, D.S.: 1967, The inheritance of liability to diseases with variable age of onset, with particular reference to diabetes mellitus. Ann.Hum.Genet. 31, pp. 1-20.
Finnegan, M.J.: 1972, Population definition on the north-west coast by analysis of discrete character variation. Ph.D. dissertation, University of Colorado, Boulder.
Grüneberg, H.: 1963, The Pathology of Development. Blackwell, Oxford.
Howells, W.W.: 1966, The Jomon population of Japan. A study by discriminant analysis of Japanese and Ainu crania. Peabody Museum Papers LVII/1, pp. 1-30.
Howells, W.W.: 1973, Cranial Variation in Man: A Study by Multivariate Analysis of Pattern of Difference Among Recent Human Populations. Peabody Museum Papers, vol. 67.
Leamy, L.: 1974, Heritability of osteometric traits in a random-bred population of mice. J. Heredity 65, pp. 109-120.
Martin, R.: 1928, Lehrbuch der Anthropologie in Systematischer Darstellung. Band 2. Gustav Fischer, Jena.
Merkel, F.: 1871, Die Linea Nuchae Suprema. Engelmann, Leipzig. Cited by Corruccini (1974).
Mizoguchi, Y.: 1977a, Genetic variability in tooth crown characters: Analysis by the Tetrachoric Correlation method. Bull.Natn.Sci.Mus., Ser. D (Anthrop.) 3, pp. 37-62.
Mizoguchi, Y.: 1977b, Genetic variability of permanent tooth crowns as ascertained from twin data. J.Anthrop.Soc. Nippon 85, pp. 301-309.
Pearson, K.: 1900, Mathematical contributions to the theory of evolution. VII. On the correlation of characters not quantitatively measureable. Phil.Trans. 195, pp. 1-47.
Rightmire, G.P.: 1972, Cranial measurements and discrete traits compared in distance studies of African Negro skulls. Hum. Biol. 44, pp. 263-276.
Rightmire, G.P.: 1976, Metric versus discrete traits in African skulls. In: Giles, E. & J.S. Friedlander (Eds.): The Measures of Man, pp. 383-407. Peabody Museum Press.
Rösing, F.W.: 1974, Auswahl von Schädelmassen. HOMO 25, pp. 117-121.
Self, G.S. & L. Leamy: 1978, Heritability of quasi-continuous

skeletal traits in a randombred population of House mice. Genetics 20, pp. 109-120.

Sjøvold, T.: 1973, The occurrence of minor non-metrical variants in the skeleton and their quantitative treatment for population comparisons. HOMO 24, pp. 204-233.

Sjøvold, T.: 1984, Das Vorkommen von verwandter Individuen unter den bemalten Schädeln aus Hallstatt, Oberösterreich. In press.

Suzuki, M. & T. Sakai: 1960, A familial study of torus palatinus and torus mandibularis. Am.J.Phys.Anthrop. 18, pp. 263-272.

Torgersen, J.: 1951a, Hereditary factors in sutural patterns of the skull. Acta Radiologica 36, pp. 374-382.

Torgersen, J.: 1951b, The developmental genetics and evolutionary meaning of the metopic suture. Am.J.Phys.Anthrop. 9, pp. 193-205.

Williams, J.: 1959, Regression Analysis. Wiley, New York.

IMPROVING THE SENSITIVITY, SPECIFICITY, AND APPOSITENESS OF MORPHOMETRIC ANALYSES

J.G. Rhoads

Department of Anthropology, Yale University, New Haven, Connecticut

The finesse and the communicativeness of papers in the physical anthropology Journals using multivariate techniques have improved quite a lot over the last ten or fifteen years. The best are excellent and it is not my intention here to write a jeremiad decrying the remaining shortcomings of the weakest papers. Still, there is value in reminding ourselves of some criteria of success in data analysis that are simple and familiar but which constantly need re-emphasis.

I will label these ideas by generalizing the sense of some fundamental terms of epidemiology: sensitivity and specificity. I will call an analytical method in morphometrics sensitive when it responds reliably to precisely the question you wish to ask of your data. It is specific when it reponds to nothing else (that is, it is not affected by irrelevant factors). These characteristics imply that units of analysis and the variables observed must be carefully chosen to be apposite. These simple notions are easy to lose sight of amid the technical details that command attention in a multivariate study.

METHOD VERSUS SUBSTANCE

It is a common notion that the lack of mathematical facility is what limits the quality of the applications of statistics that the "average" physical anthropologist makes. This seems to me to be far from the truth. The most serious shortcomings of our applications of multivariate statistical methods to physical anthropology come not from the difficulty of developing suitable mathematical treatments for asking questions that are well-defined in biological terms, but from the impoverished set of metaphorical ideas or

G. N. van Vark and W. W. Howells (eds.), Multivariate Statistical Methods in Physical Anthropology, 247–259.

conceptual archetypes with which we think about variation in and among populations of organisms. We are captives of a small, stereotyped set of procedures for going from the biological phenomena under study, to numbers, to pictures, and finally back to the interpretation of the biology again. These procedures are used as metaphors: they aid in relating the raw multidimensional observations, which are extremely difficult to think about, to simple picture-ideas like 'distance' that the physical anthropologist can operate on intuitively.

Maintaining the distinction between the model/metaphor and the actual domain of study is of critical importance. The Pythagoreans are said to have believed that the cosmos has a mathematical basis, and that it was reasonable and proper to think of things as actually being made up of numbers. There is a curious neo-Pythagoreanism in much morphometric discourse, a kind of confusion of mathematics with biology that has held us back. We often talk and behave as though the biological phenomena we study are actually as they are represented by the highly simplified (though not, to be sure, simple) mathematical representations programmed into our computers.

This is a serious overvaluation of mathematical tools above biological subject matter, and it condemns not a little morphometric work to sterility. To state the matter explicitly: there is nothing natural about linear statistical methods or about their relation to the analysis of animal structures. Like all forms of mathematics, the vector models that we manipulate exist in the universe of abstract mathematical objects. The correspondence between the abstract mathematical system and empirical reality is not something that inheres in the mathematics. It is something that we, as biological scholars, create in the course of our work, hoping to produce by this means a quantitative treatment that will increase our understanding of the biological phenomena under study.

It is a misuse of the statistical procedures, and a form of neo-Pythagoreanism, to suppose that nature conforms to inductively discoverable arithmetical ideals, and that statistical methods are active agents in an inductive search. We are the agents and must acknowledge the responsibility. The application of a particular method is, or ought to be, a thoughtful act of substantive scholarship. To the extent that the question the method answers is not identical to the best form the biological question could be posed in, we are being deflected from the questions we ought to ask, to questions we have an easy way of asking.

SPECIMENS AND MEASUREMENTS: AGAINST MORPHOMETRIC INDUCTIVISM

There has been some misplacement of emphasis by morphometricians on technique, at the cost of attention to critical decisions we may pass over almost unconsciously about units of analysis, attributes to measure, and principles of interpretation for relating our quantitative results once again to the biological world.

A specific example of the kind of fallacy reffered to can be

relating our quantitative results once again to the biological world.

A specific example of the kind of fallacy referred to can be seen in the numerical taxonomic literature in the discussion of the 'nonspecificity hypothesis'. The idea, which survives in a veiled form in some current morphometric work, is that it scarcely matters what is measured and put into a taxonomic distance model. If enough data is provided, the true biological relations of the taxonomic units will become manifest.

Clearly, though, the apparent relations emerging from a particular study are a product of the way the measurement process and the metric used for comparison were designed. It is an extraordinary leap of Platonic idealist faith to suppose that all that is necessary to derive a natural scale of taxonomic distance is to measure enough arbitrarily chosen parts of the animal. Mahalanobis distances are not given us of the gods to determine natural relations of the "total morphological pattern" in populations; obviously more is required of us as morphologists than this.

We are deluded if we think we are seeking some essential _eidos_ which exists in nature, using more and more refined mathematical tools. It is a delusion to think that in mathematics is the salvation of biology, and if this technique doesn't work, then surely the statistics or engineering journals can be ransacked for another better one. This is mere numerology. Mathematics is a way of ordering and expressing our opinions about the way the world is organized. Our theories are either oriented toward a world view, or they are sterile. A theoretical data exploration can be useful, but until we improve the quality of our viewing the hope that we will inductively arrive at new concepts is a vain hope.

EMPHASIZING RESIDUALS IN ADDITION TO REGULARITIES

Linguists speak of the elusive goal of a complete discovery procedure, a systematic and unfailing method for uncovering all of the structure and subtleties of a language under study. If our techniques of statistical analysis are not discovery procedures for biological phenomena, then what are they? I submit they are most valuable as a way of summarizing and then suppressing repetitive detail. We chose a model of a greater or lesser degree of elaboration, determine what events, objects, and circumstances it describes to a subjectively chosen degree of refinement, then try to explain the residuum, perhaps adding to our store of concepts in the process.

To look at the essentials of most morphometric studies, the universe of discourse is made up of one or more collections of organisms or parts of organisms, and the procedure is to, by some process of property selection and measurement, map them onto corresponding sets of vectors of numbers. These are then operated on by a variant of the general linear model, or a special sort of variance components analysis such as principal components analysis. Because very many published analyses are based on a small set of prototypical analyses, the sequence of procedures used usually follows a stereotyped channel, and often ends with reducing the problem to a pseudomap of one of two kinds: a cross-plot of means of artificial variates (discriminant functions or canonical variates), or a pseudo-phylogeny from a cluster analysis. These are metaphors for two appealing and understandable picture-ideas to biologists. But we need to be able to move freely from the biology to the pseudomaps and back to the biology again; because of a tendency to concentrate on the pictures themselves, the last step is often all but missing. Little effort is given to discovering or explaining why the analysis arrayed the populations as it did.

The techniques themselves are valuable and worth keeping, but unless supplemented they are limited in utility as instruments for discovering the biologically significant fact. I claim that this is because in using them we use only a severely limited set conceptual archetypes (in the phrase of the philosopher Max Black) with which to think about variation in and between populations of animals.

The error is one of overcondensation. The matter is reduced to a few summary statistics or a graph, only describing the simplest properties, location and perhaps a simplified measure of dispersion for a discriminant function or two. Because summary statistics are so effective as a way of suppressing detail and ignoring bewildering variation, we can readily commit a new form of the typological fallacy and collapse a population into a type, represented by a mean vector (or a single point in a plot).

I have called the relation between the mathematical treatment and the biological explanans metaphorical. As always with metaphors, the analogy between the multivariate method and the simple picture-idea is imperfect and there is danger of drawing false deductions when the analogy is not used carefully. Commonly the mathematical development of the model, with its necessary simplifications and compromises, has introduced distortions and irrelevancies that make the model diverge from the ideal embodiment of a desirable picture-idea that the nonmathematical user would like to think that it is. The user, and indeed the mathematician, may not be aware of what these are and how they affect the practical use of the technique.

Poor correspondence between the mathematics and the biological ideas can come from several sources. We often thoughtlessly follow inappropriate mathematical precedent, as when we use derivations based on precise but narrow definitions of optimality, optimizing with respect to criteria that may have little substantive importance. We regularly confuse biological variation, error variation, and selection idiosyncrasies of both variates and units. Or we may read too much into a mathematical abstraction, as when we uncritically take "variance" to be a complete solution to the biological problem of measuring variation, or use a multivariate technique that limits our attention to the mean vectors and variance-covariance matrices to the nearly complete neglect of careful individual study of units of analysis, variates, and the the empirical details of distributions. This is an area of practice where we have much to learn from the precepts of the "exploratory data analysis" school of statisticians, in particular that one of the most valuable properties a data-analytic strategy can have is a good likelihood of exposing to us the unexpected detail in our data.

We pay insufficient attention to the individual organisms or aggregates of organism. We pay insufficient attention to the variates, how they suit or fail to suit precisely the purpose at hand, and most importantly the processes underlying them and underlying the statistical relations among them.

We often do not examine or explain whole distributions adequately (as opposed to just their means and variances) even in one variate, let alone in a multiresponse problem. We are mentally prepared to take in only a few patterns. This is an area where statistical and computational progress is being made and will be helpful (see for example the computer graphics presented in Gnanadesikan 1978, Wang 1978).

The responsibility of the analyst and interpreter extends to enumerating the mathematical assumptions of the analysis and to describing the measures taken to assure that violations of the assumptions do not make the results useless (Were the distributions passably close to Gaussian if that matters in the analysis strategy we have used? Could some variates have been transformed to conform better to normality or to linear relations with other variables? Were the group dispersions really commensurate?).

SUGGESTED NEW EMPHASES IN MORPHOMETRICS

As morphometricians become more comfortable with multivariate statistical methods, it is to be hoped that attention can be turned from mere technique back to the biological matter that motivated the trend to quantification in the first place: to phenetics, particularly the relation of the traits we study to their genetic and evolutionary bases. This means modelling process, rather than just describing the results in some particular space of discriminant coordinates. Though like all other mathematical techniques they are tools and not philosopher's stones, such models as path analysis and state variable system descriptions can be used to begin to describe the relations and dynamic characteristics of the morphological systems, systems behind our multiple variables. They have the marked advantage of forcing us to make our model explicit, to expose the model's relation to biological concepts and display it in a form open to the critical examination and discussion of its appositeness. We then have a basis for evaluating whether our data and methods make our results responsive to the biological phenomena we have proposed for study and to no other, irrelevant factors, that is, how sensitive and specific our research strategy is. Within such a carefully-constructed substantive framework such old biological ideas as morphogenetic fields, scaling and allometry, plasticity, and functional equivalence need to be given multivariate quantitative interpretations and investigated.

There is need for a science of measurement (in a broad sense). In this is included both the mundane but important business of deciding what to put our calipers to and what the results actually mean, and also everything else that intervenes between the specimens and vectors we operate on in our models: what does it mean to select such and such a measurement set? What can we learn from comparing the multivariate distances so obtained with ones coming from other variable sets?

There is also a need for a science of what is usually called sampling, but might here be better called "accidence" ("that which happens") because physical anthropologists are so seldom in a position to do probability sampling. What are the quantitative effects of our sometimes necessarily arbitrary decisions on what specimens and measurements to include in making a particular analysis?

And finally, there is a need for further development of a science of <u>idiography</u>, of individual description of the interesting cases and groups in an analysis. It is not paradoxical to ask that we go back to a certain amount of individual description, because although we gain a great deal by the suppression of repetitive detail that is the great virtue of statistical data analysis, the final summary statistics can almost always be revealingly supplemented by a discussion of particual cases where perhaps the simple model did not fit so well.

COMPUTATIONAL TOOLS

The rather laborious procedures being advocated here would not have been seriously suggested even a few years ago. To suggest recomputing a multivariate analysis, possibly many times, was obvious folly in the days when matrices were inverted on desk calculators, or even in the days of the first generations of statistical computing program packages.

But computing resources are no longer much of a limitation. Data analysis of this kind must unfold interactively at a computer terminal to be practical, though, and there is still some programming to be done to bring the capabilities of the interactive statistical packages in the area of multivariate analysis up to their other capabilities. But the ingredients are there for such packages as SAS and GENSTAT, in the form of a macro-coding facility to allow users to make their own extensions to the built-in procedures, a matrix manipulation language, and a driver program for generalized plotting on a graphics terminal or hardcopy plotter.

By programming small amounts of macro code, it is now readily possible to examine multivariate data very efficiently and flexibly while sitting at a terminal, rapidly producing such tabular and graphical information as:

- Distances relative to any chosen metric
- Histograms, normal plots or other quantile-quantile plots for any variate or principal component
- Scatter plots of variates, components, or discriminant functions with the option of using different graphic symbols to represent groups
- Displays of residuals from any general linear model
- Tukey 'Exploratory data analysis' graphics such as stem-and-leaf diagrams or smoothed quantiles of one variate plotted against another, to allow the exposure of more complicated sorts of bivariate dependencies than are depicted by ordinary regression analysis.
- Quantitative comparison of the configurations of metric or ultrametric distance generated by different measurement sets or clustering methods (cf. Gower 1971).

Existing technological innovations which will be extremely valuable for this work when they become more widely available in university computing centers include interactive graphics terminals with high-resolution displays, preferably in color to allow more information to be represented, and the facility for using a light pen or cursor to identify a particular data point (an outlier, for example) on a terminal plot and request measurement information for it, or perhaps request a re-analysis with the point omitted.

EXAMPLES: EXPOSING THE IRREGULAR AND THE USEFULNESS OF SIMPLE INDICATORS

The recent trend in statistics to paying more attention to robust methods and to simple methods useful in data exploration has been guided by a principle that can be valuable in morphometrics: the philosophy of dividing up the data set into the regular and the irregular, and reducing the impact of the irregular on methods intended to seek generalizations. For concreteness, I will briefly discuss two small examples where this idea has helped me in my own work.

In a study of patterns of gene frequency variation in human populations in Melanesia (Rhoads 1983), I first used principal components analysis to try to isolate the major axes of covariation in allele frequencies in 134 localities sampled. Later I discovered by a simple plot of Mahalanobis distances of each population from against the expected value for their rank among the set of distances if the populations had simply a multinormal distribution around the centroid (Q-Q plot against gamma quantiles, Gnanadesikan 1978; see figure 1) that there were very many large distances and that in particular some small island populations likely to have been affected by founder effect and random genetic drift were very dissimilar to all other populations. Using an algorithm due to N.C. Campbell (1980) for robust principal components analysis which reduces the influence of extreme outliers on the orientation of the principal components axes, I obtained a set of components quite different from those given by ordinary principal components analysis and which seemed to represent general trends in the bulk of the data more faithfully.

Another principle of the "exploratory data analysis" school is that very simple displays designed to separate residual exceptions from general trends can be much more valuable for substantive interpretation than the results of elaborate analyses. In an attempt to understand the biological implications of anthropometric means in a set of Solomon Islands populations, I used the very simple-minded notion that means of different measurements in different groups might be expected to reflect a simple scaling up or down of a common shape. The mean of, say, arm length in one group would then be expected to be the same multiple of arm length in a second group that stature in the first group would be of stature in the second. When all measurements are transformed to a logarithmic scale, this corresponds to an additive model with group effects and measurement effects. A simple iterative method based on row and column medians ("median polish", Tukey 1977) gave estimates for these effects which could then be translated back to percentage deviations from the expected value under the simple multiplicative (log-additive) model (table 1). This allows quick recognition of such things as that the row effects show the largest means overall are in the Ontong Java and Lau populations, exceeding the overall averages by something like two to two-and-a-half per cent, and that even with this general large size taken into account, the largest deviations from fit are in the heaviness of those same populations, a fact with ecological and possibly with genetic implications.

The table can be still further simplified by displaying the residuals from fit with single symbols, "." for residuals in the middle 50% of the distribution of residuals, "+", "#" and "P" for progressively greater positive deviations from fit and "-", "=" and "M" representing greater and greater negative deviations (see table for details). Though a great deal of information has been collapsed into a few symbols here, the table is rich in interpretable detail. (The Lau and Ontong Java populations are also

Table 1. Log-transformed Solomon Islands anthropometry means: residuals from median polish fit

	WT	HT	CHBR	STHT	HDLN	HDBR	MNFR	BIZY	BIGN	TOTF	NSHT	NSBR	Row Effect
Aita	14	26	-268	-13	-180	-128	39	19	-40	-27	-110	89	40
Nasioi	-14	49	- 24	4	24	-108	3	30	-76	57	37	162	-33
Nagovisi	-24	-6	- 64	- 4	- 12	19	3	78	22	- 2	-106	-28	-47
Lau	332	-64	150	-19	- 17	- 19	58	- 1	-37	16	272	- 9	90
Baegu	-20	51	4	72	15	39	95	1	-22	-39	- 37	71	-58
Kwaio	-138	6	-4	104	12	37	80	-19	26	-92	-102	-11	-35
Ulawa	67	-28	68	-71	20	-141	91	-21	49	2	114	9	33
Ontong Java	295	-39	144	37	- 78	-139	191	-115	39	13	289	-257	111

Unit: base 10 logarithm, x 10^4 Measurements = WT = weight, HT = stature, CHBR = chest breadth, STHT = sitting height, HDLN = head length, HDBR = head breadth, MNFR = minimum frontal breadth, BIZY = bizygomatic breadth, BIGN = bigonial breadth, TOTF = total fairheight, NSHT = nose height, NSBR = nose breadth

Table 2. Coded table representation of Solomon Islands anthropometry data

	WT	HT	CHBR	STHT	HDLN	HDBR	MNFR	BIZY	BIGN	TOTF	NSHT	NSBR
Aita	.	.	M	.	=	+	.	.	–	.	–	+
Nasioi	.	+	.	.	.	–	.	.	–	+	.	#
Nagovisi	.	.	–	.	.	.	.	+	.	.	–	.
Lau	P	–	+	.	.	.	–	.	.	.	P	.
Baegu	.	+	.	+	.	+	+	.	.	–	–	+
Kwaio	–	.	.	+	.	.	+	.	.	–	–	.
Ulawa	+	.	+	–	.	–	–	.	+	.	+	.
Ontong Java	P	–	+	.	–	–	=	–	+	.	P	=

Symbols:

.	in middle 50% of distribution of residuals from "median polish" fit, i.e., between 25th and 75th percentiles
–,+	below 25th/above 75th percentile
=,#	"outside" below/above point 1.5 X mildspread (interquartile range from 25th/75th percentile
M,P	"far outside" below/above point 3.0 X midspread from 25th/75th percentile.

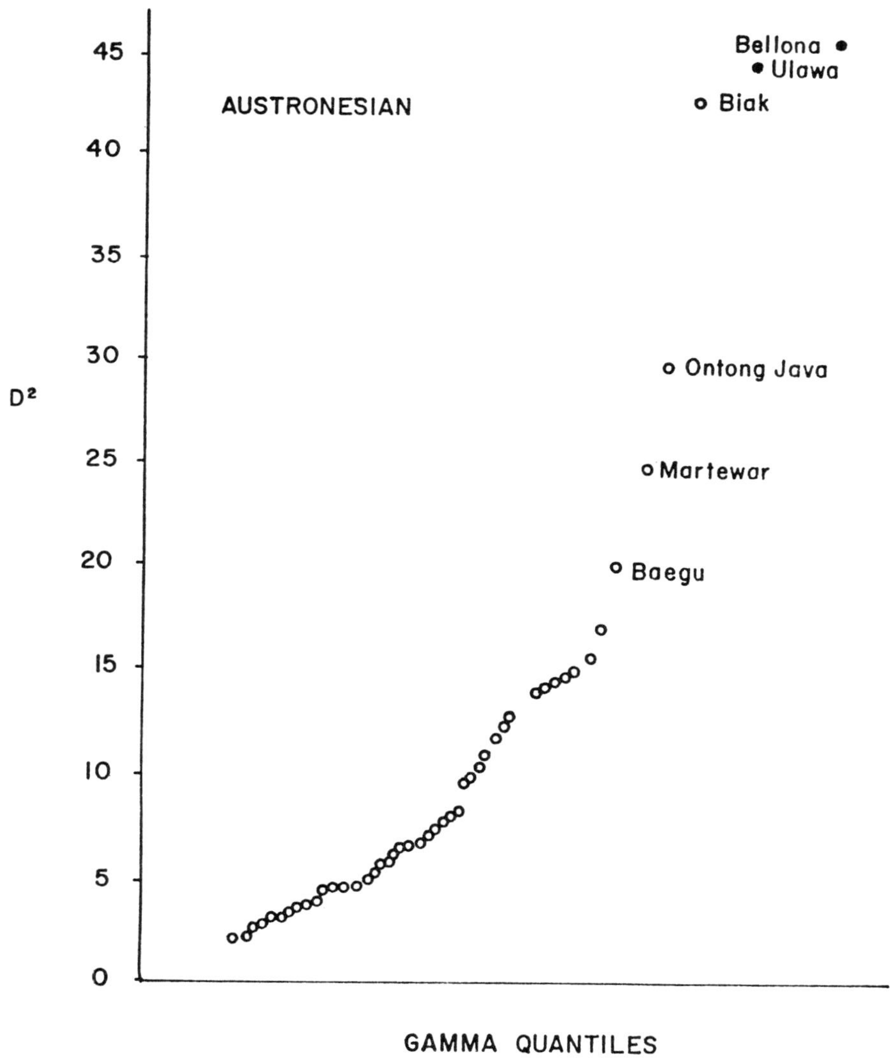

Figure 1. Quantile-quantile plot of Mahalanobis D^2 from centroid for 51 Melanesian populations based on ABO, MNSs, and Rh gene frequencies data. Shows presence of extreme multivariate outliers.

relatively short in stature, have in common relatively long noses, and so on.)

These examples do not of course amount too much in relation to the rather grand proposals for changes in approach I am making in this paper, but are intended to give the flavor of how we can supplement the generalizing and data-aggregating methods of classical multivariate morphometrics with simple means for being attentive to details of variables and units.

CONCLUSIONS

Let there be experimentation with concepts as well as with techniques - the statisticians cannot provide procedures until we have devised and articulated apposite biological questions. We can also do much better with the procedures that exist. We need to pursue a more flexible view of how we should wield our tools; we can afford to use more, and more diverse ones, with detailed explicit attention to their appositeness, requiring them to respond sensitively and specifically to the most thoroughly thought-out view we can devise of the biological processes that might underly the observed data.

Particular mathematical models are metaphors which form and limit our view of nature. It follows that we should study carefully the properties, wanted and unwanted, of the ones that we have, and that we should search (in concert with the statisticians) for a richer variety of them, to be combined flexibly and pluralistically to give a variety of views of nature. For the best and most biologically apposite results (as I think no reasonable statistician would deny) this enterprise must be guided by biological concepts and values, not mathematical ones. We must avoid the situation where there is under the surface some Procrustean transfiguration of the question to fit the technique. I commend to us all a principle that was very well put by J.W. Tukey (1962): "Far better an approximate answer to the right question, which is often vague, than an exact answer to the wrong question, which can always be made precise."

References

Black, M.: 1962, Models and metaphors. Ithaca, N.Y.: Cornell University Press.

Campbell, N.A.: 1980, Robust procedures in multivariate analysis I: Robust covariance estimation. Journal of the Royal Statistical Society C 29: 231-237.

Gnanadesikan, R.: 1978, Methods for statistical data analysis of multivariate observations. New York, Wiley.

Rhoads, J.G.: 1983, Melanesian gene frequencies: a multivariate data-analytic approach. Journal of Human Evolution 12.

Tukey, J.W.: 1962, The future of data analysis. Annals of Mathematical Statistics 33: 1-67.
Tukey, J.W.: 1977, Exploratory data analysis. Reading, Mass.: Addison-Wesley.
Wang, P.C.C., ed.: 1978, Graphical representation of multivariate data. New York: Academic Press.

TOWARDS AN UNDERSTANDING OF DATA IN PHYSICAL ANTHROPOLOGY

Susan R. Wilson

Department of Statistics, IAS, The Australian National University, PO Box 4, Canberra

1. INTRODUCTION

Of recent years it has become popular in the analysis of physical anthropological data to eschew the yesteryear commonly-done, univariate, and simple comparative analyses of such data, and to launch into complex multivariate analyses that are commonly available in popular statistical packages on high-speed computers. Used with sufficient care and understanding, these multivariate approaches to the analysis of data have given additional insight, besides that given by the univariate analyses, into the complex underlying structures of the data. However the formal numerical techniques of data analysis (as described in basic statistics texts, and developed in journals of statistics) are too often designed to yield specific answers to rigidly defined questions. In general, the objectives of anthropological data analysis are neither so narrow nor so formal as described and implied by some statistical theories of estimation and hypothesis testing.

In many situations involving the analysis of physical anthropological data, the techniques which are needed are informal exploratory techniques rather than formal confirmatory techniques, such as tests of specific hypotheses. Graphical techniques are well-suited for this purpose since they tend to be less formal and confining, and enable the investigator to gain 'insight' into the structure of his data, without imposing restrictive constraints. The use of diagrams may often reveal aspects of the data that might have escaped notice were the data tabulated only in numerical form. The diagrams may help to uncover features of the data that were totally unanticipated prior to the analysis, or to suggest hypotheses which may be further investigated.

However basic univariate and multivariate analyses of data

G. N. van Vark and W. W. Howells (eds.), Multivariate Statistical Methods in Physical Anthropology, 261–282.

are merely a "push button" away in their ease of execution, and so they tend to be used first by the majority of researchers. Hence the first part of this article will be devoted to exploring the underlying structure (inherent in the assumptions) that is being assumed, by the basic multivariate analyses, to have generated the data. Some of the more commonly asked questions for which such analyses might supply an answer will be discussed in some detail, as well as questions which are related to, or follow from, these basic questions. Although, in this article, there appears to be little emphasis placed on univariate analyses, this should not be interpreted as their being unimportant. Instead it is stressed that such analyses do from an integral part of obtaining a complete understanding of the data.

In the final section there is a brief review of some of the visual techniques that have been proposed for multivariate data. In the future, graphical procedures will be more readily available than they are at present. Even so, researchers should be encouraged now to use graphical techniques at the commencement of their analysis. As well, in this article, an audio-visual technique that is being developed by the author, exploratory-and-sound data analysis, will be discussed. This technique has a particular advantage for physical anthropological data, and this aspect will be emphasised.

Finally, it is noted that in the publication of their conclusions, researchers should outline the complete process they underwent to obtain their results, and the reasons they chose that particular approach, and not any alternative ones. Too often, the obtained results depend completely on the statistical analysis chosen. This analysis is usually not given in sufficient detail for a statistician to make complete comments, and further correspondence with the author is necessary to elucidate the exact method of analysis, so as to complete these; (see, for example, Campbell, 1980; Wilson, 1979). Journal editors should insist, despite the cost, that all necessary details be published. As well, articles wherein the conclusions depend on the statistical approach taken, should have had a critical statistical review by a trained statistician. The collection of the data is usually costly and time consuming. Since the findings from the data collection are so often dependent on the results of the statistical analysis, as much care and time should be given to this analysis, as to the collection.

2. INTERACTIVE MULTIVARIATE DATA ANALYSIS: SOME QUESTIONS AND THEIR ANSWERS

In this section, and the next, part of the data set given in Ashton, Healy and Lipton (1957) will be used to illustrate the techniques, and approaches, to be discussed. It is stressed that it is not the aim of this article to give a complete analysis of these data.

The data set consists of eight measurements (see Table 1) on the permanent first lower premolar for three types of modern man (British, West African native, Australian aboriginal) and both sexes of three groups of living great apes (gorilla, orang-outang, and chimpanzee). These measurements were also taken on a series of fossils (see Table 1). The aim was to compare each of the fossils with the nine groups. The measurements taken were designed to describe the main features of the tooth, and it is noted that analyses of different measurements may yield quite different conclusions. (As described by Ashton et.al., two previous analyses appeared to give different results because the dimensions used in one analysis did not describe the differences in shape which were emphasised in the other analysis.) This article does not address the question of which measurements are to be used in the analysis. However, it is noted that the inclusion of unnecessary measurements may well hide important aspects in the data that the anthropologist would wish to elucidate. So it is necessary for the anthropologist to give a great deal of careful thought at the commencement of the project to his choice of measurements, rather than to measure everything possible.

Having chosen the measurements to be analysed, it is necessary, in choosing the form of analysis, to keep in mind both the basic question/s being explored and all the information one has. For these data, the basic aim is to describe the differences and the similarities between each of the fossils and the nine groups. However, as will be discussed in this section, none of the 'standard' multivariate techniques appear to be a completely satisfactory approach to the problems that the fossils pose. This is because of the additional information given by Ashton et.al., namely "it is almost certain that these particular creatures were neither human beings nor members of the existing species of great ape".

Now many of the basic multivariate techniques assume that an unknown specimen actually belongs to one or other of the groups, and they provide a method for allocating a specimen to its proper group. If one does have the additional information that the specimen does belong to one of the groups, g, (g=1,...,G), then one can calculate the specimen's allocation score, S_g, for each group g, and this is given by

$$S_g = -[\Pi_1 P_1(\underset{\sim}{U})r_{1g} + \dots + \Pi_G P_G(\underset{\sim}{U})r_{Gg}] \ , \ g=1,\dots,G,$$

where $P_g(\underset{\sim}{U})$ is the probability density for a given set of measurements $\underset{\sim}{U}$ on an individual in group g, Π_g is the prior probability of belonging to population g, and r_{ij} represents the loss in identifying an individual of the ith population as a member of the jth population. The specimen is assigned to that population for which his allocation score, (also sometimes known as 'discrimination cation score for the gth group is proportional to the Mahalanobis this technique are given in Rao (1965).

In the very special case when one can assume not only that the specimen does belong to one of the groups, but also that $P_g(\underset{\sim}{U})$ is a multivariate normal probability density with variance-covariance matrix independent of g, and that $r_{ij}=1$ for $i \neq j$ and $r_{gg}=0$ for all g, and that the prior probabilities are equal, then the allocation score for the gth group is a decreasing function of the squared Mahalanobis distance,

$$D_g^2 = (\underset{\sim}{U}-\underset{\sim}{u}_g)^T \Sigma^{-1}(\underset{\sim}{U}-\underset{\sim}{\mu}_g)$$

where $\underset{\sim}{\mu}_g$ is the mean vector for group g and Σ is the (common) dispersion matrix for the G groups. A specimen is assigned to that group for which the allocation score is the highest, or the Mahalanobis distance has the lowest value. As well, in this special case, multivariate-normal statistical theory gives us the probability of observing a Mahalanobis distance which is less than or equal to the observed value. This probability is obtained by referring the value, F, where

$$F = n_g(n_g - p)p^{-1}(n_g^2-1)^{-1}D_g^2$$

to tables for an F-distribution on p and (n_g-p) degrees of freedom, (where n_g is the number of observations in group g and p is the number of measurements, and is taken here to have the same value for each group).

Although, as mentioned above, classical theory suggests one should allocate the individual to the group for which the allocation score is largest, or here equivalently, the Mahalanobis distance has the lowest value, it is good practice to allow for regions of doubtful allocation. For an individual may be allocated to another group even though the individual is also within a reasonable probability ellipse with respect to its original group. Further considerations on the role of individual Mahalanobis distances in discrimination and allocation are given in Campbell (1984).

In a series of articles by Schaafsma and Van Vark (1977, 1979) and Schaafsma (1984), the assumption of requiring equal prior probabilities of belonging to each of the groups is relaxed, and it is shown how one can adjust the (known or estimated) prior probabilities to obtain confidence-intervals for the posterior probabilities of the specimen belonging to each of the groups.

Often the additional information one has concerning the specimens under consideration is softer in kind than the above knowledge. Namely, one does not know whether the specimen does belong to one of the comparative groups or not. Then, the sort of question one may wish to answer is the following: How likely is it that the specimen belongs to each of the groups (considered separately)? The answer to this question may be that it is not likely that the specimen belongs to any of the comparative groups, or that it is likely that the specimen belongs to one group but it is not likely to belong to any of the other groups, or that it is likely that the specimen could belong to two, or more, of the groups. A method to answer this sort of question is that based on the likelihood approach to the analysis of data, as presented in detail by Edwards (1972). The use of this approach in physical anthropology was first proposed by Thorne and Wilson (1977), who were comparing prehistoric skeletal remains from Australian sites with a comparative series drawn form the osteological collections of the National Museum of Victoria. The evidence was not at all clear as to whether the skeletal remains might have belonged to the comparative series, or not. Assuming (after appropriate transformation of the variables) multivariate normality for the underlying distribution, it was shown that twice the increase in support given by the alternative hypothesis (that the specimen does not belong to the comparative group under consideration) over the support for the null hypothesis (that the specimen may belong to this comparative group) is given by the Mahalanobis distance. It was suggested, by reference to the asymptotic theory, that one could compare this value with a χ^2-distribution with p degrees of freedom.

Recently, Wilson (1981) has shown that the assumption of multivariate normality implies, essentially, a sample size in the comparative group that is far too large, (of order of magnitude proportional to a power of p). Instead, one should assume an underlying multivariate-t distribution to accommodate the finite sample size in the comparative group. (This is analogous to the use of the t-distribution instead of the normal distribution when p=1, and sample sizes are small.) Assuming this distribution, twice the increase in support, $2\Delta S_g$, is given by

$$2\Delta S_g = n_g \ell n\{1+n_g(n_g^2-1)^{-1}D_g^2\}.$$

Approximate (and asymptotic) values for the support limits are obtained from values of the χ_p^2-distribution. It is of interest to compare the conclusions from the two forms of additional information, namely that of <u>not</u> knowing whether the specimen does come from one of the groups, or not, with that of knowing the specimen does come from one of the comparative groups. From this comparison it is found that the 'harder' additional information, of the latter kind, yields 'harder' conclusions, while the 'softer' additional information, of the formel kind, yields 'softer' conclusions. As noted above, the 'softer' additional information leaves open the possibility of the specimen possibly belonging to more than one group, or may be not belonging to any of the groups.

For the data we are using illustratively, the size of the comparative groups is not given. In the following we have assumed that the sample size is identical in all the groups and has the value 50. The conclusions may change if the sample sizes actually differ markedly.

To show how the above technique can be used we consider some of the data in Table 1. It is noted that the measurements had been transformed to logarithms before analysis to ensure normality, and homogeneity of the variance-covariance matrices within groups. Also, the canonical variates were standardised to have unit standard deviation within groups.

Table 1. Permanent first lower premolar (Ashton et.al., 1957)

	coefficients of canonical variates							
trans.-bas.lab.	+10.2	+ 3.7	- 6.6	-18.1	- 6.7	+19.3	- 6.4	+ 9.9
trans.-mid.lab.	- 1.4	+ 0.9	+ 0.1	+14.7	-16.4	- 0.8	+13.8	+19.7
trans.-max.lab.	+24.6	- 3.8	+ 2.7	+24.2	+18.0	- 6.0	- 0.7	-30.5
ht.-lab.	+ 0.8	+ 0.9	- 8.1	- 1.8	+ 5.0	- 9.6	- 5.9	+ 7.5
ht.-prox.seg.	- 5.0	+ 2.4	+ 2.8	+ 4.6	+ 2.1	+ 8.9	- 2.8	0.0
thick.-bas.	- 2.0	+14.0	- 5.3	- 8.4	+17.7	+ 3.3	+25.6	- 1.3
thick.-mid.lab.	-13.1	+11.5	-19.9	+ 2.7	-10.3	- 1.2	- 7.6	-13.5
thick.-max.	- 1.2	+ 0.6	+42.0	-10.7	-15.5	-15.5	-19.4	+ 7.7
(constant term	-31.78	-54.53	-16.55	-14.23	+10.98	+ 7.97	+ 7.76	+ 3.16
	means of groups on eight canonical variates							
A. West African	-8.09	+0.49	+0.18	+0.75	-0.06	-0.04	+0.04	+0.03
B. British	-9.37	-0.68	-0.44	-0.37	+0.37	+0.02	-0.01	+0.05
C. Australian aboriginal	-8.87	+1.44	+0.36	-0.34	-0.29	-0.02	-0.01	-0.05
D. gorilla: male	+6.28	+2.89	+0.43	-0.03	+0.10	-0.14	+0.07	+0.08
E. female	+4.82	+1.52	+0.71	-0.06	+0.25	+0.15	-0.07	-0.10
F. orang-outang: male	+5.11	+1.61	-0.72	+0.04	-0.17	+0.13	+0.03	+0.15
G. female	+3.60	+0.28	-1.05	+0.01	-0.03	-0.11	-0.11	-0.08
H. chimpanzee: male	+3.46	-3.37	+0.33	-0.32	-0.19	-0.04	+0.09	+0.09
I. female	+3.05	-4.21	+0.17	+0.28	+0.04	+0.02	-0.06	-0.06
	fossils							
J. Pithecanthropus	-6.73	+3.63	+1.14	+2.11	-1.90	+0.24	+1.23	-0.55
K. pekinensis	-5.90	+3.95	+0.89	+1.58	-1.56	+1.10	+1.53	+0.58
L. Paranthropus robustus	-7.56	+6.34	+1.66	+0.10	-2.23	-1.01	+0.68	-0.23
M. Paranthropus crassidens	-7.79	+4.33	+1.42	+0.01	-1.80	-0.25	+0.04	-0.87
N. Meganthropus palaeojavenicus	-8.23	+5.03	+1.13	-0.02	-1.41	-0.13	-0.28	-0.13
O. Proconsul africanus	+1.86	-4.28	-21.4	-1.73	+2.06	+1.80	+2.61	+2.48

First, considering the fossil <u>Paranthropus Crassidens</u>, it is found to be most like the Australian aboriginal group, with

$$2\Delta S_c = 12.2$$

which can be compared with $\chi^2_{8;0.05} = 15.5 (\chi^2_{8;0.10} = 13.4)$. So one can entertain the null hypothesis that the fossil may belong to this group, (with respect to <u>all</u> the measurements). No other value of $2\Delta S_g$ for this fossil lies within these support limits. Second, considering the fossil <u>Proconsul africanus</u>, it is most like the chimpanzee with $2\Delta S_H = 24.7$, $2\Delta S_I = 24.5$. So, it would appear that it is unlikely that this fossil belongs to any of the comparative groups. Similarly, one could continue with likelihood comparisons of the fossils with the groups.

Ashton et.al. (1957) proceeded in the following way for their analysis. They did a 'discriminant analysis', (a canonical variate analysis), on the nine groups, and this is now a standard multivariate statistical technique, readily available in popular packages. The basic aim of this analysis is to determine a <u>linear</u> combination of the original variables which 'best discrimates' between the groups. Here, 'best discriminates' means maximising the between-groups variation relative to the variation within groups. This linear function is called either the first 'discriminant' function, or to avoid confusion with the allocation score discussed above, it is called the first canonical variate. Successive linear combinations are chosen to be uncorrelated within groups, as well as between groups. A total of k canonical variates are determined, where k=min(p,G-1). A clear description of this technique can be found in Tatsuoka (1971). When the canonical variates are used for descriptive purposes only, no distributional assumptions are necessary. To test appropriate hypotheses, distributional assumptions become necessary, and these are usually multivariate normality and homogeneity of the dispersion matrices. For sensible interpretation, it is desirable that the latter condition be satisfied, and Ashton et.al. took logarithmic transformations to achieve this. Also, for meaningful description, there is an inherent assumption that the groups be 'non-overlapping'. This term is not used in the geometric sense (as misinterpreted by Campbell, 1980), but is used in the following sense. It is necessary, based on external criteria, that the individuals who compose the comparative groups can be <u>clearly</u> allocated to their correct comparative groups. So, for example, it is necessary that one can clearly distinguish the males from the females in the ape groups, and so on.

Since all but two of the eigenvalues, (corresponding to the canonical variates), were small in magnitude, Ashton et.al. represented the groups and the fossils in those two dimensions, and this is shown in Figure 1. The circles represent the approximate 90% confidence limits for each of the populations. The positions for each of the fossils with respect to these two dimensions are also shown. In comparing each fossil with the groups, there is an implicit assumption, (which is more explicit in the above comparisons), that the covariances for (each of) the populations from which each fossil came are similar to those for the groups. It is of interest to compare some of the conclusions one may form from this diagram with those from the likelihood analysis above.

First, <u>Proconsul africanus</u> lies in the 90% confidence regions of the chimpanzee populations, yet from above, one would conclude that this fossil does not appear at all likely to belong to either

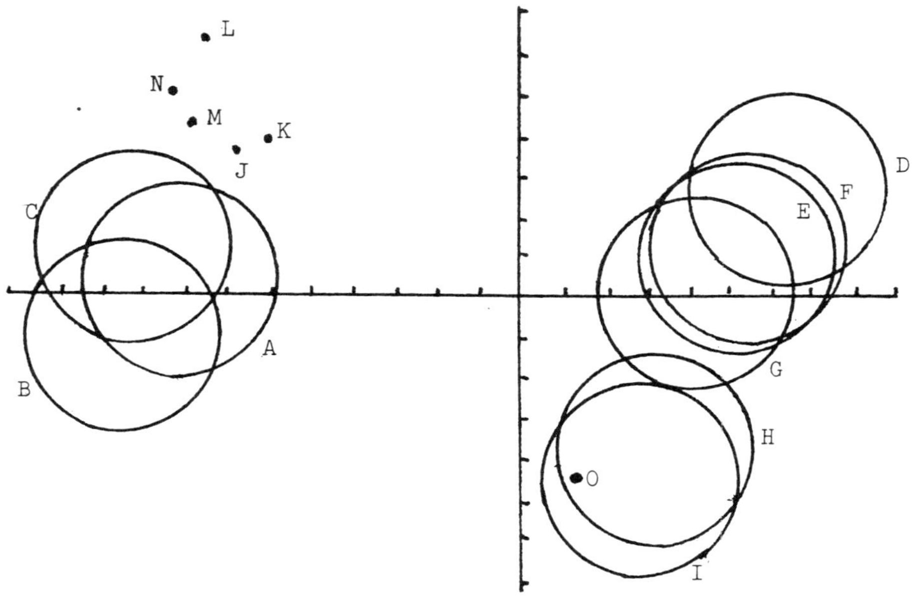

Figure 1. Representations of the comparative groups and the fossils in the space of the first two canonical variates (Ashton et.al., 1957)

A - West African
B - British
C - Australian
D - Gorilla, male
E - Gorilla, female
F - Orang-outang, male
G - Orang-outang, female
H - Chimpanzee, male
I - Chimpanzee, female
J,K - Pithecanthropus pekinensis
L - Paranthropus robustus
M - Paranthropus crassidens
N - Meganthropus palaeojavanicus
O - Proconsul africanus

of these populations. This apparent discrepancy can be explained, heuristically, by the covariances for the population to which this fossil belongs being different to the covariances for the comparative populations. So all the canonical variates should be considered for comparing this fossil.

On the other hand, Paranthropus crassidens lies outside the confidence regions, on the first two canonical variates, for all the groups. Yet, in the above analysis considering all the canonical variates, we would entertain the hypothesis of this fossil belonging to the Australian aboriginal group. Again, this suggests that this fossil belongs to a population with a different covariance structure to that for the comparative populations.

These are examples of a common problem that occurs with multivariate analyses, namely which dimensions should one be considering? The problem is even more complex if one is searching among the original, correlated, variables to find that set which 'best discrimates', say, particular groups. Usually this search is restricted to linear combinations of the variables and 'best' is defined in terms of a mathematical criterion. Various aspects of variable selection procedures are discussed further in Van Vark (1982).

For these data, since the original set of measurements was judged to be meaningful and all measurements were necessary for the comparison, and since the aim was to compare the fossil with the groups, then the relative position of the fossil on each (canonical) variate gives the most insight into the nature of the similarities to and differences from the comparative groups. In other words, only the two canonical variates need be considered in making inferences concerning the comparative groups. But all variates should be considered in making inferences concerning the fossils which "it is almost certain ... were neither human beings nor members of the existing species of great ape". Ideally, if the original data were available, the values on the original variates, and biologically meaningful combinations of these, would be investigated. Since these data are not available, in the following we will use the canonical variates for illustrative purposes, to describe the graphical techniques.

Some questions which should be asked by the researcher, as an integral part of canonical variate analysis, are the following. How stable are the results? Do the conclusions differ if some of the observations are removed? It is noted here that the use of robust techniques for the estimation of means and of dispersion matrices guards against the situation of atypical observations unduly affecting results (see Campbell, 1983). Also, are the computer program algorithms good ones from a numerical computation point of view? It is not always realised that successive versions of a package may have undergone extensive changes to their algorithms. Some researchers have rerun the same analyses for the same data on a later version of a package, and obtained quite different results and interpretations.

One way of guarding against the many pitfalls into which a researcher can tumble using multivariate statistical techniques, is to use, with care, multivariate graphical techniques. Such techniques are also useful for exploratory data analysis, that is for generating hypotheses concerning the data, compared with confirming (or not) prior hypotheses (which has been the main emphasis in this section). The use of these techniques may often reveal aspects of the data that might have escaped notice were the data tabulated only in numerical form; they may help to uncover features of the data that were totally unanticipated prior to the analysis, or to suggest hypotheses which may be investigated further. Ideally, what appears to be desirable in the exploration and analysis of multivariate data is far more interaction between the two approaches of 'formal' inference (as addressed in this section) and 'informal' (to be addressed in the next section).

3. SOME RECENTLY PROPOSED GRAPHICAL TECHNIQUES FOR MULTIVARIATE DATA

It is not a particularly straightforward procedure to gain insight into multiresponse data configurations. Yet some graphical procedure should be used for this purpose. Some of the more basic plots, such as 2- and 3-dimensional scatter plots of bivariate and trivariate subsets of the original data, and probability plots of the observations on each response separately, and so on, are described in texts like Gnanadesikan (1977) and Everitt (1978). Also, it can be useful to do quantile-quantile (Q-Q) plots of the individual Mahalanobis distances (in association with the analyses described in the previous section) to examine the degree of agreement with an underlying multivariate Gaussian distribution (see Campbell, 1983). The wide class of general multidimensional scaling techniques and their associated graphical techniques are extremely useful for prividing insight into multivariate data (see Gower, 1983).

Rather than describe the total set of plots that could be used for these illustrative data, instead our attention will be restricted to a few of the suggestions that have been made recently.

The first technique to be discussed has been used on anthropological data by Feldesman, Howells, and Oxnard, among others. The high dimension technique was proposed by Andrews (1972) and is a method of mapping each multiresponse observation into a function, f(t), of a single variable t. f(t) is defined to be a linear function of orthonormal functions in t with the coefficients in the linear combination being the observed values of the responses. More specifically, given the 8-dimensional observation, $\underset{\sim}{U} = (U_1, U_2, \ldots, U_8)^T$, Andrews chose to map this observation into

$$f_{\underset{\sim}{U}}(t) = \frac{1}{\sqrt{2}} U_1 + U_2 \sin t + U_3 \cos t + U_4 \sin 2t + U_5 \cos 2t$$
$$+ U_6 \sin 3t + U_7 \cos 3t + U_8 \sin 4t \,, \quad -\Pi < t < \Pi \,.$$

The group means $\underset{\sim}{u}_g = (u_{g1}, u_{g2}, \ldots, u_{g8})^T$ can also be mapped, by replacing U_i by u_{gi}. Using this plotting technique on these data, Andrews came to interesting conclusions that differ in their emphasis compared with the conclusions given by the analyses of the previous section.

The resulting curves for the nine groups are shown in Figure 2. This diagram shows that the curves for the three human groups are quite well separated from the curves of the six groups of apes, and among the apes, the chimpanzees stand out from the other two types. The sexes within each ape type tend to have relatively closely spaced curves, especially in the left hand part of the graph. Also, at specific values of t the separations among the groups are very pronounced in relation to the separation within groups. For example, at t_2 and t_4, the human groups are very cohesive and more clearly distinguished from the values for the curves for any of the apes. At t_1 and t_3 the chimpanzees seem to stand out more clearly from the remaining two groups of apes. So this shows how the technique can be used for detecting directions of clusterings among the groups.

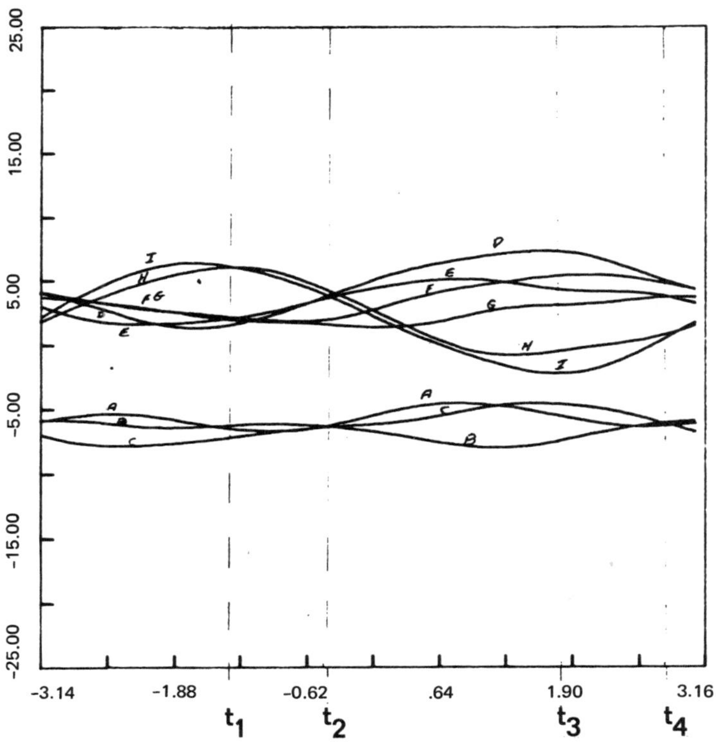

Figure 2. High-dimensional plot of the centroids of the comparative groups (Andrews, 1972)

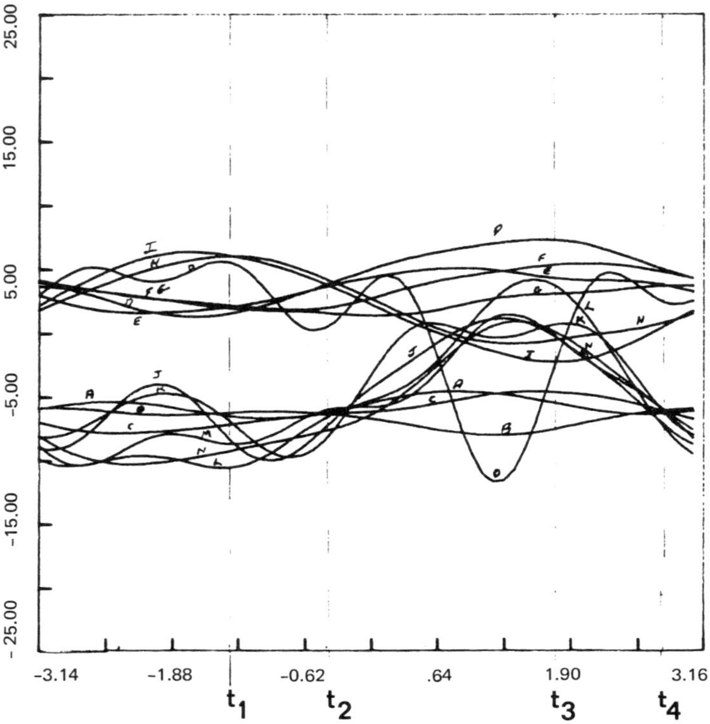

Figure 3. High-dimensional plot of the centroids of the comparative groups, and the fossils (Andrews, 1972)

In Figure 3, the curves for the six fossils have been plotted, as well as the curves for the groups. Immediately, one fossil, Proconsul africanus shows up as being different, although for specific (but different) values of t it comes close to each of the groups. In particular, in the left hand part of the plot and especially at t_1, it is most like the chimpanzee. The remaining fossils tend to have curves that are more like those for the human groups, especially at t_2 and t_4, and in the left hand part of the plot. It requires experience in using the technique, and comparing the relative behaviour of the curves at each value of t, (and not simply just regarding the functions as curves in their own right), before the researcher finds the representation useful.

Fienberg (1979) also uses two further graphical techniques on these data. First he uses the STARS technique for the nine groups, and this is shown in Figure 4. Each canonical variate is located along one of the eight rays, beginning with variable 1 located at the 3 o'clock position, and then going counterclockwise. So, in Figure 4, the ray at 3 o'clock corresponds to the value in row A,

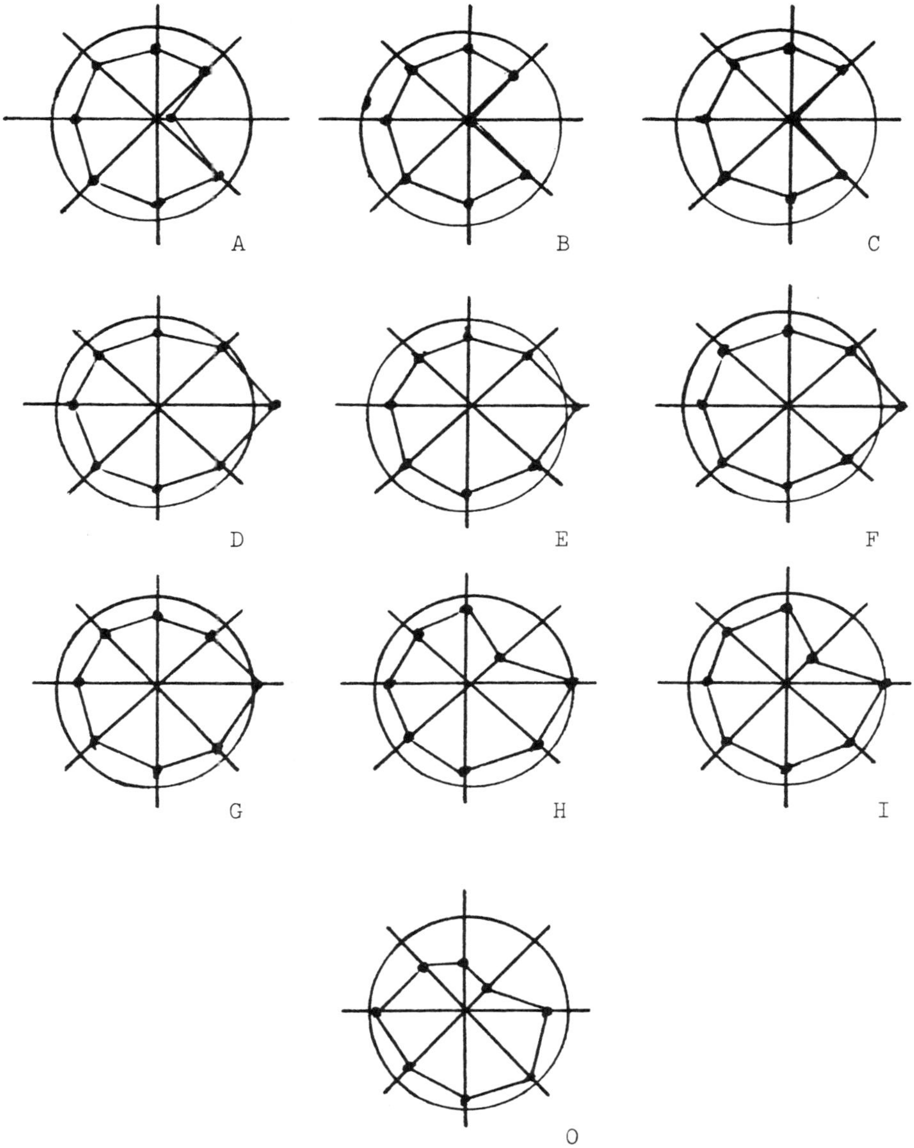

Figure 4. STARS for the canonical variates of the comparative groups and Proconsul africanus

column 1 of Table 1, and the other seven rays going counterclockwise correspond to columns 2 to 8 of this row, respectively. Each of the data components has been normalised by using the overall minimum and maximum of all the data components.

Examination of the STARS for the nine groups suggests that A, B, and C (the humans) form one group; D, E, F, and G (the gorillas and orang-outangs) form a second; and H and I (the chimpanzees) form a third. Most of the separation into groups is based on the values of the first two canonical variates. Examination of the STARS for the fossils shows how, for example, Proconsul africanus is similar to the chimpanzees on the first two canonical variates, but very different from all the comparative groups on the remaining variates.

Noting that people grow up studying and essentially classifying faces, Chernoff (1973) proposed representing a point, in space of up to 18 dimensions, by drawing a face whose characteristics (including shape of face, length of nose, curvature of mouth, and so forth) are determined by the coordinates or position of the point. Using the same normalisation as for STARS, the FACES corresponding to the nine groups are given in Figure 5 (taken from Jacob, 1980). Again the same basic clustering of the nine groups as observed in Figure 4 is seen here.

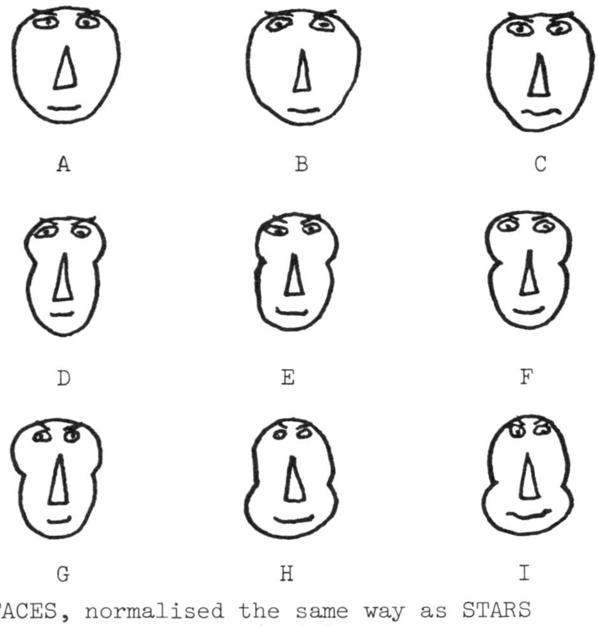

Figure 5. FACES, normalised the same way as STARS Adapted from Jacob (1980).

The use of the FACES technique requires more experience than the above techniques (see Fienberg, 1979), and both this technique and Andrews' plots are affected by changing co-ordinates. A variety of displays need to be tried before one can arrive at the best one for a given data set. On the other hand, the STARS technique is more strongly influenced by which components are adjacent to which other ones (because large variations in adjacent components result in sharp points). The visual effects of the parameters of FACES, by contrast, are inter-related in much more complex, and less well-known ways. The technique is distinctly culture bound. A Japanese or Chinese investigator, for example, might see possibly different groupings of faces. Certainly some of the groupings of faces that we see are biased by the forms of the faces. Assigning, for example, one set of variables to the mouth position so that one is comparing differences produced by different groups of smiles may lead to a different grouping than that produced by assigning that set of variables to the position of ears (which play much less of a part in our emotional view of faces). (Recently Flury and Riedwyl (1981) have extended the FACES technique to accommodate 36 variables by varying the face parameters of the left and the right side separately. This construction is somewhat more realistic.) Both the STARS and FACES representations may also be altered quite considerably by the form of normalisation used (Jacob, 1980). The choice of an appropriate representation depends, of course, on what one is studying. Here it might be the main division into clusters, or the structure within them.

All of the above, visual, techniques require some degree of training by the user if they are to be used beneficially. Essentially, this is because our eye has become conditioned to be alerted to certain aspects of the pictorial representation we are viewing, and highlighting such aspects, in certain situations, may well be misleading. Also one's eye tends to remove a lot of features that it judges to be 'noise', from the pictorial representation, and these may well be of importance for our interpretation.

Our different senses detect different messages, in different ways, from any global surrounding in which the individual is placed. One sense whose faculty could be used with success, for understanding data, is our aural sense. Aspects of the data that are highlighted by what we hear are very different to those that are highlighted by what we see. The author has used this tenet to develop a system of exploratory and sound data analysis (Wilson, 1982). The application of this system to our illustrative data set will be explained.

It is noted that there are two major obstacles hampering any explanation of this system. First, the readers of this article will differ quite considerably in their degree of musical training, and hence musical conditioning, to which they have been exposed, compared with, say, the degree of conditioning for viewing graphics to which they have been exposed. Also the system is based on a tonic triad representation which forms the basis of classical western music. So those who are tone deaf, or have been raised in a different (musical) culture will encounter difficulties with the technique. Second, the only medium at the author's immediate disposal for demonstrating how this aural system assists the researcher is a visual one.

The aural system is based on the following observation. If one gives an individual a sequence of digits, say the first row of Table 1, (and this is a particularly simple set), then on recall, some of these will be forgotten, as would the details of the Andrews' plot, STARS or FACES representations. On the other hand, give the same individual the corresponding 'tune' (Figure 6) this will be recalled more easily and more accurately. (Training will be required to achieve this easily.) When it comes to comparing, say, two groups, then it is easier to remember the tunes and compare the regularities and detailed irregularities in these, than to do this comparison for any of the visual representations.

To keep the normalisation of the data the same as that for STARS and FACES, it was decided to convert all the canonical variate values using the same scale, given by the following:

Canonical variate value	-10		-8		-6		-4		-2		0		2		4		6		8		10
Scaled value		-4		-3		-2		-1		0		1		2		3		4		5	

In other words, coefficients within the range (-8,-10) take the scaled value -4, and so forth. Usually it is only necessary to give the range for one scaled value, say 0, and the remaining values can be readily determined. The scaled values have the corresponding notes shown in Appendix 1. The scaled values for Table 1 are given in Table 2. Now the values in Table 2 can be transcribed using Appendix 1, and written out for each of the 9 groups and 6 fossils. (The underlying reasons for choosing this system are given in Wilson, 1982). An example of group A is shown in Figure 6. (The similarity to the profiles clustering technique (Hartigan, 1975) can be seen in this representation.) A form of (indirect) weighting has been introduced by scaling the canonical variates identically. This can be overcome by weighting the canonical variates in proportion to their corresponding eigenvalues in the following way. A time signature is introduced. The time signature chosen was $\frac{4}{4}$. The upper figure shows the number of divisions (accents) in a bar, here 4, and the lower specifies their quality, or value, and here is a crotchet (see Macpherson and Payne, 1970). The 9 group means and 6 sets of fossils are shown in Figure 7. First, it is remarked that the notes and the

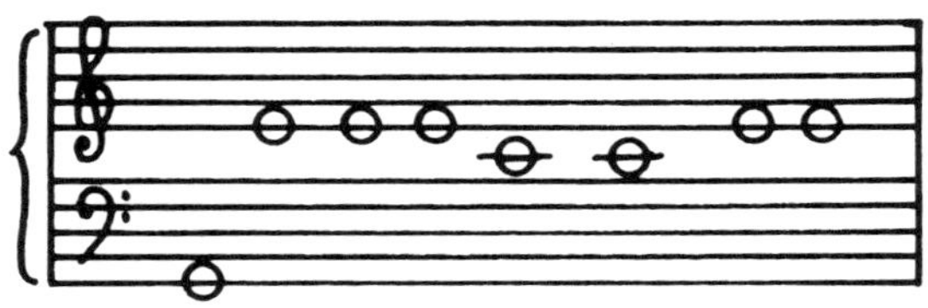

Figure 6. One sound representation of Group A

Table 2. Rescaled values for data on Table 1

	group means							
A.	-4	1	1	1	0	0	1	1
B.	-4	0	0	0	1	1	0	1
C.	-4	1	1	0	0	0	0	0
D.	4	2	1	0	1	0	1	1
E.	3	1	1	0	1	1	0	0
F.	3	1	0	1	0	1	1	1
G.	2	1	0	1	0	0	0	0
H.	2	-1	1	0	0	0	1	1
I.	2	-2	1	1	1	1	0	0
	fossils							
J.	-3	2	1	2	0	1	1	0
K.	-2	2	1	1	0	1	1	1
L.	-3	4	1	1	-1	0	1	0
M.	-3	3	1	1	0	0	1	0
N.	-4	3	1	0	0	0	0	0
O.	1	-2	-1	0	2	1	2	2

corresponding canonical variates have been reversed, so the first note represents the eighth canonical variate. The time signature has been chosen, first to be an approximation to the relative size of the eigenvalues, and second to incorporate the effect of the strongest accents falling on the eighth and first canonical variate values; the less strong accents occur on the third and second canonical variate values.

Playing these cadences shows the grouping of the humans together, the gorillas and orang-outangs together, and the chimpanzees together. The ending of the cadence (in particular the last three notes) determines this grouping. (It is noted that in practice this sort of grouping is usually made with respect to the first two canonical variates: last two notes.) If one plays the last five notes of each cadence, then one hears that both sexes of gorillas are similar, and different from the sexes of orang-outangs which are similar to each other. Fienberg (1979) remarked that one of his Chernoff faces representations seemed to recover the sexes. The last three canonical variates, (first three notes) does this here; the last four canonical variates, though, would group the Australian aboriginal and female orang-outang groups together, and also the West African and male chimpanzee groups are similar, and opposite to the female gorilla and female chimpanzee groups which are similar.

The original aim of analysing these date was to compare the fossils to the populations. Now considering the cadences for the fossils, the two *Pithecanthropus pekinensis* fossils are similar to each other, and to a lesser extent, they are similar to

Figure 7. One sound representation of the centroids of the comparative groups and the fossils

Paranthropus robustus and crassidens which are similar to each other (last four notes). Now comparing these four to the population groups, they are most like the human groups, and this is determined by the last two notes of the cadences.
Meganthropus palaeojavanicus is also like these four fossils in its cadence ending (last three notes), and the early part of its cadence is just as for the Australian aboriginal group. (In fact, this fossil sounds just like this group, but with a greater 'jump' at its ending.)
Proconsul africanus seems markedly different from the other fossils and, comparing to the nine groups, its cadence ending is more like the chimpanzees (last three notes). These aspects of the data would be worth further consideration. In other words use of the musical representation, in an interactive way, would lead the anthropologist to further explore certain aspects of the comparison.

A major aim for this technique is to provide a tool for exploratory data analysis, in the sense of generating hypotheses rather than confirming (or not) preconceived hypotheses. The value of this can be well appreciated for these data from the following consideration. It is well-known in anthropology that populations change in shape and size over time. So a fossil may be from a population which is a direct ancestor of the modern population to which one is making the comparison. But due to shape and size changes, a significance test for whether the fossil 'belongs' to the modern population may strongly reject this hypothesis. However, this form of data representation may show similarities which are of interest to the anthropologist. For example, the beginning of the cadence for Proconsul africanus has the same pattern, but at a higher level, as that for the male gorilla group. Is this due to changes over time to form today's male gorillas? Or is it (remembering the comment on sexing above) because Proconsul africanus is a (large) male and this group is a male population, and so on. Further similarities may also become evident by considering the original measurements for the groups and the fossils in this way. Of all the multivariate "graphical" techniques considered, this technique has given the greatest insight into the multivariate structure of these data.

Finally, it is stressed that this is not a complete analysis of the Ashton et.al. data and that the data have been used for illustrative purposes only. Also, it is noted that since the canonical variate functions have been used to illustrate these various graphical representations, the representations have been useful only for comparative purposes. Little back interpretation to the original measures is possible. In applying these techniques to data sets, both the original measures (possibly transformed), and the suitably combined measures (an example of which has been used here) should be (separately) considered. This would allow for a full description of the final comparisons.

Acknowledgements: Comments made by Charles Oxnard and Willem Schaafsma considerably improved this manuscript

Appendix 1.

The diagram accompanying this appendix, Figure 1.1, shows the correspondence between the pianoforte keyboard (in upper part of diagram), the numbering system used for the rescaled measurements, (the numbers being written on the corresponding notes of the keyboard), and the written notation (in the lower part of diagram). The English names of the notes and the Italian names (Sol-fa system) are also given for the corresponding numbers, and also for the immediate notes for the octave (8 notes) commencing on "Middle C".
For those who have not learnt to play the piano and/or read music, this diagram should suffice for them to play the cadences for the example in this article.

Figure 1.1

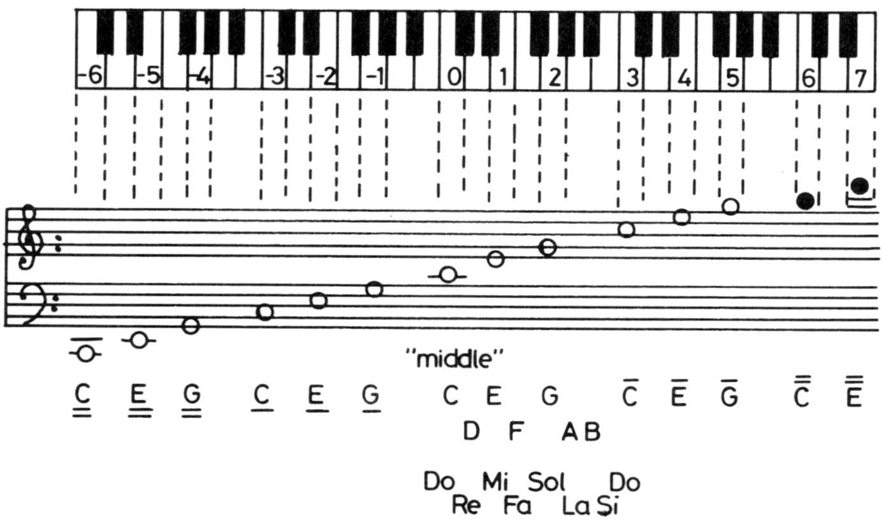

References

Andrews, D.F.: 1972, Plots of high-dimensional data. Biometrics 28, pp. 125-136.

Ashton, E.H., Healy, M.J.R., and Lipton, S.: 1957, The descriptive use of discriminant functions in physical anthropology. Proc. Roy.Soc. B 146, pp. 552-572.

Campbell, N.A.: 1980, On the study of Border Cave remains: statistical comments. Curr. Anthropol. 21, pp. 532-535.

Campbell, N.A.: 1984, The role of the individual Mahalanobis distance in discrimination and allocation. IUAES Intercongress 1981 Amsterdam.

Chernoff, H.: 1973, Using faces to represent points in K-dimensional space graphically. Journal of the American Statistical Assoc., 68, pp. 361-368.

Edwards, A.W.F.: 1972, "Likelihood", Cambridge University Press.

Everitt, B.S.: 1978, "Graphical Techniques for Multivariate Data", Heinemann Educational Books.

Fienberg, S.E.: 1979, Graphical methods in statistics. The American Statistician 33, pp. 165-178.

Flury, B., and Riedwyl, H.: 1981, Graphical representation of multivariate data by means of asymmetrical faces. Journal of the American Statistical Assoc., 76, pp. 757-765.

Gnanadesikan, R. : 1977, "Methods for Statistical Data Analysis of Multivariate Observations", John Wiley and Sons, New York.

Gower, J.C.: 1984, Some recent advances in multivariate analysis applied to anthropometry. IUAES Intercongress 1981 Amsterdam.

Hartigan, J.: 1975, "Clustering Algorithms", Wiley-Interscience.

Jacob, R.J.K.: 1980, Letter to the Editor, The American Statistician 34, pp. 252-253.

Macpherson, S., and Payne, A.: 1970, "Rudiments of Music", Galliard, New York.

Rao, C.R.: 1965, "Linear Statistical Inference and Its Applications", John Wiley and Sons, New York.

Schaafsma, W.: 1984, The asymptotic variance estimators for the posterior probabilities: A useful tool which should not be neglected. IUAES Intercongress 1981 Amsterdam.

Schaafsma, W., and Van Vark, G.N.: 1977, Classification and discrimination problems with applications, part I. Statistica Neerlandica 31, pp. 25-45.

Schaafsma, W., and Van Vark, G.N.: 1979, Classification and discrimination problems with applications, part IIa. Statistica Neerlandica 33, pp. 91-126.

Tatsuoka, M.M.: 1971, "Multivariate Analysis", John Wiley and Sons, New York.

Thorne, A.G., and Wilson, S.R.: 1977, Pleistocene and recent Australians: a multivariate comparison. Journal of Human Evolution 6, pp. 393-402.

Van Vark, G.N.: 1984, On the determination of hominid affinities. IUAES Intercongress 1981 Amsterdam.

Wilson, S.R.: 1979, Comment on Rightmire, G.P., Implications of Border Cave skeletal remains for later pleistocene human evolution, Current Anthropology 20, p. 32.

Wilson, S.R.: 1981, On comparing fossil specimens with population samples, Journal of Human Evolution 10, pp. 207-214.

Wilson, S.R.: 1982, "Exploratory and Sound Data Analysis". In preparation.

DATA BANKS AND MULTIVARIATE STATISTICS IN PHYSICAL ANTHROPOLOGY

I. Schwidetzky

Anthropologisches Institut der Johannes Gutenberg-Universität, Mainz

In recent decades, the fields of administration and economy, the press and - last but not least - the sciences have been characterized by an "explosion of knowledge", and, as a consequence, by the problem of managing the rapidly increasing mass of information. It has been estimated that knowledge doubles each five years, and even that the interval of doubling seems to decrease. The main response to this challenge are computerized and structured data collections called data banks. "Data banks are systems of data collections which are organized according to logical and/or formal criteria; they should make it possible to reproduce the data of the total collection arranged according to different aspects ..." (REICHERTZ 1975, p. 916). Data banks began to develop in the midst of the sixties (MERTENS 1970), and if we see their present importance for instance in medicine they themselves demonstrate the "explosion of knowledge". Medical documentation already includes thousands of publications and a dozen or more special periodicals (KOLLER and WAGNER 1975). Not only universities but also important and less important hospitals have their own documentation and computer centers. The main fields of data banks, however, are economy and administration, with still larger information systems. In this landscape physical anthropological computerized data collections of course represent a very small section. It is remarkable, however, that the idea of data banks arose rather early in physical anthropology, and that it had already been realized for prehistoric physical anthropology at the end of the sixties.

During the congress of French speaking anthropologists in Brussels, November 1967 (its main subject: Physical anthropology of the Neolithic period), M.-A. Sauter, Geneva, proposed the creation of a

G. N. van Vark and W. W. Howells (eds.), Multivariate Statistical Methods in Physical Anthropology, 283–288.

"banque néolithique" in Mainz. He proposed Mainz because the Anthropological Institute of the University had organized there in 1966 an international symposion on the physical anthropology of the Neolithic. Besides, preparatory work had been done by establishing a bibliography and a collection of metrical data - averages on Randlochpunching cards - which had been used for multivariate comparisons. A matrix of Penrose-distances for 95 Mesolithic, Neolithic and Bronze Age population samples was one of the materials offered to the participants of the symposion. This was of course not yet a data bank but a first step towards it. Another data collection of Neolithic as well as Early Bronze Age samples already existed in Geneva. The proposal of M.A. Sauter was readily accepted. In 1969 the Working Group of European Physical Anthropologists met in Mainz and accepted the program of the data bank; it decided to extend the program to more recent periods and to collect on punched cards individual skull and skeleton measurements from the Mesolithic up to the present. Information on the samples as a whole, e.g. culture, dating, author and localisation of the site, should be noted in special "serial cards" so that the data could be reproduced for computer programs according to several criteria. The first report was published in 1971 (SCHWIDETZKY and CREEL 1971). By this time the data from 1117 samples with 9015 individuals had been stored. The data exchange with Geneva for the Neolithic and Early Bronze Age helped to build up a certain data stock rather quickly (meanwhile the data bank in Geneva also enlarged its program in exchange with Mainz; s.MENK 1979).

It seems that the data banks in Mainz and Geneva are the most comprehensive ones in physical anthropology. Other, smaller computerized data collections for special purposes and studies exist but no listing of these is available. At the beginning of the Human Adaptability program the project of a data bank for living populations was discussed but could not be realized because of the high costs. The data bank in Mainz had been organized with a minimum of cost (5000 Dollars from the Wenner Gren-Foundation as the starting sum was the only grant it received, and, consequently, with a minimum of collaborators. However, it now stores skull measurements (+ the length of 4 long bones) for about 43500 individuals (many of course incomplete) from ca. 3910 series and additional postcranial measurements for about 6300 individuals (1.9.1981). There may be two reasons why the project could be realized. In comparison with the HA-project: it is easier to collect skeletal measurements since the number of variables and the number of indiviuals which have been studied and published is smaller than in the biology of living populations; and in comparison with medical documentation there is an additional reason: the main variables of prehistoric physical anthropology, the measurements, have been standardized for a long time. With regard to the immense and rapidly growing literature on medical documentation this seems to be the most important reason that the data bank in Mainz could begin to work rather early. It was for example the basis for a series of multivariate "comparative studies" (s.e.g., ROSING and SCHWIDETZKY 1977) which studied the geographical differentiation of the European populations (and those of the non-European but Europoid borderlands) in a modern way, that is without any a priori type system (s. HOWELLS 1973, 1978). This had been done for different time levels so that diachronic changes could also be considered. Other papers using data bank materials are e.g. HENKE 1979, KLUG 1982 and MENK 1977.

But although it is relatively easy to collect human skeletal data, many problems remain. We shall give some examples.

a. It has been said that one advantage of skeletal anthropology - in comparison with medicine but also with some sections of the biology of living populations - is that the measures have been standardized long ago. But, a limitation of this statement is necessary. There are not only some differences in the measuring techniques e.g. between MARTIN and the system of the English biometricians which have to be considered (and in early papers the measures are sometimes not identifiable) but also the interpretation of the individual researcher of the measuring system adapted may lead to systematic differences.

b. A second problem is formed by the collection and storage of morphological, i.e. non-metrical characters. These characters are apparently more important for sex determination than are measurements (cf. FEREMBACH, STLOUKAL & SCHWIDETZKY 1980). But non-metrical characters are not yet stored in data banks. This is true also for the so-called epigenetic characters which are much appreciated (perhaps too much) in present day physical anthropology. Many research workers think that these variables more precisely reflect genetical relationships between populations than do measures, since the latter may be more liable to modification by environmental factors.

It would indeed be desirable to include non-metrical characters in a data bank but this could give rise to problems of standardization and comparability which already fill so many volumes in the field of medical documentation. Only the information on the epigenetic characters may be comparable in as far these are (or should be) presence-absence characters. For these characters there exists a certain standardization which was proposed by BERRY & BERRY (1967). Age determination in skeletons is based on morphological characters only. Therefore the same problem exists as to storing and computerizing. And there is another big handicap for storing: most frequently the morphological characters on which sex and age determination is based are not published. It may be mentioned, however that in the Anthropological Institute of Hamburg a multivariate statistical program has been developed for the age determination according to the NEMESKERI "complex-method" (NEMESKERI et.al., 1960; ACSADI and NEMESKERI 1970; FEREMBACH,

STLOUKAL and SCHWIDETZKY 1980). In summary: big tasks remain for the further development of data banks in physical anthropology.

c. Referring once more to the problem of the sex diagnosis of skeletal parts, we recall to mind that we do not as yet know which sets of measures provide the "maximum" sex discriminatory capacities, and, in connection with this, where these maxima lie. This holds true for skulls as well as for post-cranial bones. The determination of the sex of e.g. prehistoric individuals to whom the skeletal material under investigation belongs is indispensable for obtaining valuable results regarding the degree of morphological similarity of the prehistoric populations concerned. As a consequence, this gap in our knowledge has up to now formed a serious limitation in the investigation of inter alia prehistoric human populations.

Procedures for the sex diagnosis of bones which derive from prehistoric individuals are discussed in VAN VARK (1971, 1974). A further elaboration together with practical examples is presented in UYTTERSCHAUT (1983). However, an optimal use of these procedures requires that they be preceeded by variable ordering and selection procedures (cf. VAN VARK 1982) which are applied to extensive reference material. This material should ideally consist of skeletal data taken from a "very large" number of male as well as female specimens, and should cover the scores on a "very large" number of variables. Only if calculations of this nature may be carried out will questions be settled as to which sets of variables are optimum for sex diagnostic purposes, or how well samples of, e.g. prehistoric skull samples may be sexed. Doubtless, this will at the same time imply that a major step is taken in the investigation of prehistoric populations.

Similar considerations hold for age determinations and the diagnosis of skeletal remains of poor quality. In all these instances, the usefulness of the results is to a large extent determined by the availability of extensive reference material.

These are among the reasons why G.N. VAN VARK requests a new anthropological data bank or a new section of the existing one with a comprehensive data collection of skeletons of known sex and age. According to VAN VARK (CIRCULAR LETTER of 16.1.1980), this reference series should consist of a sample of a homogeneous population, and the sample should not be selective.

In general, an anthropological data bank should serve the following purposes:

a. the development and the testing of the practical merits and demerits of new statistical and data-analytical techniques; this includes the determination of the practical range of competitive methods,

b. the standardization of the measuring technique,

c. the provision of material to investigators who do not have the time and/or the financial support to collect this material themselves,

d. the development of a retrieval system for the literature,
e. the provision of easily convertible computer programmes, and accessory information,
f. the enlargement of the scope, the variety, the number, and the diagnostic power of anthropological investigations.

In summary, one may state that the "new methodology", if we may so call advanced multivariate methods such as those presented in this volume, can only function satisfactorily in conjunction with a well-equiped anthropological data bank. Without such a bank, a major part of the possibilities that are opened by this methodology will never be realized and will remain the subject of theoretical discussions only. Meanwhile, the data banks in Mainz (and Geneva) have there limited importance: to collect the data scattered over so many periodicals and to make them more easily available to those who are interested in this kind of data; and to offer these in a form which is needed by computer programmes of uni- or multivariate statistics.

References

Acsádi, Gy., J. Nemeskéri: 1970, History of human life span and mortality. Budapest.
Ferembach, D., M. Stloukal, I. Schwidetzky: 1980, Recommendations for age and sex diagnoses of skeletons. J.Hum.Evol. 9: 517-549.
Henke, W.: 1979, Beitrag zur Optimierung der metrischen Geschlechtsbestimmung am postcranialen Skelett. Mém.Soc.Hell.Anthrop. 48, 112-122.
Howells, W.W.: 1973, Cranial variation in man. A study of multivariate analysis. Peabody Mus. Papers 67.
Howells, W.W.: 1978, Quantitative systematics in human populations: an overview. Homo 29, 217-228.
Klug, S.: 1982, Discriminant function analysis on Slavic and Germanic populations. Homo 33, 237-248.
Menk, R.: 1977, La néolithisation: impact de l'innovation culturelle sur la biologie et la dynamique des populations. Arch.Suisses Anthrop. Gén. 41, 31-35.
Menk, R.: 1979, Data banks in historical anthropology: the material infrastructure for interdisciplinarity. Arch. Suisses d'Anthrop.Gén. 43, 331-342.
Koller, S., G. Wagner: 1975, Handbuch der medizinischen Dokumentation und Datenverarbeitung. 1485 pp. Stuttgart-New York.
Mertens, P.: 1970, Datenbank-Organisation. DV-Praxis 13.
Nemeskéri, J., L. Harsány, Gy. Acsádi: 1960, Methoden zur Diagnose des Lebensalters von Skelettfunden. Anthrop. Anz. 24: 70-95.
Perscheid, M.: 1974, Das Mainzer Lochkartenarchiv für postkraniales Skelettmaterial prähistorischer Populationen. Homo 25: 121-124.
Reichertz, P.L.: 1975, Datenbanken und Informationssysteme. pp. 913-939 in S. Koller, G. Wagner: Handbuch der medizinischen Dokumentation und Datenverarbeitung.

Rösing, F.W., I. Schwidetzky: 1977, Vergleichend-statistische Untersuchungen zur Anthropologie des frühen Mittelalters (500-1000 n.d.Z.) Homo 28: 65-115.
Schwidetzky, I.: 1979, Stand der Mainzer Datenbank für die prähistorisch-historische Anthropologie. Arch.Suisses d'Anthrop. Gén. 43, 327-330.
Schwidetzky, I., N. Creel: 1971, Das Mainzer Lochkartenarchiv für die prähistorische Anthropologie. Homo 22: 1-2.
Uytterschaut, H.T.: 1983, Affinities of Phillipine populations. An application of multivariate techniques to human skull data. Thesis, Groningen University.
Van Vark, G.N.: 1971, Eine Methode zur Geschlechtsbestimmung prähistorischer Individuen auf Grund von Skelettüberresten. Homo 22: 76-84.
Van Vark, G.N.: 1974, The investigation of human cremated skeletal material by multivariate statistical methods. Ossa 1: 63-96.
Van Vark, G.N. & P.G.M. van der Sman: 1982, New discrimination and classification techniques in anthropological practice. Zeitsch. Morph.Anthrop. 73/1: 21-36.

BIOLOGICAL EVOLUTION AND HISTORY IN 19th CENTURY PORTUGAL

Jean Pierre Bocquet-Appel

Centre National de la Recherche Scientifique,
Laboratoire d'Informatique pour les Sciences de l'Homme.
54, Boulevard Raspail, 75270 Paris Cédex 06, France.

Studies of interpopulation differences in non-metric morphological characters (also termed "discrete" and "epigenetic") of the skeleton raise problems which are both biological and statistical.

Generally considered to have an "important" component of heredity, these characters have been used in many investigations that compare worldwide populations (Wood-Jones, 1930-31, Riesenfeld, 1956; Berry and Berry, 1967; Kellock and Parsons, 1970, Knip, 1971, Muller, 1977, Pietrusewsky, 1977; Sjøvold, 1977; Kaul, Anand, Corruccini, 1979; Rathbun, 1979; Thoma, 1981). However, strictly speaking nothing is known as to the genetic base of these traits. While it is entirely legitimate to compare distributions of a morphological character in different populations, it is not legitimate to make such comparison in the belief that one is making genetic comparison only, in disregard of possible environmental factors. This belief nevertheless seems to be frequent in craniological studies. What, actually, are we comparing when we examine series of skulls of eskimos and Bantus? Genotypic structures, or those related to periglacial versus jungle climates, or both together? What is the true nature of these non-metric variations of the human skeleton? How sensitive are these characters to rapid modifications of the environment? How do their frequency distributions fluctuate, among sub-units of a single population, when these are subjected to different environments?

The present study is meant to ask certain questions about the biological nature of minor morphological traits of the human skeleton. It will be pursued through three steps. The first, of no theoretical importance, is devoted to selecting a set of traits that are independent of the effects of age and of pathology present at death; it consists of the use of simple statistical procedures. The second step comprises an analysis of the statistical definitions

G. N. van Vark and W. W. Howells (eds.), Multivariate Statistical Methods in Physical Anthropology, 289–321.

usually given for particular morphological traits of the skeleton: are they discrete or continuous? This question is not merely an academic one: it leads into the general conception of variation, and of the definition of metric scales in quantifying the difference between groups. Furthermore, the task of the anthropologist does not consist solely of describing the size or the scale of the biological variation, or the significance of the difference between groups. It consists primarily of attempting to explain these things, that is, of trying to reveal the mechanism of every sort (genetic, environmental, demographic, geographical, cultural) which may influence the distribution of characters within a population and the redistribution of the same in the succeeding generation.

For these reasons the third step will be addressed to the variation of the characters themselves. Going beyond simple comparisons of interpopulation frequencies, or of means of the morphological traits, we ask: where does the origin of the differences lie? In attempting to answer I shall use a simple model for breaking down the intergroup morphological variation into matrices for different contributing factors. To every matrix of biological distance, D (or S for similarity, where 1 on the principal diagonal instead 0), there can be fitted a certain number of matrices of postulated contributions or hypothesis matrices, H, such as matrices of the environment, of geographical distances, of coefficients of isonymy, of migration, and so forth. The residual after a convenient scaling, E = D - H, can measure the difference between the observed sample and the explanatory model. Attempts in this direction have already been performed by Howells (1966), Workman, Harpending, Lalouel, Lynch, Niswander, Singleton (1973), Friedlaender (1975). However, the degree of congruence between D and H may be weak. Then what is the fraction of D (or S) correlated to H? When the whole dimensionality (or a part of it) of D and H is conserved, the first point can be resolved by Procrustean analysis and statistics; the second point can be resolved by multiple correlation analysis (or even canonical correlation analysis) between the eigenvectors from the D and unmodified H matrices. To consider the concrete case, we shall consider the variation in samples of crania representing generations of Portuguese born in the 19th century. The method of decomposition allows us to subdivide the total morphological variation into categories of rapid and slow modification. The analysis (below) shows that the group of characters showing rapid response is linked to the immediate external environment to which each generation is subjected, while the characters showing slow modification are independent of the immediate environment and fluctuate only over long periods. For example, we shall see that the frequency distribution of incisura frontalis is seen to be associated with that of the price of wheat over generations (such trait being more frequent in times of less favorable nutriture) rather than showing a secular effect over time. Certain other traits exhibit the opposite connection.

1. MATERIAL

1.1. Biological variables

The Institute of Anthropology of the University of Coimbra possesses some 2300 skulls of individuals, of both sexes, who died during the 19th century and at the beginning of the 20th. Collection was begun originally by B. Machado at the end of the last century, and continued by E. Tamagnini. The crania derive from individuals born from 1803 to 1917 and interred in the cemeteries of Coimbra and Oporto from 1897 to 1936, as the inhumation rights of 5 years expired. For each individual the age at death, sex, date at death, probable cause of death, place of birth and occupation are all known. For reasons related to historical documentation (see below) a subseries of crania was selected which comprised only individuals born from 1810 to 1909 and with an age at death of at least 18 years (see table 1).

Table 1. Distribution of crania sample from Coimbra by birth generation and age at death (M : males, F : females)

Age at death	Birth generation											
	1820		1840		1860		1880		1900		Total	
	M	F	M	F	M	F	M	F	M	F	M	F
18-39	0	0	0	0	50	30	61	52	247	178	358	260
40-59	0	0	61	22	114	66	176	176	23	21	374	285
60+	42	33	95	127	125	157	22	18	0	0	284	335
Total	42	33	156	149	289	253	259	246	270	199	1016	880

Note that cranial sample is not longitudinally distributed (i.e. in a given birth generation all the age-classes at death are equally distributed) but semi-transversely (in the oldest birth generations there only are oldest age-classes at death, i.e. the age-classes at death are negatively correlated with the birth generations). This fact introduces a constraint specially in the selection of morphological traits which should be independent of age at death; this will be considered below. Although it is not easy to classify individuals by occupation under the sociological designations in use today, nonetheless the terms given do allow recovery of appropriate groupings (for example "small artisan" from toothbrush-maker, or "unskilled laborer" from porter) to summarize the occupational categories making up the sample as shown in table 2. The female sample is composed mainly of domestica, or housewives. The term must not be misinterpreted: it indicates a station much closer to that of a female farmworker than that understood by the same term today in western Europe. To summarize, the series of crania is drawn essentially from the ordinary classes of Portuguese society.

Table 2. Occupational categories of individuals from the anthropologic collection of Coimbra. (Individuals aged at least 18 years at death and born between 1810-1909.)

Categories	Males	Females	Total
Farm-labourer	84	12	96
Soldier	49	-	49
Sailor	16	-	16
Mill hand	47	3	50
Small artisan	224	13	237
Upper class	51	3	54
Unskilled laborer	294	2	296
Servant	49	59	108
Houswive	-	733	733
Lower class (Hobu, prostitute)	52	36	88
Diverse	112	7	119
Peasant proprietor	32	3	35
Total	1010	871	1881

To study genetic and environmental influences on morphological variation in this population, 15 craniometric and 35 cranioscopic variables were collected, between 1977 and 1980. These characters are listed in table 3, with the frequencies of each category.

1.2. Environmental variables

The geographic or climatic environment in Portugal has hardly changed in the last century. Contrariwise other variables, for example the level of availability of good, fluctuated, sometimes considerably, in the course of the 19th century. Years of food shortages, or privation, followed years of relative abundance. Such fluctation was certainly not without effect on the physical development of segments of the population. Accordingly I have endeavored to characterize the fluctuations by a classically used measure to indicate the nutritional condition of a population: the current price of articles of diet. In the 19th century and earlier, high prices for foodstuffs corresponded with periods of adversity and vice versa. Inflation, on the other hand, is a contemporary phenomenon which made its appearance only with the

Table 3. Distribution of the non-metric traits in the crania sample from Coimbra University (Individuals aged at least of 18 years at death. The sign * designs a frequency < 5% in the two sexes). Characters damaged or crania with closure suture cause the differences in sample size.

		Males						Females					
			Occurence						Occurence				
Trait		N	0	1	2	3	4+	N	0	1	2	3	4+
Sutura frontalis	left	1058	1003	55				915	851	64			
	right	1058	1045	13				915	893	22			
Incisura frontalis margo orbitalis	left	1057	213	797	47			913	193	688	35		
	right	1056	194	816	46			915	144	717	54		
Foramen supra-orbitale	left	1057	790	236	28	3		914	665	222	26	1	
	right	1057	798	226	30	3		915	679	213	23		
Foramen ethmoidale	left	1014	9	84	646	265	10	880	18	65	517	269	16
	right	1007	12	86	631	264	14	884	14	67	536	258	9
Os epipterium	left	951	900	51				863	799	64			
	right	952	900	52				858	795	63			
Os epipterium anterius	left	950	929	21				862	840	22			
	right	952	932	20				859	847	12			
Os epipterium posterius	left	950	882	68				862	771	91			
	right	952	891	61				859	782	77			
Pterion:"H"	left*	1015	7	1008				896	14	882			
	right*	1010	10	1000				888	13	875			
Pterion:"I"	left*	1017	1010	7				896	882	14			
	right*	1012	1001	11				889	877	12			
Os interparietale		1010	896	114				890	793	97			
Ossa suturae lambdoideae	left	860	530	156	78	48	48	812	552	124	61	41	34
	right	848	570	122	77	39	40	805	562	121	56	28	38
Os bregmaticum*		1057	1048	9				914	910	4			
Ossa suturae coronalis	left*	1035	1015	17	2	1		894	874	19	1		
	right*	1035	1022	11	1	1		894	13	3	1		
Ossa suturae saggitalis*		1028	997	23	5	3		893	857	24	12		
Foramen parietale	left	1057	437	607	13			912	392	514	6		
	right	1057	394	643	20			912	354	541	17		
Canalis hypoglossi	left	1057	6	835	212	4		913	4	695	212	2	
	right	1057	7	864	183	3		915	4	752	153	6	
Condylus occipitalis	left	1041	943	98				889	876	13			
	right	1040	947	93				892	875	17			
Torus palatinus		1056	995	61				902	841	61			
Canalis basilaris medianus superior		1057	725	332				913	692	221			
Ligamentum pterygospinale	left*	1058	1027	31				915	900	15			
	right*	1058	1023	35				915	899	16			

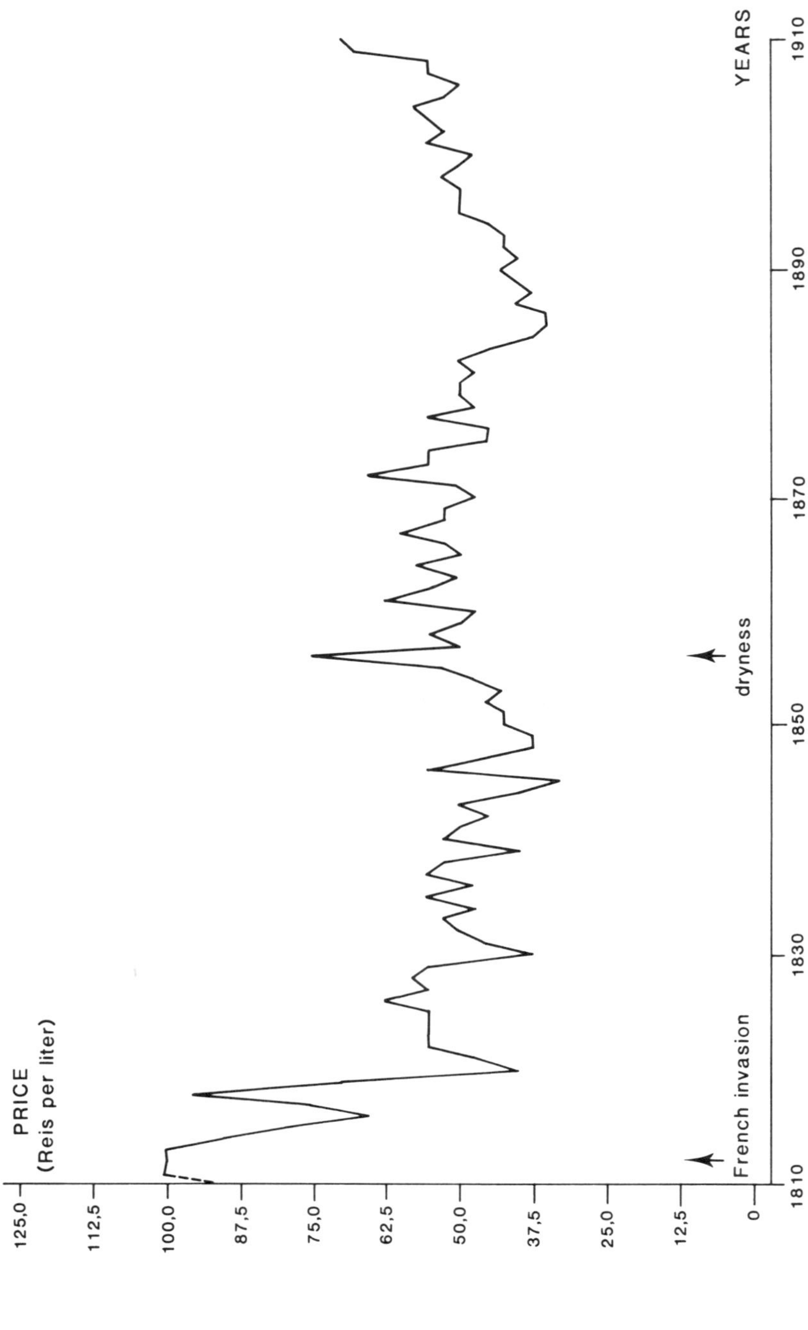

Figure 1. Wheat price trend for the city of Oporto (1810-1909)

1914 war. The price we shall employ here is that of wheat, a dietary staple playing an important role in the agricultural societies of pre-industrial Europe. We shall use only the price of wheat in August in the south and in September in the north, because, traditionally, Portuguese peasants judged the quantity of produce available for the year by its volume during those months (Soares, 1859, p. 437). It is, furthermore, the August or September prices of foodstuffs that are found most often in the archives of municipalities. Figure 1 shows the movement in the price of wheat for the city of Oporto.

2. METHOD

Accordingly, the object of this study is to examine whether the frequencies of non-metric traits of the skull remain stable from one cohort of births to another over a period of 100 years. If not, I will seek to determine the importance which may be assigned to factors of nutritional origin on the one hand or "secular" factors on the other; that is to say, factors associated with chronology but indeterminate as to origin. Specifically, we ask whether morphological variation between birth cohorts, V(D) is related to the variation in the price of wheat V(P) and/or to the variation in chronology V(T), by:

$$V(D) = V(P) + V(T) + E$$

where E is an uncontrolled residue.

The statistical procedure I shall introduce taking into account the nature of the variables themselves is therefore intended to

- i select a set of characters independent of systematic influences not accounted for by the model (i.e. traits influenced by age and pathology prevailing at death are eliminated);
- ii measure satisfactorily the morphological differences between birth generations;
- iii measure the relationship between a biological <u>structure</u> (morphology) or a simple part of it and causal hypotheses (nutrition and time).

2.1. Select a set of characters: construct a filter

The selection of characters with regard to different criteria under investigation (low frequencies, age and pathology present at death) can be compared to a filter. The criteria chosen to construct this filter are the following:

<u>Step 1</u> Eliminate the characters presenting low frequencies of occurrence ($f < 5\%$ in both sexes). Intergeneration fluctuations in such low frequencies might make interpretation difficult.

Is the frequency of the s_{th} occurrence of the k_{th} character, f_s, lower than 5% in both sexes?

yes: s = s - 1. If s = 0 (or 1, depending of the distribution under consideration) eliminate the character. Otherwise return to step 1.
no: continue.

The results are given in table 3. The characters in question are already recognized in the literature as rare.

Table 4. Test of symmetry of the distribution of bilateral characters. McNemar χ^2_{MC} test of independence (Individuals aged at least 18 years at death. Figures in parentheses under the name of the trait design what states were used to form contingency table).

Trait	Males			Females		
	χ^2_{MC}	d.f.	N	χ^2_{MC}	d.f.	N
Incisura frontalis (0, 1+)	1.473	1	1055	10.195***	1	913
Foramen supra-orbitale (0, 1+)	.269	1	1056	1.010	1	914
Foramen ethmoidale (0-1, 2, 3+)	3.146	3	992	1.666	3	867
Os epipterium (0, 1+)	.033	1	931	1.347	1	848
Ossa suturae lambdoideae (0, 1, 2+)	13.377*	6	843	2.287	6	799
Foramen parietale (0, 1+)	5.945**	1	1057	5.121*	1	912
Canalys hypoglossi (0-1, 2-3)	3.719*	1	1057	13.176***	1	913
Condylus occipitalis (1, 2)	.563	1	1034	.800	1	881

* :P ⩽ .05; ** :P ⩽ .01; *** :P ⩽ .001

Step 2 Test the symmetry of the distribution of the bilateral characters. Next to be eliminated were characters which are correlated with age or pathology present at death. It must be borne in mind that 14 characters have a bilateral distribution as against 7 with unilateral. Not to test independence of age and pathology, particularly in bilateral distributions which may be redundant (7 unilateral + (14 x 2) = 35) runs the risk of needlessly increasing time spent in computation without obtaining new information. We can take a shortcut by first testing the symmetry of the distributions of the bilateral characters. Where these are symmetrical, the test of independence of age or pathology may be carried out for one side only (Following craniological tradition I have used the left side). Where the two distributions are asymmetrical the test must be made for each side.

Is the bilateral distribution of the k_{th} character symmetric in both sexes? Following McNemar (Bishop, Fienberg and Holland, 1975), in a S x S table, with a_{ij} being the totals observed in the i_{th} row and the j_{th} column:

$$\chi^2 = \sum_{i<j}\sum \frac{(a_{ij} - a_{ji})^2}{a_{ij} + a_{ji}} \quad \text{with d.f.} = S(S - 1)/2$$

Is $\chi^2_{obs.}$ significant?
yes : continue on the left side cranium only.
no: continue on both sides.
The results are given in table 4.

Step 3 Eliminate the characters which are correlated with age at death. The semi-transversal distribution of crania (birth generations x age at death) makes necessary to take simultaneously into account the age at death and the birth generation. Since Yule in the 1900's there is a number of association coefficients for qualitative data as the Yule's Q and Y, Pearson's C and Ø, and Tschuprow's T. It is also possible to seek an association between two variates A, B, in a sub-population based on a third criterion C; this defines the partial association coefficient $Q_{AB.C}$. Nevertheless with s different attributes under consideration (s > 2), the number of partial associations becomes very large. In case of categorized variables, a partial rank-order statistics (e.g. the Spearman's r_s, the Goodman and Kruskal's G, the Kendall's t_c and Stuart's t_b) or the product moment correlation based on rank order of the indices can be simply used. The respective advantages of this different coefficients are discussed by Kendall and Stuart (op.cit.). In the present study G, r_s and r gave the same results. The product moment correlation based on the rank order, r, would be perhaps less sensitive.

Is the partial rank coefficient between the character k and the age at death D (birth generation, B, held constant) $r_{kD.B}$ significant? Here

$$r_{kD.B} = \frac{r_{kD} - r_{kB}r_{BD}}{[(1-r^2_{kB})(1-r^2_{BD})]^{\frac{1}{2}}}$$

where r represents the coefficient G, r_s or r, k is defined as a set of occurrences of the k_{th} character (0, 1, 2,...,s_k), and B and D are defined as a set of indices of birth generations (1, 2,...,p) and the set of indices of the age classes at death (1, 2,...,q), respectively.

Table 5. Partial association coefficients (Goodman and Kruskal's G., Spearman's rank $r_{s.}$ and product moment correlation based on rank order of indices $r_.$) between a character and the age-class at death, birth generation held constant. Ratio of the statistics to its estimated asymptotic standard error ($r/s_o(r)$) is given below when only significant.

Traits		Males				Females			
		N	$G_.$	$r_{s.}$	$r_.$	N	$G_.$	$r_{s.}$	$r_.$
Sutura frontalis (0, 1)		1018	-.082	-.016	-.019	881	-.115	-.041	-.031
Incisura frontalis (0, 1+)	left	1018	.065	.027	.023	879	-.037	-.010	-.015
	right	1017	.059	.022	.022	881	.037	.023	.011
Foramen supra-orbitale (0, 1+)	left	1018	.008	-.002	.003	880	.126	.041	.052
Foramen ethmoidale (0-1, 2, 3+)	left	990	.087	.037	.045	857	-.007	-.004	-.003
Os epipterium (together)	left	946	-.307	-.103	-.104	846	-.151	-.061	-.054
			-3.22***	-3.19***	-3.22***				
Ossa suturae lambdoideae (0, 1, 2+)	left	887	-.029	-.030	-.015	813	-.102	-.065	-.050
	right	879	-.080	-.044	-.043	809	-.218	-.111	-.104
							-2.98**	-3.18**	-2.98**
Foramen parietale (0, 1+)	left	1018	-.185	-.089	-.089	879	.118	.048	.054
			-2.86**	-2.87**	-2.86**				
	right	1018	-.032	-.012	-.013	879	.182	.069	.081
							2.41*	2.05*	2.41*
Canalis hypoglossi (0-1, 2-3)	left	1018	.072	.023	.028	879	.021	.005	.007
	right	1018	.007	-.001	.002	881	.134	.040	.046
Condylus occipitalis (1, 2)	left	1018	-.168	-.049	-.048	864	.266	.022	.023
Torus palatinus		1017	.373	.077	.081	873	.204	.047	.045
			2.60**	2.49**	2.60**				
Canalis basilaris		1018	.133	.058	.058	879	.249	.095	.123
			1.88	1.87	1.88		3.67***	2.83**	3.67***

* : $P \leqslant .05$; ** : $P \leqslant .01$; *** : $P \leqslant .001$

yes : eliminate the character in both sexes.
no : continue.
The results are given in table 5.

Step 4 Eliminate the characters which are correlated with pathology present at death. Is the k_{th} character linked to the causes of death, i.e. is $\chi^2_{obs.}$ for homogeneity, significant?

yes: eliminate the character
no : continue.
The results are given in table 6.

Table 6. Distribution of characters and pathology present at death. χ^2-test of homogeneity. Pathology present at death is distributed in the 11 following categories: Tuberculosis, cancer, gastro-intestinal diseases, respiratory diseases, cardio-vascular and kidney diseases, stroke, no-tuberculous epidemic diseases, septicemia, violent death, death in childbirth, diverses (unknown, cachetia, cirrhosis, hemological diseases, endocrine diseases).
Note that if the number of individuals was < 20 in a category it was classified as diverses.

Characters		Males			Females		
		χ^2	d.f.	N	χ^2	d.f.	N
Sutura frontalis (0, 1)		4.623	9	1023	11.505	10	883
Incisura frontalis (0, 1+)	left	11.634	9	1022	16.549	10	881
	right	9.146	9	1021	9.440	10	883
Foramen supra-orbitale (0, 1+)	left	4.361	9	1022	13.412	10	882
	right	14.100	9	1022	4.930	10	883
Foramen ethmoidale (0-1, 2, 3+)	left	19.743	18	980	22.897	20	848
	right	14.726	18	972	17.010	20	852
Ossa suturae labdoideae (0, 1, 2+)	left	29.057*	18	825	18.192	20	782
Canalis hypoglossi (0-1, 2-3)	left	4.598	9	1022	4.850	10	881
	right	8.398	9	1022	12.912	10	883
Condylus occipitalis (1, 2)	left	12.714	9	1007	8.692	10	857
	right	1.770	9	1006	5.067	10	860

* : $P \leqslant .05$.

Thus we arrive at a final subset of 11 traits: Incisura frontalis right and left, foramen supra-orbitale right and left, foramen ethmoïdale right and left, Canalis hypoglossi right and left, Condylus occipitalis right and left, Sutura frontalis. The study will now be carried out on this set.

2.2. The measure of morphological differences

During the 1960's the supposed genetic origin of minor non-metric variations of the skeleton (e.g. Berry and Berry, op.cit.) influenced the mathematical treatment of such variables as well as the choice of a measure of biological distances. It was perhaps through analogy with allelic frequencies that the nature of such variables, and consequently the form of distributions and the biological distance, was perceived, leading to this train of thought: discrete variable, distribution bi- or multinomial, independence of two distributions; angular transformation of frequencies in order to stabilize the variance. The simplified form of the distance for discrete variables, following bio-statistical reasoning, between populations i and j is:

$$D_{ij}^2 = \Sigma(\theta_i - \theta_j)^2 \quad \text{with } p_i = \frac{x_i}{n_i}, \quad \theta_i = \text{arc sin}(1-2p_i).$$

where n_i is the sample size of the i_{th} population, x_i is the number exhibiting the trait, and p_i is the frequency thereof.

Gruneberg (1952), the principal instigator of studies bearing on the variation of non-metric morphological characters, calls this variation "quasi-continuous". The model formulated by Gruneberg, the only one available, relates to variation in a threshold character, that is, a character with discontinuous outward manifestation but with a continuous underlying scale.

Based on the apparent discontinuity, two conceptions of the statistical nature of the variation are possible: discrete or continuous. It is discrete variation which until now has prevailed in quantifying a biological phenomenon as described by Gruneberg, but this introduces certain constraints which I propose to consider now.

The first constraint relates to the difficulty of handling correlated discrete distributions. Authors in fact acknowledge that their distance measures do not avoid consequent redundancies in the information derived from the characters. In spite of apparent independence in their distributions, correlated characters are numerous; this is so with all bilateral morphological characters (22 out of 30, for example, in the list of Berry and Berry, op.cit.). Doubtless it is this difficulty that has led to the variety of statistical treatments of characters with bilateral distributions, which include:

- summing the right and left characters (Berry and Berry, op.cit.)
- treating the right and left characters independently (Sjøvold, 1977)
- using right and left characters while taking account of their correlation (Green, Suchey, Gokhale, 1979).

The second constraint lies in the obligation to deal only with dichotomous variables (presence/absence) excluding characters exhibiting multiples states. An attempt made by myself in this direction, using formulae proposed by Sjøvold (op.cit., p. 110) yielded a number of negative distances. Put otherwise, the procedure for the study of morphological variation between groups may demand a drastic and artificial reduction of the variation itself. This is hardly acceptable.

I was led to discard distance measures resting on explicitly discontinuous variables for the following three reasons:

i characterization as quasi-continuous of these skeletal traits by Gruneberg,

ii difficulty of handling bilateral traits and, more generally, intercorrelated traits, by means of distance measures for discrete variables,

iii the true range of the biological variation being reduced, in this case to a simple dichotomy,

and to turn instead to the alternative approach of treating graded but non-metric traits of the human skeleton as if they were continuous variables.

Let us remind that the device of treating ordered categorical variables as if they were quantitative is common. The problem raised is the scoring methods of the traits. This question has already been considered by Yates (1948), Williams (1952) and summarized by Kendall and Stuart (1973). Optimum transformation by canonical analysis accordingly to the underlying conception of distributions have also been proposed by Kendall and Stuart (op.cit.). The simplest scoring system is the crude observed score: the character k can take all the integer values $m = \begin{cases} 0, 1, 2, \ldots, s_k \\ \text{or} \\ 1, 2, \ldots, s_k \end{cases}$.
This is the system first proposed by Yates - the use of the sets of natural number - which I shall use here. Thus with p populations $I = (1, \ldots i, \ldots, p)$ and r characters $K = (1, \ldots k, \ldots, r)$, for which characters k takes all the integer values $m = 0, 1, \ldots, s_k$ or $m = 1, 2, \ldots, s_k$, we associate with the i_{th} population a vector of means $\bar{m} = (\bar{m}_{i1}, \ldots, \bar{m}_{ik}, \ldots, \bar{m}_{ir})$ and a covariance matrix S_i in a manner identical to that for continuous variables. It then becomes possible to calculate, without requiring that the r characters have a multinormal distribution and that $S_i = S$ for all i's, a coefficient of morphological dissimilarity among groups, of which the algebraic form is identical to the D^2 of Mahalanobis (Rao, 1980):

$$D^2_{ij} = \delta'_{ij}\ S^{-1}\ \delta_{ij}$$

with $\delta_{ij} = (\bar{m}_i - \bar{m}_j)$, the prime sign (') denoting the transpose of the vector, and S^{-1} the inverse of the covariance matrix. No test of significance can be associated with this coefficient, but we shall see that this is a minor inconvenience. Contrariwise, this coefficient allows us to answer two objections we have formulated above when faced with the concept of discontinuity in non-metric characters of the skeleton and the implications thereof, viz.

- it accomodates the true range of variation of the characters (0, 1, ..., s), in place of a simple dichotomy (0, 1)
- it takes account of the intercorrelations between characters, including those that are bilateral.

Having defined the statistic for measuring morphological distance, we may now approach the object set out at the beginning of this work: analysis of the morphological variation in Portugal in the 19th century.

3. MORPHOLOGICAL EVOLUTION IN PORTUGAL (1810-1909)

Recall that the model we wish to test is:

$$V(D) = V(P) + V(T) + E$$

Put otherwise in a defined historical and geographical setting, what is the total influence of state of nutrition on morphological variation?

Considering the relatively short period covering the births of the individuals composing the sample (100 years), as well as the relatively small numbers of crania for the earliest period. I have subdivided the century involved into 5 generations of "births", each of 20 years: 1810-1829, 1830-1849, 1850-1869, 1879-1889, 1890-1909. The 20 years period was chosen instead of a 5 or 10 years period since a small number of crania is observed in the oldest birth generations. If $p_{i1}, p_{i2}, \ldots, p_{ir}$ is the mean price of wheat in the i_{th} generation for the 1, 2, ..., r_{th} region, and $\pi_{i1}, \pi_{i2}, \ldots, \pi_{ir}$ are the a priori probabilities of each region in the i_{th} generation, then the mean price of wheat in the i_{th} generation is: $P_i = p_{i1}\pi_{i1} + p_{i2}\pi_{i2} + \ldots + p_{ir}\pi_{ir}$. As the distribution of generations and regions are not the same in the two sexes, the mean prices are different. Table 7 gives the totals of individuals born in each generation and each region, separately by sex. Table 8 gives the mean prices of wheat for each generation in the two sexes. Table 9 give the matrices of intergeneration coefficients of dissimilarity $D^2 = (D^2_{ij})$ for the two sexes. Along with the matrix D^2 it is also possible to form, using Tables 7 and 8, two other matrices for the p generations: a matrix of squared intergeneration differences in the price of wheat on the one hand, and a matrix of squared time intervals between the generations on the other.

The matrix of squared price differences between generations i and j is:

$$P^2 = (p^2_{ij}) = (p_i - p_j)^2 \quad \text{if } i \neq j \qquad i = 1, \ldots, p-1$$

$$(p^2_{ii}) = 0 \qquad j = 2, \ldots, p.$$

The elements p_{ii} on the principal diagonal are equal to 0 since $(p_i - p_i)^2 = 0$; the off-diagonal elements represent the squared price differences between generations i and j.

The matrix of squared time intervals between generations i and j is:

$$T^2 = \log(t_{ij}^2) = \log(t_i - t_j)^2 \quad \text{if } i \neq j \qquad i = 1, \ldots, p-1$$

$$(t_{ii}^2) = 0 \qquad j = 2, \ldots, p.$$

The elements t_{ii} on the principal diagonal are equal to 0, since $(t_i - t_i)^2 = 0$. The other elements represent the logarithm of the squared time elapsed between generation i and generation j.

Table 7. Distribution of the crania sample used in the analysis, from Coimbra University, by birth generation and region. (North: District of Porto, Braga, Bragança, Viana do Castelo, Vila Real: Middle: District of Coimbra, Aveiro, Castelo Branco, Guarda, Leiria, Porto Alegre, Vizeu; South: District of Lisboa, Beja, Evora, Faro, Santarem, Setubal.) (M = males, F = females.)

	North		Middle		South		Total		
Generation	M	F	M	F	M	F	M	F	
1820	4	5	18	9	16	18	38	32	70
1840	13	13	87	92	38	27	138	132	270
1860	23	27	175	181	51	24	253	228	481
1880	23	16	188	193	22	11	233	220	453
1900	19	4	208	167	18	9	245	180	425
							907	792	1699

Table 8. Mean price of wheat by generation (Reis per liter). Explanation see p. 306, 307.

Birth generation	Mean price of wheat	
	Males	Females
1820	57.33	59.00
1840	36.17	35.16
1860	43.83	42.99
1880	40.78	40.46
1900	49.37	49.32

Table 9. Intergeneraton morphological dissimilarity matrix, D^2 (11 non-metric characters of the skull.)

		Males				
		1820	1840	1860	1880	1900
	N =	38	138	253	233	245
1820	38	.000				
1840	138	.473	.000			
1860	253	.390	.091	.000		
1880	233	.315	.073	.083	.000	
1900	245	.336	.110	.033	.043	.000

		Females				
		1820	1840	1860	1880	1900
	N =	32	132	228	220	180
1820	32	.000				
1840	132	.737	.000			
1860	228	.583	.176	.000		
1880	220	.472	.317	.109	.000	
1900	180	.646	.261	.083	.112	.000

For each sex, we thus have a matrix of intergeneration morphological dissimilarity D^2, a matrix of price differences P^2 and a matrix of time differences T^2. The level of congruence of the three matrices should indicate the acceptability or otherwise of the explanatory model of morphological variation we have proposed. This is the purpose of the Procrustean analysis and statistics. As Sibson says (1978): "Consider two configurations of points in k-space with a preassigned correspondance between the points of the two configurations: how much do the configurations differ?" Such a question has been investigated by Green (1952), Schönemann and Carrol (1970), and Gower (1971, 1975). A review was performed by Sibson (1978). The procedure I shall use here is that which was published by Lalouel (1973). The measure of goodness of fit of two matrices A and B is found by least squares, estimating different parameters, notably e_s (a symmetric measure of fit, varying from 0 (A = B) to 1 (no relation between the elements of A and B), and r_c (correlation coefficient between the elements of the final fit). No test of significance has been proposed. This procedure of fitting between two matrices minimizes the squared distances between the two configurations while leaving invariant the distances within each configuration (program MATFIT, Lalouel op. cit.). Criteria of the goodness of fit of the three matrices are given in table 10.

Table 10. Measure of congruence between intergeneration morphological dissimilarity matrix (D^2) and the intergeneration differences in price of wheat (P^2) and the differences in location in time (T^2) (5 birth generations: 1820, 1840, 1860, 1880, 1900).
Upper triangular part: symmetric measure of fit (e_s).
Lower triangular part: correlation between coordinates (r_c) of the two matrices.

Males

	D^2	P^2	T^2
		e_s	
	0.	0.087	0.171
r_c	0.812	0.	0.120
	0.743	0.819	0.

Females

	D^2	P^2	T^2
		e_s	
	0.	0.176	0.239
r_c	0.625	0.	0.117
	0.641	0.825	0.

Males and Females

	D^{*2}	P^2	T^2
		e_s	
	0.	0.076	0.055
r_c	0.631	0.	0.081
	0.743	0.623	0.

In males intergeneration morphological differences are fairly correlated to price differences (r_c = .812) and less correlated to time differences (r_c = .743). In females matters have a rather different aspect: the association of morphology in successive generation with the price of wheat and with chronological distance are of the same order of size (r_c = .625 and .641 respectively). Compared to males females exhibit a certain morphological inertia in relation with nutritional differences and time.

It is also interesting to examine fluctuation of all the biological structure (male and female) together, with the same statistical procedure. This leads us to include the sexual dimorphism during this period. The corresponding intergeneration matrix, $D^{2}*$, including the two sexes, is given in table 11.

Table 11. Intergeneration morphological dissimilarity matrix, D^{2*}, including the two sexes (11 non-metric charaters of the skull. M : males, F :females.)

			1820	1840	1860	1880	1900	1820	1840	1860	1880	1900
			M	M	M	M	M	F	F	F	F	F
		N =	38	138	253	233	245	32	132	228	220	180
1820	M	38	.000									
1840	M	138	.491	.000								
1860	M	253	.395	.098	.000							
1880	M	233	.310	.085	.087	.000						
1900	M	245	.338	.128	.036	.046	.000					
1820	F	32	.866	.812	.965	.865	.934	.000				
1840	F	132	.447	.212	.229	.232	.284	.750	.000			
1860	F	228	.358	.292	.234	.287	.279	.600	.137	.000		
1880	F	220	.378	.337	.278	.344	.298	.481	.245	.103	.000	
1900	F	180	.481	.228	.177	.245	.250	.632	.179	.081	.103	.000

The principal diagonal element D^{2*}_{ii} represents the sexual dimorphism for generation i, while the off-diagonal element D^{2*}_{ij} represents dimorphism between male generation i and female generation j, D^{2*}_{ji} dimorphism between male generation j and female generation i.
The model I attempt to fit is the following: dimorphism in generation i, D^{2*}_{ii}, is a function of the mean price of wheat in generation i (p_i) and of time ($t_i = \log(t_i - 1800)$); the dimorphism between generations i and j (D^{2*}_{ij}, D^{2*}_{ji}) therefore depends on the squared difference of price $(p_i - p_j)^2$ during generations i and j together with the squared difference of time of i and j, $\log(t_i - t_j)^2$. Accordingly we have the following three matrices (of which only the lower part is given):

D^{2*} =

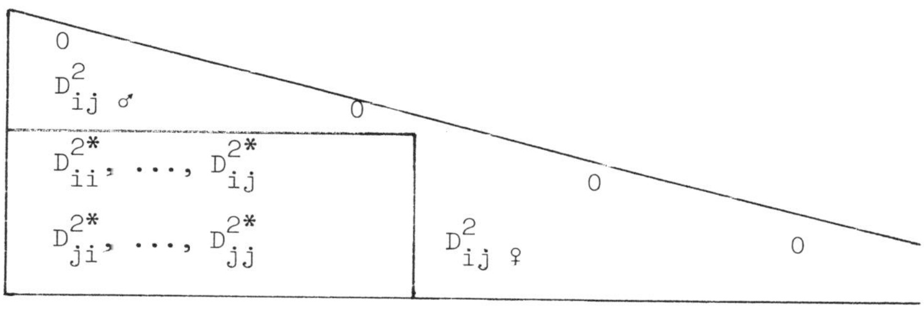

P^2 =

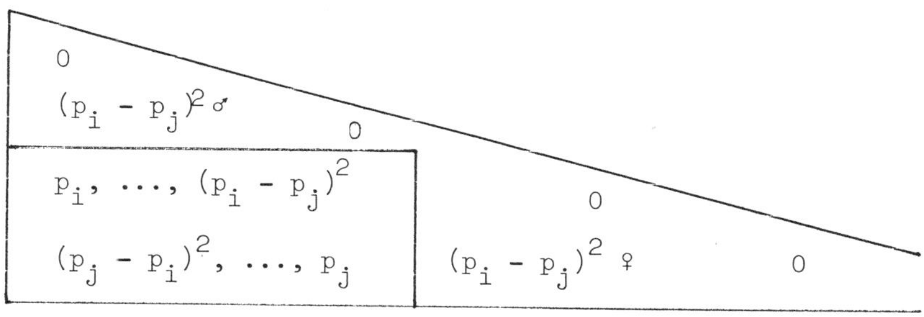

T^2 =

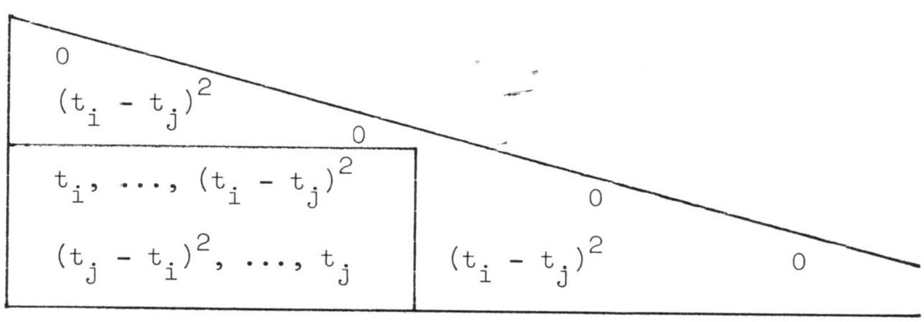

The parameters measuring the goodness of fit of these three extended matrices are also given in table 10 (lower part). In a whole, the morphological intergeneration differences seem to be more correlated to time than to nutritional differences (r_c = .743 and .631 respectively).

However, Procrustean analysis and statistics display some disadvantages if relationships between biological distance matrix and causal underlying hypotheses are sought. The fit between structures can only be performed pairwise (B fitted to A) whereas usual hypotheses involve a higher number of structures (B and C and D fitted to A). Moreover, adjustment measures (e_s or r_c) give discordant results, for instance with the same r_c = .743 two different e_s values are obtained: .171 with 5 x 5 matrix, and .055 with 10 x 10 matrix. Adjustment measures seem to be sensitive to the matrix dimension making interpretation of the results difficult.

Once the whole biological structure and causal hypotheses have been examinated by Procrustean analysis, the following question can be raised: which part of the biological structure (and not all) is correlated to the causal hypotheses?

A suitable transformation for Jacobi's method of decomposition permits to transform a dissimilarity or distance matrix D (a_{ii}=0) into a similarity matrix S (a_{ii} = 1), by: $a_{ij} = 1 - a_{ij}/\max a_{ij}$. Table 12 shows the S matrix, as derived from the D^2 matrix in Table 9.

Table 12. Intergeneration morphological similarity matrix, S (11 non-metric characters of the skull.)

	Males				
	1820	1840	1860	1880	1900
	38	138	253	233	245
38	1.000	0.000	.175	.334	.290
138		1.000	.808	.846	.767
253			1.000	.824	.930
233				1.000	.909
245					1.000

	Females				
	32	132	228	220	180
32	1.000	0.000	.209	.360	.123
132		1.000	.761	.560	.646
228			1.000	.852	.887
220				1.000	.848
180					1.000

As is well known, diagonalization of a symmetric p x p matrix S gives an orthogonal p x p matrix U, where the columns $u_1, u_2, \ldots, u_p$ are the eigenvectors (possibly standardized such that $u_j'u_j = 1$), and Λ the diagonal matrix of order p whose diagonal elements are $l_1, l_2, \ldots, l_p$, the eigenvalues. The spectral decomposition of S can be expressed as: $S = U\Lambda U'$. The relative contribution of each eigenvector to the total dispersion of S, for the i_{th} eigenvector is: $l_i \,/\, \sum_{k=1}^{p} l_k = l_i \,/\, \mathrm{tr}\, S$, where tr is the trace. One can also form a p x r hypothesis matrix H, whose columns $h_1, h_2, \ldots, h_r$ are simply the coordinates of hypotheses under investigation. An approach suggested by Gower gives the possibility to measure relationship between all or a simple part of S, and H: perform a multiple correlation analysis (or even a canonical correlation analysis) in an ordinary way between the set of eigenvectors extracted from S (U), and the column vectors from H. In this way all the dimensionality of S can eventually be kept (i.e. p - 1). However there is a restriction to this. Let us remind that as the number of U_q (q = 1,2,...,p-1) in the regression system approaches the matrix dimension, the multiple correlation coefficient tends to 1, even when its population value is zero, as shown by Wishart:

$$\text{if } \rho = 0 \quad E(R^2) = \frac{q}{N-1} \quad (\text{here } N = p-1)$$

An empirical observation shows that only about half of the eigenvectors extracted from S are of practical use. As the distribution of $\rho(u,h)$ is unknown a sample of it should be generated to test the null hypothesis. Such a sample can be obtained by means of random simulations on U, via S issued from random distributions of characters in the population. As an example, figure 2 shows the distribution of $\rho(u_1, h_2)$ obtained from 1,000 random simulations.

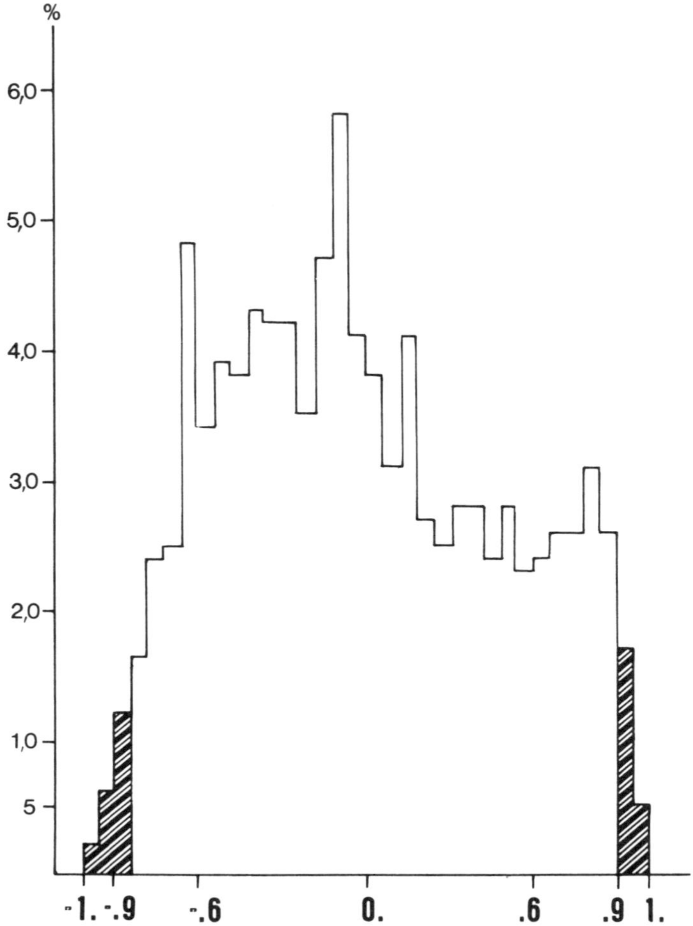

Figure 2. Distribution of $\rho(u_1, h_2)$ obtained from 1,000 random simulations. Shaded lines represent the rejection region at the 5%-level (The procedure is described above.)

This distribution looks irregular. However, the rejection region of the null-hypothesis at the 5%-level, constructed on this distribution is similar to the one observed in a bivariate normal distribution. In fact this close relation reaches the Girshick (1939) and Anderson's conclusions (1951, 1963): Principal component scores are approximatively distributed according to a multi-normal distribution, independently of their eigenvalues, as the population size becomes very large. Therefore the ordinary tests of the null-hypothesis $\rho(u, h) = 0$ are approximatively relevant. R^2 distribution

Table 13. Canonical correlation between eigenvectors (u_i) (extracted from the intergeneration morphological similarity matrix S) and the location in time of the birth generations (t), the price of wheat (p) (χ^2_o is a Bartlett's test of sphericity). (Note that probability level is given without proof, see next page)

Vector	Fraction of the dispersion of S	t				p				t and p		
		r	R^2	F	d.f.	r	R^2	F	d.f.	R^2	F	d.f.
						Males						
u_1	.721	.892*	-	-	-	-.756	-	-	-	.976*	40.41	2, 2
u_2	.202	-.646	-	-	-	.894*	-	-	-	.891	8.25	2, 2
u_1 and u_2	.923	-	.974*	39.24	2, 2	-	.824	4.70	2, 2	.990	$\chi^2_o = 1.93$ d.f. = 1	
						Females						
u_1	.671	.907*	-	-	-	-.649	-	-	-	.948*	18.44	2, 2
u_2	.208	-.717	-	-	-	.842	-	-	-	.804	4.14	2, 2
u_1 and u_2	.879	-	.874	6.94	2, 2	-	.707	2.43	2, 2	.948	$\chi^2_o = .80$ d.f. = 1	

* : $P \leqslant .05$; **: $P \leqslant .01$

was not simulated but it can be thought, at least roughly, that it is a simple generalisation of the previous $\rho(u, h)$ distribution.

In the Portuguese sample of crania we have for each sex a p x p matrix U, and a p x r hypothesis matrix $H = (h_1, h_2) = (p,t)$ where h_1 is a vector column representing the price (the set of mean price of wheat by generation, $h_1' = (p_1, p_2, \ldots, p_p)$), h_2 is a vector column representing the time (the set of chronological location of the successive generations, with the transformation $t_i = \log(t_i - 1800)$, $h_2' = (t_1, t_2, \ldots, t_p)$). Table 13 gives the correlation coefficients between U and H.

In the male population, the first eigenvector of S, u_1 (72,1% of the dispersion) is first correlated to the chronological location of generation t ($P \leq .05$) but also to the chronological location of generations t and price p ($P \leqslant .05$). The second eigenvector u_2 (20,2% of the dispersion) and price p are correlated ($P \leqslant .05$). In the female population although the correlation structures are similar to those in males, only u_1 and chronological location T are significantly correlated. The main difference between the two sexes is that an eigenvector of the male intergeneration morphological similarity matrix (and its associated eigenvalue) is correlated with the price of wheat, which is not the case in females. The biological significance of this eigenvector, assuming other things to be equal (genotypic structure being stable in the population) is that the nutritional status influences the behavior of morphology in the generations of males, while those of the females, although differing morphologically, manifest a degree of inertia in the face of this particular factor. Here again we encounter the general observation as to a greater sensitivity in males to external factors (Ferembach, 1975). Figure 3 gives for each sex simultaneous rendering of the bidimensional approximation of S and the hypothesis matrix, H. This is clearly visible in the plot.

Similarly an analysis can be carried out on the intergeneration morphological similarity matrix S^* in cluding the two sexes, as derived from the D^{2*} matrix. Table 14 shows the S^* matrix, table 15 the correlation coefficients between U and H for this case. u_1 and u_5 are each correlated with time and price (73,5% of the dispersion). They are also the best subset of eigenvectors, U^*, for canonical correlation analysis (i.e. $R^2_{(H,U^*)}$ significant at the 5%-level). Since the matrix U is orthogonal ($u_i'u_j = 0$) the fraction of the dispersion of S^* explained by H could be approximatively taken as (without proof):

$$(R^2_{(H,U^*)} \sum_i^q l_i^*)/ \operatorname{tr} S^*$$

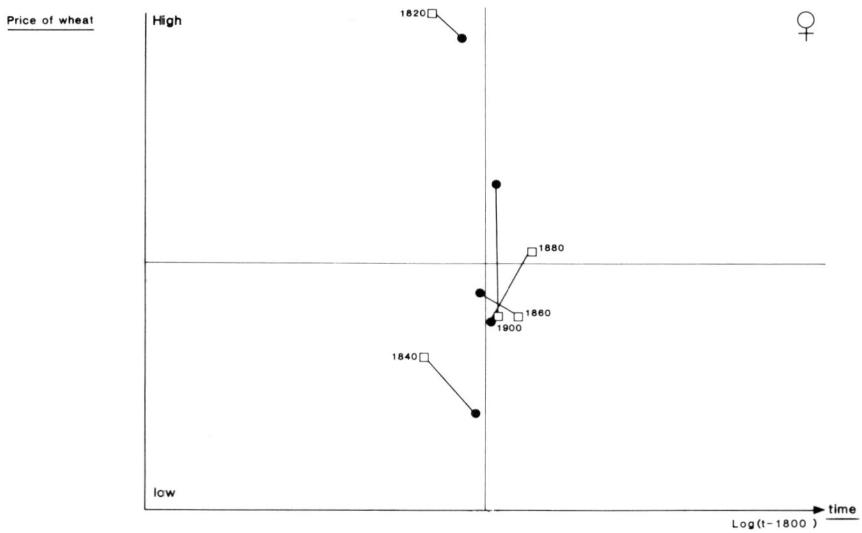

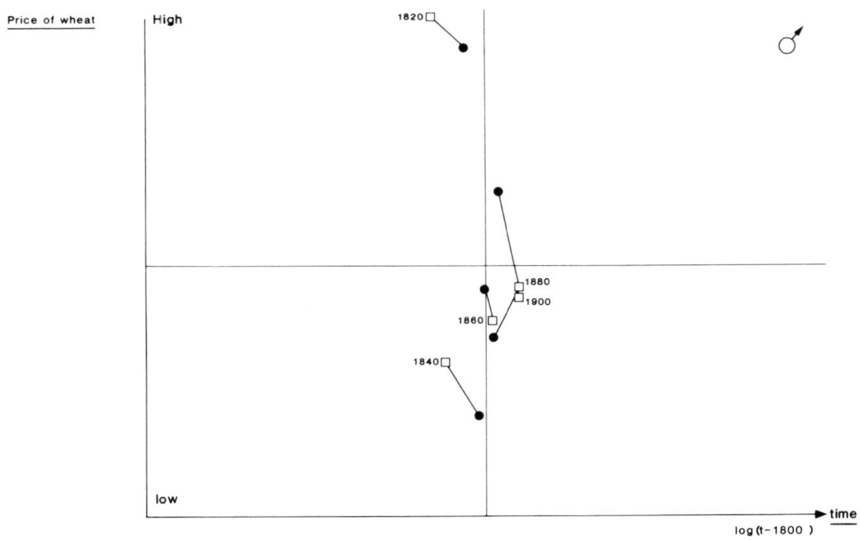

Figure 3. Simultaneous rendering of the bidimensional approximation of the intergeneration morphological similarity matrix, S (points designed by □), and the hypothesis matrix H (points designed by •)

Table 14. Intergeneration morphological similarity matrix, S^*, including the two sexes (11 non-metric characters of the skull. M : males, F : females.)

	1820	1840	1860	18880	1900	1820	1840	1860	1880	1900
	M	M	M	M	M	F	F	F	F	F
N =	38	138	253	233	245	32	132	228	220	180
M 138	.491									
M 253	.590	.898								
M 233	.678	.912	.910							
M 245	.649	.867	.962	.951						
F 32	.102	.158	.000	.103	.031					
F 132	.536	.781	.763	.760	.706	.221				
F 228	.629	.698	.758	.703	.711	.377	.858			
F 220	.608	.650	.718	.644	.692	.501	.746	.893		
F 180	.501	.764	.817	.746	.741	.344	.815	.916	.893	

Table 15. Canonical correlation between eigenvectors (u_i) (extracted from the intergeneration morphological similarity matrix. S^*) and the location in time of the birth generation (T), the price of wheat (P) (χ^2_o is a Bartlett's test of sphericity). (Note that probability level is given without proof, see p. 312.

Vector	fraction of the dispersion of S*	t	p	t and p		
		r	r	R^2	F	d.f.
u_1	.709	.744*	-.701*	.734**	9.67	2, 7
u_2	.133	-.367	.385	.198	.87	2, 7
u_3	.062	-.445	.556	.362	1.98	2, 7
u_4	.045	-.328	.367	.170	.72	2, 7
u_5	.026	-.548	-.281	.624*	5.81	2, 7
u_6	.009	.064	-.183	.033	.12	2, 7
(u_1, u_2, ..., u_6)	.983	R^2 = .861 F = 3.11 d.f. = 6, 3	R^2 = .915 F = 5.39 d.f. = 6.3			
		Best subset of vectors				
u_1	.709	unchanged	unchanged	unchanged		
u_5	.026	unchanged	unchanged	unchanged		
(u_1 and u_5)	.735	R^2 = .743** F = 10.10 d.f. = 2, 7	R^2 = .650 F = 6.53 d.f. = 2.7			

χ^2_o between the two set of variates: 14.81** d.f. = 4

Canonical correlation = .866 χ^2_o = 5.71* d.f. = 1

* : P ⩽ .05; ** : P ⩽ .01

where $R^2_{(H,U^*)}$ is the squared canonical correlation coefficient, U^* is the subset of the q eigenvectors selected from U for the analysis (q < p) and ordered in an increasing order of indices, H is the hypothesis matrix, l_i^* is the eigenvalues associated to the i_{th} eigenvectors $u_i^* \in U^*$, tr is the trace of S^*. Then 55.2% of the dispersion of the intergeneration morphological similarity matrix is due to the variation of time and price.

Finally, table 16 gives, for each sex, coefficients of Spearman's rank correlation (r_s) between means of each character by generation ($\bar{m}_{ik}$) and price of wheat (p_i); table 17 gives again the same coefficients but using time. 3 out of 11 traits in the two sexes are individually linked with fluctuations in the price

Table 16. Coefficients of Spearman's rank correlation (r_s) between means of each character by generation and price of wheat. (5 birth generations)

Characters		Males	Females
Incisura frontalis	right	1.0**	-.40
	left	.40	-.10
Foramen supra-orbitale	right	.0	.30
	left	.20	.30
Foramen ethmoidale	right	.70	.40
	left	.70	.90*
Canalys hypoglossi	right	-.80	.20
	left	.10	.90*
Condylus occipitalis	right	.70	-.70
	left	.0	.10
Sutura frontalis		.30	.70

* : P ⩽ 0.05; ** :P ⩽ 0.01

Table 17. Coefficients of Spearman's rank correlation (r_s) between means of each character by generation and location in time (5 birth generations)

Characters		Males	Females
Incisura frontalis	right	-.10	.70
	left	.80	1.0**
Foramen supra-orbitale	right	-.90*	-.90*
	left	-.50	-.60
Foramen ethmoidale	right	-.10	-.70
	left	-.10	-.30
Canalys hypoglossi	right	.50	-.50
	left	-.70	.0
Condylus occipitalis	right	-.10	.10
	left	.90*	.97*
Sutura frontalis		-.60	-.10

* : P ⩽ 0.05; ** : P ⩽ 0.01

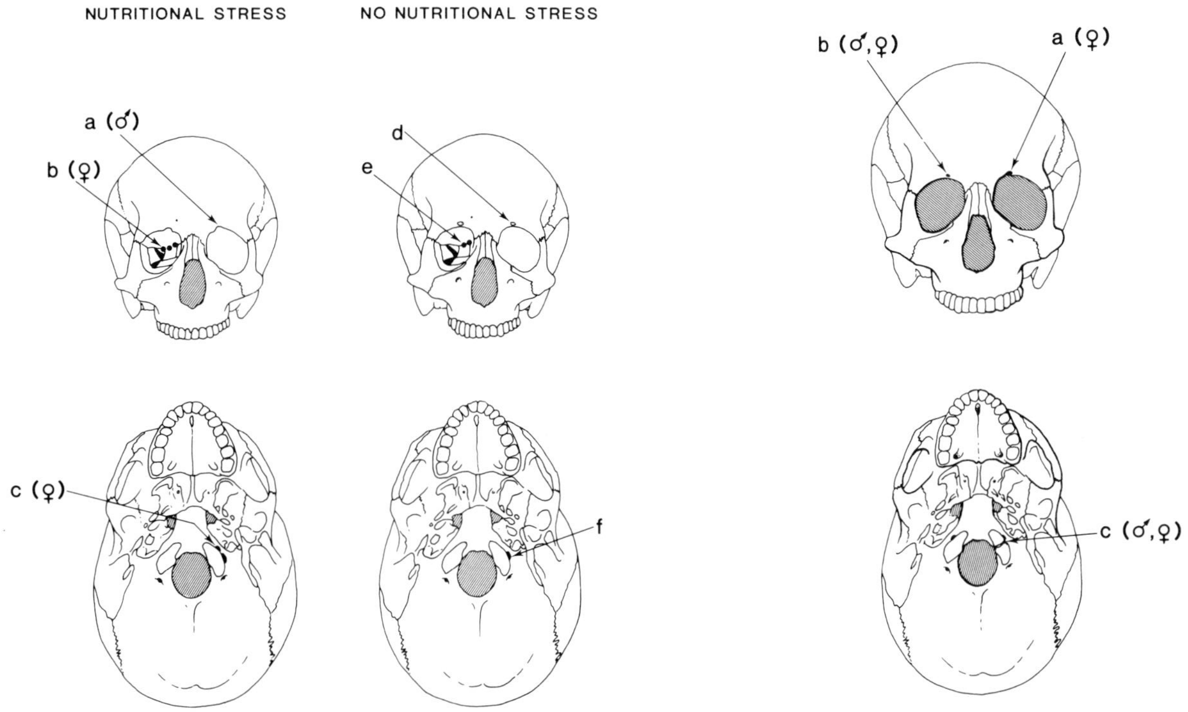

Figure 4. Distribution in schematic fashion of some traits in crania as a function: 1) of the most contrasting conditions of dietary (left), 2) of the secular change (right).
For left: a) Incisura frontalis margo orbitalis, b) Foramen ethmoïdale multiple, c) Canalys hypoglossi multiple, d) Foramen supra-orbitale, e) Foramen ethmoïdale double, Canalys hypoglossi simple.
For right: a) Incisura frontalis margo orbitalis frequency increase, b) Foramen supra-orbitale frequency decrease, c) Condylus occipitalis double: frequency increase.

of wheat (Incisura frontalis right in males, Foramen ethmoidale and Canalys hypoglossi left in females), while 5 out 11 are individually linked to time (in males: Foramen supra-orbitale right, Condylus occipitalis left: in females: Incisura frontalis left, Foramen supra-orbitale right, Condylus occipitalis left). Figure 3 represents in schematic fashion the distribution of these traits in crania, as a function of the most contrasting conditions of dietary, sufficiency or insufficiency, and as a function of the secular change obtaining in Portugal in the 19_{th} century. Nevertheless we should note that traits previously eliminated through the filter are strongly linked with fluctuations in the price of wheat. This is especially observed for all the sutural bones whether they are apparently linked to age (Ossa epipterium) or pathology (Ossa lambdoideae) : such ossicles being more frequent in times of less favorable nutriture.

4. CONCLUSION

This study has developed an essentially quantitative conception of the nature of variation in non-metric characters of the human skeleton. This conception is in strict conformity with the model proposed by Gruneberg: variation in quasi-continuous characters.

I have wished to test, in a precisely defined historical and geographical framework, a model of variation in morphology between generations which is a function in part of their dietary situation and in part of their location in time. Two ways of measuring the relationship between a biological structure and causal hypothesis were used: Procrustean analysis and canonical correlation analysis on causal hypotheses and vectors generated from a similarity matrix. This latter method permits to consider possibly a simple fraction of the similarity matrix and not its whole dimensionality. Furthermore, whereas only one hypothesis can be considered by Procrustean analysis, various hypotheses can simultaneously be analyzed by this method.

It is rather reassuring that the main fraction of the inter-generation morphological similarity matrix and the causal hypothesis of the model are correlated. This could not have been the case. However, what does t really mean? t measures underlying variables which are undefined and not immediately environmental. What does p mean? Undoubtedly it represents the price of wheat and more generally the current price of articles of diet. p can also measure correlated factors such as pathological ones. In the past the sanitary situation was dependent on the dietary situation. Underlying and undefined variables can simultaneously act, although this does not appear in formal variables (i.e. $\text{cov}(t,p) \simeq 0$). The magnitude of the foodstuff price may induce local or regional migrations which may in turn modify the genotypic structure of local populations. The diversity of possible situations is tremendous.

Except for the males which show a morphological component actually linked to environmental situations the main part of the intergeneration variation is linked with both time and price. This points out that non-metric characters as other morphological characters are polyfactorial, i.e. depending on inextricably mingled genetic and environmental factors.

ACKNOWLEDGMENT

I am very grateful to M.A. Maia Neto, N.H. Xavier de Morais, M.A. Tavares da Rocha, and to the council members of the Institute of Anthropology from Coimbra University, Portugal (chairman: Prof. Dr. F. Ferrand de Almeida). I would like also to thank P. Darlu, D. Ferembach, J.C. Gower, M. Hainsworth, W.W. Howells, J.M. Lalouel, J.P. Lecoutre for their helpful comments on different aspects of this paper.

This paper was translated from French by W.W. Howells and F. Demenais.

The drawings were done by Ch. Theureau and G. Delamare.

Computing facilities of the Laboratoire d'Informatique en Sciences Humaines (L.I.S.H., C.N.R.S., Paris) were used for the analysis.

References

Anderson, T.W.: 1951, The asymptotic distribution of certain characteristic roots and vectors, Proceeding of the Second Berkeley Symposium on Mathematical Statistics and Probability, University of California Press, Berkeley, 103-130.

Anderson, T.W.: 1963, Asymptotic theory for principal component analysis, Annals of Mathematical Statistics, 34, 122-148.

Berry, A.C., Berry R.J.: 1967, Epigenetic variation in the human cranium. Journal of Anatomy, 101, 2, 361-379.

Bishop, Y., Fienberg, S.E., Holland, P.W.: 1975, Discrete Multivariate Analysis, MIT Press, Cambridge.

Cheverud, J.M., Buiska, J.E., and Twichell, E.:1979, Relationship between non-metric skeletal traits and cranial size and shape. American Journal of Physical Anthropology, 50, 2, 191-198.

Ferembach, D.: 1975, Les effets de l'alimentation sur l'expression du phénotype chez l'homme. Arquivo de Anatomia e Antropologia, 1, 271-305.

Friedlaender, J.C.: 1975, Pattern of Human Variation, Harvard University Press. Cambridge, Mass.

Girshick, M.A.: 1939, On the sampling theory of roots of determinantal equations, Annals of Mathematical Statistics, 10, 203-224.

Gower, J.C.: 1971, Statistical methods of comparing different multivariate analysis of the same data. In:Mathematics in the Archaeological and Historical Sciences. Edinburgh, University Press, 138-149.

Gower, J.C.: 1975, Generalized Procrustean analysis. Psychometrika, 40, 33-51.
Green, B.F.: 1952, The orthogonal approximation of an oblique structure in factor analysis. Psychometrika, 17, 429-440.
Green, R.F., Suchey, J.M., Gokhale, P.V.: 1979, The statistical treatment of correlated bilateral traits in the analysis of cranial material. American Journal of Physical Anthropology, 50, 4, 629-634.
Gruneberg, H.: 1952, Genetical studies on the skeleton of the mouse. IV. Quasi-continuous variations. Journal of Genetics, 51, 95-114.
Howells, W.W.: 1966, Population distances biological, linguistic, geographical, and environmental. Current Anthropology, 7, 531-540.
Kaul, S., Anand, V., Corruccini, R.S.: 1979, Non-metric variation of the skull in samples of four indian populations. Journal of Human Evolution, 8, 693-697.
Kellock, W.L., Parsons, P.A.: 1970, A comparison of the incidence of minor non-metrical cranial variants in Australian aborigenes with those of Melanesia and Polynesia. American Journal of Physical Anthropology, 33, 2, 235-240.
Kendall, M.G., Stuart, A.: 1973, The advanced Theory of Statistics, 2, Charles Griffin ed., London, 2nd ed.
Knip, A.S.: 1971, The frequencies of non-metrical variants in Tellem and Nokara skulls from the Mali Republic I. Koninkl. Nederlandse Akademie van Wetenschappen, Amsterdam, serie C, 74, 5, 423-430.
Lalouel, J.M.: 1973, Topology of population structure. Genetic Structure of Populations. N.E. Morton ed., University of Hawaii Press, Honolulu, 139-149.
Muller, H.: 1977, Caractères non-métriques du squelette de la tête chez les populations médiévales de Thoiry (Ain, France) et de Bavois (Vaud, Suisse). Archives Suisses d'Anthropologie générale, 41, 2, 123-164.
Pietrusewsky, M.: 1977, Etude des relations entre les populations du pacifique par les méthodes d'analyse multivariée appliquées aux variations crâniennes. L'Anthropologie, Paris, 81, 1, 67-97.
Rao, C.R.: 1980, Diversity and Dissimilarity coefficients: An unified approach. Technical report, no. 80-10, University of Pittsburgh.
Rathbun, T.A.: 1979, Metric and discrete traits variation among southwest asian populations. American Journal of Physical Anthropology, 50, 3, 473.
Riesenfeld, A.: 1956, Multiple infraorbital, ethmoidal and mental foramina in the races of man. American Journal of Physical Anthropology, 14, 1, 85-100.
Sibson, R.: 1978, Studies in the Robustness of Multidimensional Scaling: Procrustres Statistics. Journal of Royal Statistical Society B, 40, 2, 234-238.

Schönemann, P.H., Carroll, R.M.: 1970, Fitting one matrix to another under choice of a central dilation and a rigid motion. Psychometrika, 35, 245-255.

Sjøvold, T.: 1977, Non-metrical divergence between skeletal populations, Ossa, 4, 1.

Soares, de Morais, R.: 1859, Dos preços de cereas e outros generos alimentares no continente do reino. Archivo Rural, II, 436-440.

Thoma, A.: 1981, The pattern of quasi-continuous variation in Homo sapiens. Journal of Human Evolution, 10, 303-310.

Williams, E.J.: 1952, Use of scores for the analysis of association in contingency tables. Biometrika, 39, 274-289.

Workman, P., Harpending, H., Lalouel, J.M., Lynch, C., Niswander, J., Singleton, R.: 1973, Population studies on Southwestern Indian tribes VI. In Genetic Structure of population. N.E. Morton ed., University of Hawaii Press, Honolulu, 166-194.

Wood-Jones: 1931-32, The non-metrical morphological characters of the skull as criteria for racial diagnosis. I, II, III. Journal of Anatomy, 65, 179-195, 368-378, 438-445.

Yates, F.: 1948, The analysis of contingency tables with grouping based on quantitative characters. Biometrika, 35, 176-181.

ON THE DETERMINATION OF HOMINID AFFINITIES

G.N. van Vark

Department of Anatomy and Embryology, Groningen University, Oostersingel 69, 9713 EZ Groningen, The Netherlands

INTRODUCTION

When comparing populations or classifying unknown specimens on the basis of their morphology, one can either rely on visual comparison or follow an approach which involves the use of the mathematical techniques of discriminant analysis.

Although both approaches are essentially of a multivariate statistical nature, it is only the latter which fully takes into account the correlations between the variables used. This advantage over the visual approach has raised high expectations. Hence, discriminant analytical techniques have become very popular in physical anthropology.

Unfortunately, many applications of discriminant analysis to anthropology yielded results which seem less convincing than those obtained by means of visual inspection. This holds in particular for the classification of hominid specimens, where the "visualists" so far seem to have obtained the more consistent results(1).

It is our belief that in this case the poor performance of the mathematical approach is mainly caused by the uncritical use of standard multivariate techniques.
The anthropological problems at issue require that procedures be developed which are specifically designed for the purpose.

In some recent papers (2, 3), we have presented an example of such a procedure which was primarily developed for the investigation of hominid specimens. Its application to hominid skull material collected by C.B. Stringer produced some striking results which might lead to new insights into hominid phylogeny, and possibly even into the process of evolution as such.

This has led to reexamine these results by investigating their stability with other variable sets, and by carrying out

G. N. van Vark and W. W. Howells (eds.), Multivariate Statistical Methods in Physical Anthropology, 323–349.

some experiments. In the present paper we shall, after a recapitulation of earlier findings, present results obtained with these further investigations.

Furthermore, we shall point to some other possibilities for the investigation of hominid (and other) specimens issuing from our approach.

PRECEDING METHODOLOGICAL CONSIDERATIONS (AND SOLUTIONS)

The advantages of a mathematical approach result primarily from the fact that the correlations between the variables are fully taken into account. This leads, at least theoretically, to results which are more reliable.

An advantage of visual comparison over current mathematical procedures is that the variation of the present world population is implicitly used as a standard for determining whether or not a specimen falls within the range of a certain subspecies. This implies that information is used which the usual mathematical techniques lack.

A number of earlier investigations led us to find how essential this extra information is for the classification of hominid specimens.

The starting-point may be said to lie in the early seventies, when we carried out some simulation experiments with discriminant functions. On screening the results, we found that beyond a certain bound, p*, the effective discriminatory power decreases if variables are added to the preceding selection. It turned out that the pattern recognition and mathematical statistical literature contained similar experiments (4). Meanwhile, mathematical statistical explanation has been found: the phenomenon is essentially caused by sampling fluctuations (5).

The nature of this phenomenon is demonstrated in figure 1. This figure contains five curves referring to three different concepts of discriminatory value in a series of discriminant functions which develop when variables are added one by one to the preceding selection.

The solid line represents the curve of true discriminatory values (Mahalanobis distances), Δ. This is a well-defined concept but, for a given set of variables, its value is an unknown parameter which has to be estimated from samples.

The two broken lines denote the curves of the usual estimates D for Δ calculated from two samples. The upper line refers to relatively "small" samples and the lower one to relatively "large" samples. On comparing these lines, we see that the bias in the usual estimates is positive and tends to increase as the sample sizes are decreased and/or the number of variables is increased. (The exact bias of D^2 as an estimator for Δ^2 can be derived, and an unbiased estimator for Δ^2 can be constructed. If the root of the unbiased estimator for Δ^2 is plotted as a function of p, a corres-

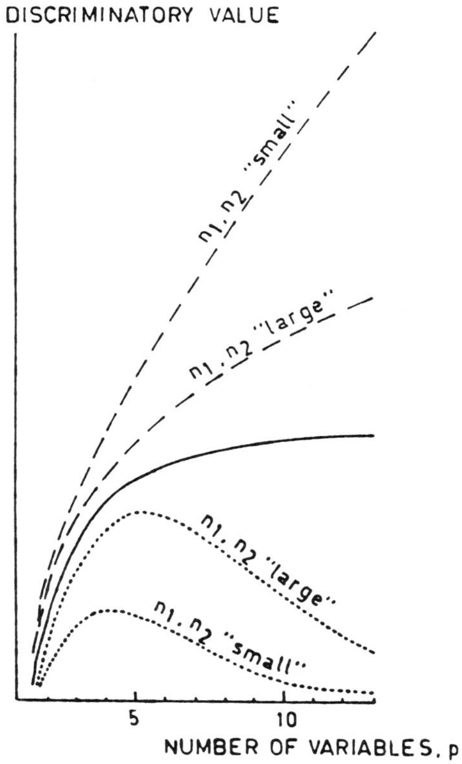

Figure 1. For explanation see text. Originally published in the "Zeitschrift für Morphologie und Anthropologie", Vol. 73/1, p. 23: G.N. van Vark and P.G.M. van der Sman, "New discrimination and classification techniques in anthropological practice.

ponding line is obtained which will be close to the (unknown) line of true Δ's (solid line)).

The two lower lines are the most intriguing ones. They refer to the true discriminatory value of the sample-based discriminant function, that is, the discriminatory value which applies to individuals that are to be diagnosed. These curves suggest that there exists an "optimum" number of variables, p*, which depends on the sizes of the samples from which the functions are calculated: as these sizes decrease, the number of effective variables also decreases.

The obvious next step was to try to estimate p*. To that end, we proposed a number of variable selection techniques for k-group (k>2) discrimination problems (6). We here mention a data-analytic variant which is based on Lachenbruch's so-called "leaving-one-out" (LOO) method (7). In the two-population case, given the set of variables, the following steps are taken:

A first discriminant function is calculated with the first individual of the first sample omitted from the calculation. The discriminant function score of this omitted individual, y_{11}, is registered. (Note that this one individual takes the position of an individual to be diagnosed.)

A second discriminant function is calculated by restoring the first and omitting a second individual from the first sample. The score of this individual, y_{12}, is also registered. This procedure is continued until n_1+n_2 scores $y_{11},\ldots,y_{2n_2}$ are obtained. Next, mean scores $\hat{\mu}'_h = n_h^{-1} \sum_{j=1}^{n_h} y_{hj}$ (h=1,2) and a standard deviation $\hat{\sigma}'$ are obtained, and the "discriminatory value in samples to be diagnosed" is calculated according to

$$D'_t = |\hat{\mu}'_1 - \hat{\mu}'_2|/\hat{\sigma}'.$$

Using this as a measure for the quality of the given set of variables, we propose to use the set of variables for which D'_t is as large as possible. It turns out that D'_t as a function of the number p of variables shows the required behavior.

For k-group discrimination problems the method can be generalized by maximizing the sum of

$$\sum_{1<h<g<k} D'^2_{t;gh}$$

of squared Mahalanobis distances,

$$D'^2_{t;gh} = (\hat{\mu}'_g - \hat{\mu}'_h)\, \hat{\Sigma}'^{-1}(\hat{\mu}'_g - \hat{\mu}'_h) \tag{2.1}$$

in the *min*(k-1,p) dimensional space of canonical variate vectors $y_{11},\ldots,y_{k\,n_k}$ obtained by successively omitting individuals analogous to the way mentioned above.

It is obvious that for assigning unknown individuals one proceeds as follows: By adding variables in a stepwise fashion a number p* is sought for which D'_t or $\Sigma D'^2_{t;gh}$ is as large as possible. An unknown individual with score vector y on the set of discriminant functions using p* variables is then assigned to the population for which the squared Mahalanobis distance,

$$D'^2_t = (y - \hat{\mu}'_g)\, \hat{\Sigma}'^{-1}(y - \hat{\mu}'_g)$$

is smallest (8).

The conclusion from experiences with these techniques was that, in cases where the sample sizes are as small as in hominid investigations, p* will usually not be greater than two. This makes the results so crude that many an anthropologist (the author included) would prefer results obtained by means of visual comparison of the morphology.

Quite apart from that, the number of populations (genera, species, or subspecies) in the material at hand cannot be regarded as sufficiently known. This cluster-analytic nature of the classification problem makes it even more complex.
Clearly, for a more successful classification of hominid specimens by mathematical means, a different approach had to be sought.

3. A NEW APPROACH AND A FREQUENT COMPLICATION

At various anthropological congresses, we were more impressed by the results obtained by visual comparison than the results obtained by numerical measurements and mathematical statistical evaluation. When trying to find an explanation, and mindful of the experience mentioned in the preceding section, we found the following clue: when classifying unknown hominid specimens on the basis of a visual comparison of the morphology, visualists implicitly take the variation of the present world population as a standard for judging whether or not a specimen falls within the range of a certain "subspecies" (9); information which the customary mathematical procedures do not utilise.

The next step was obvious. The mathematical statistical evaluation should incorporate similar information, i.e. a variance covariance matrix for the present world population. If such a matrix was available, many of the earlier uncertainties in the mathematical approach would disappear. Mahalanobis distances between single specimens and group averages can be calculated with respect to this standard covariance matrix. In the earlier approach, one had to estimate such matrices from scarce observations. This caused many uncertainties and led to an unrealistically small value of p*.

In principle, the required variance-covariance matrix might be estimated if one were to have a "very large" sample from the present world population. As a result, a mathematical statistical procedure can be envisaged which combines the advantages of the visual and mathematical approaches by (analogous to the visualist's approach) postulating that the populations of hominids have the same variance-covariance matrix as the present-day world population.

Our elaboration of these ideas was as follows: Skull measurements for the estimation of a variance-covariance matrix for the present world population were kindly sent to us on tape by W.W. Howells (10). These measurements were taken from samples of 30 recent subpopulations (totalling 2,471 individuals) which by and large cover the variation of the present-day human subspecies. Measurements from 105 hominid skulls, also taken according to Howells' method, and dating from about 1 million years onward, were kindly put at our disposal by C.B. Stringer.

An unbiased estimator of the variance-covariance matrix of

the sampled part of the composite present world population was derived for us by W. Schaafsma. This estimator is

$$\hat{\Sigma} = \sum_{g=1}^{k} (n_g-1)^{-1} \varepsilon_g \hat{W}_g \{1-n_g^{-1} (1-\varepsilon_g)\} + \sum_{g=1}^{k} \varepsilon_g(\hat{\mu}_g-\hat{\mu})(\hat{\mu}_g-\hat{\mu})^T,$$

where

k = number of sampled recent sub-populations,
$\hat{\mu}_g$ = estimated centroid of the g-th sub-population,
$\hat{W}_g$ = estimated within-groups SSCP matrix of the g-th recent sub-population ($\hat{W}_g = (n_g-1)\hat{\Sigma}_g$),
ε_g = size of the g-th recent sub-population divided by the size of the total recent world population,
$\hat{\mu} = \sum_{g=1}^{k} \varepsilon_g \hat{\mu}_g$ = estimated centroid of the recent world population.

Using the matrix $\hat{\Sigma}$, estimated from the scores of Howells' recent skull series, as if it were the "true" matrix, Σ, the original scores of all individuals (recent as well as older hominid individuals as far as they had a complete score vector) were replaced by principal component scores, which were standardized so that they have mean 0 and unit covariance matrix for the recent world population. Subsequently Mahalanobis distances between centroids of groups and/or single specimens were calculated.

Although the problem of the small sample sizes of the hominid populations (c.q. subspecies) has thus found a satisfactory solution, at least if one accepts the idea of using $\hat{\Sigma}$ as the true variance covariance matrix (see also Sec. 6), we are still faced with a second problem, typical for hominid skeletal material: that of the fragmentary condition of many of the skulls concerned. This implies that many hominid specimens could not be used in the above-mentioned analysis because this is based on scores on all the variables employed. (Of course, some relief can be obtained by restricting the attention to a small subset out of the total of Howells' 70 variables.)

For those cases where it is a priori known to which population the fragmentary specimens belong (for example, some Neanderthal specimens), missing data techniques were developed by De Goede (see elsewhere in this volume). However, in the majority of our cases, the problem is more complex because its above-mentioned cluster-analytic nature.

Although we do not know of a satisfactory solution of the problem in its general form, it was empirically found that the following technique (11) could, in some instances, be helpful. Essentially, it is based on the prediction of a complete set of principal component scores from the incomplete set of raw scores by applying multiple regression. The underlying assumption is that the individual whose scores are predicted is regarded as a random drawing from the present-day world population. Details are given in the diagram on page 334.

We have carried out a number of simulation studies in order to get an idea of the usefulness of this technique for our problem. Missing scores were artificially generated for all hominid specimens which actually scored completely on all variables of the set we wanted to use in subsequent D^2-calculations. Thus each completely scoring specimen led to a score vector with some missing elements.

Using the above technique, the incomplete vector of raw scores was transformed into a complete vector of principal component function scores. Consequently, each of these specimens was finally represented by two sets of principal component function scores, viz. one set of "real" principal component scores, and one obtained by means of regression. Next, Mahalanobis distances were calculated between the two vectors of scores for each of the specimens concerned.

It was found that in a number of instances similar to those in the example of the next section, the resulting D^2-values were all smaller than 2.0. This suggests that, in those instances, general interpretations (which are based on much larger D^2-values) will not be affected if missing scores are "estimated" by means of our regression technique and then used as the real ones. As a consequence, the comparability of the hominid skulls is increased.

4. APPLICATIONS; A COMPARISON WITH RESULTS BASED ON VISUAL DIAGNOSIS

In table 1, Mahalanobis squared distances between 5 groups of skulls (the grouping being mainly based on prior information) and the well-known skulls of Broken Hill (Zambia), Petralona (Greece), and Steinheim (Germany) were calculated, using the first 8 principal components extracted from 17 variables. The selection of 17 variables from Howells' original set of 70 variables was made on the basis of the incompleteness found in many of the hominid skulls we wanted to compare. 83 hominid skulls scored completely on this set. Nine skulls lacked one score, and six other ones (one of which was the Steinheim skull which will play a crucial part in our arguments) two scores. It was concluded in the previous section that for the present calculation no harm will be done if the few missing scores for these 15 skulls are replaced by their estimates.

Using the appropriate χ^2-tests with eight degrees of freedom, corresponding to Hotelling's two-sample test (in the present instance, the variance-covariance matrix is regarded as known), all distances in table 1 are significant at the 5% level, with the exception of the distances between Broken Hill, Petralona, and Steinheim.

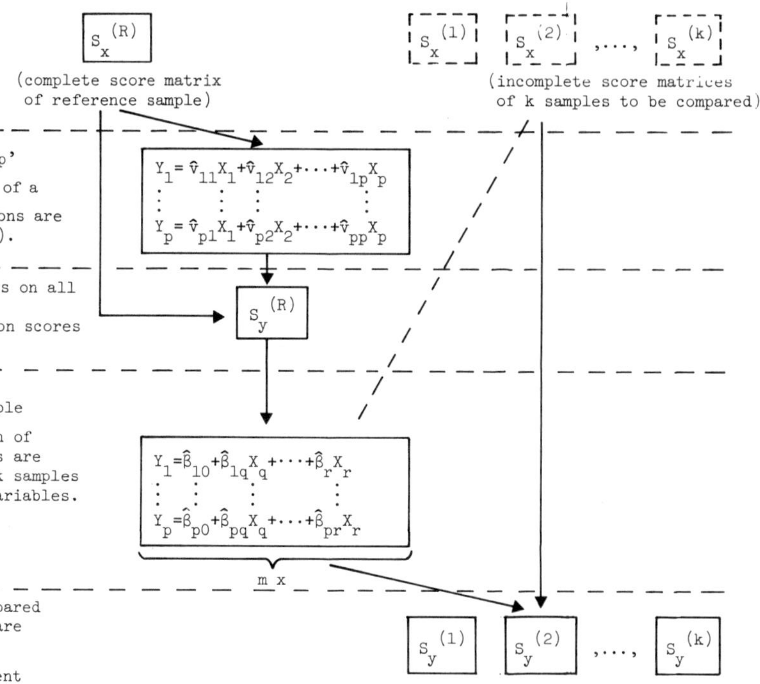

Calculation steps

1 p principal component functions, $Y_j = \hat{v}_{j1}X_1 + \hat{v}_{j2}X_2 + \dots + \hat{v}_{jp}X_p$, are calculated from the complete n'xp scorematrix, $S_x^{(R)}$, of a reference sample. (In the present instance, these functions are standardized for the present world population (see text)).

2 For all n' individuals of the reference sample the scores on all p principal component functions are calculated. Thus, a complete n'xp matrix, $S_y^{(R)}$, of principal component function scores is obtained.

3 From $S_x^{(R)}$ and $S_y^{(R)}$ m ($m \leq n$) different sets of p multiple regression functions which each use a different selection of the p original variables are calculated. The m selections are made such that at least one of the n individuals of the k samples to be compared scores on one of the m sets of selected variables.

4 For each of the n individuals of the k samples to be compared scores on the appropriate set of p regression functions are calculated. Thus, k complete complete score matrices, $S_y^{(1)}$, $S_y^{(2)}$, ..., $S_y^{(k)}$, of "estimated" principal component functions are obtained.

5 From the complete score matrices, $S_y^{(1)}$, $S_y^{(2)}$, ..., $S_y^{(k)}$, Mahalanobis distances between the k samples to be compared (which may partly or each contain 1 individual) are calculated.

Table 1.

MEASURES USED

	MEASURE NO. ACCORDING TO HOWELLS, 1973	
1	1	GOL
2	2	NOL
3	5	XCB
4	6	XFB
5	7	STB
6	11	ASB
7	36	SOS
8	37	GLS
9	39	FRC
10	40	FRS
11	41	FRF
12	42	PAC
13	43	PAS
14	44	PAF
15	48	VRR
16	49	NAR
17	54	FMR

D^2 values

		Rec.	U.P.	Ne.	So.	H.e.	Br.H.	Petr.
1	Recent (n=14)							
2	Upper Palaeolithic (n=22)	4.9						
3	Neanderthal (n=8)	39.1	32.5					
4	Solo (n=4)	41.2	34.3	10.4				
5	Homo erectus (n=2)	55.0	54.8	15.6	14.1			
6	Broken Hill (n=1)	106.6	93.4	39.6	47.1	61.7		
7	Petralona (n=1)	99.3	87.4	34.5	38.9	58.0	2.3	
8	Steinheim (n=1)	103.9	103.9	43.5	51.7	44.4	13.3	16.8

n = sample size

group 1 = Norse ♂ + ♀, Dogon ♂ + ♀, Bush ♂ + ♀, Austr. ♂ + ♀, Tasman ♂ + ♀, Buriat ♂ + ♀, Arikara (American Indian) ♂ + ♀.

group 2 = Ortucchio 1, San Teodoro 2 and 4, Oberkassel ♂ + ♀, Chancelade, Laugerie Basse 2, Le Roc du Sers 2, Kostenki 2, Predmost 3 and 4, Brno 1 and 3, Mladec 1 and 3, Dolni Vestonice 3, Engis, Cro Magnon 1, 2 and 3, Grimaldi 6 and Combe Capelle.

group 3 = Monte Circeo, La Chapelle, La Quina, Gibraltar, Le Moustier, Spy 1 and 2. Saccopastore 1.

group 4 = Solo 2, 3, 4, and 5.

group 5 = Sinanthropus 2 and E.

For measurement definition see Howells, 1973 (10).

For technical reasons, 14 recent specimens were selected in table 1, 2, 4 and 5.

For further explanation see text.

Undoubtedly, the most striking finding displayed by the table is the fact that the Broken Hill, Petralona, and Steinheim skulls while mutually being relatively close, are very distinct from all other skulls. Further mathematical statistical elaboration is as follows:

For our principal components the recent world population has mean 0 and unit covariance matrix. Let x_1 and x_2 denote the means of samples of respectively n_1 and n_2 elements taken from population 1 (here: the population to which the three skulls mentioned are assumed to belong) and population 2 (Homo erectus). Let the corresponding expectations be denoted by μ_1 and μ_2. It is in line with our earlier assumptions that $\bar{x}_h$ is the outcome of a random vector, $\bar{X}_h$, which has the $N_p(\mu_h, n_h^{-1} I)$ distribution (h=1,2). We are interested in testing the null-hypothesis

$$H_0 : \|\mu_1\|^2 = \|\mu_2\|^2$$

against the two-sided alternative because we want to investigate whether

$$\|\mu_1\|^2 < \|\mu_2\|^2 \text{ or } \|\mu_2\|^2 < \|\mu_1\|^2.$$

An appropriate procedure was derived by T. Steerneman. The test statistic

$$T = \frac{n_1 n_2(\|\bar{x}_1\|^2 - \|\bar{x}_2\|^2) + p(n_1 - n_2)}{\{4n_1 n_2(n_2\|\bar{x}_1\|^2 + n_1\|\bar{x}_2\|^2) - 2p(n_1^2 + n_2^2\}^{\frac{1}{2}}}$$

approximately has the $N(0,1)$ distribution if H_0 is true. If T > + 1.96 we conclude that there is sufficient evidence for $\|\mu_1\|^2 > \|\mu_2\|^2$, if T < -1.96 for $\|\mu_1\|^2 < \|\mu_2\|^2$ (α = .05).

The test was found to be significant at the 5% level, even if the Broken Hill, Petralona and Steinheim skulls were considered separately (T = + 2.80, + 2.68, + 2.45, respectively). Thus we may conclude that each of these three skulls is, at the significance level adopted, more "distinct" from skulls of recent man than the skulls of Asiatic Homo erectus.

Remark.

The calculations of this section were based on the first eight principal components. This number seemed to give the most plausible results. We had the impression that on using fewer principal component functions real biological information was missed whereas increasing this number seemed to produce more extra noise than real biological information.

An interesting fact was that the above test by Steerneman gave the most significant results with the same number of principal component functions. However, since the same data are used, this should not be regarded as a clear indication that the selection of the number of principal component functions was "optimum" according to some statistical criterion. A selection of variables procedure, analogous to the one discussed in Section 2, has, so far, not been developed for this test.

5. INVESTIGATING THE RELIABILITY OF THE RESULTS OBTAINED IN SECTION 4

The finding in Section 4 that Asiatic Homo erectus is closer to recent man than Petralona, Broken Hill, and Steinheim was quite unexpected. In order to get an idea of its reliability, we have checked it in various ways.

First, we have investigated the dependence on possibly unjustified assumptions concerning the underlying variance-covariance matrix. To test the stability of the results we investigated what happens if all weighing coefficients, ε_g (see p.332), are taken equal. The resulting pattern of D^2-values, given in table 2, can be compared with that in table 1. The pattern of distances remains essentially the same though the special character of the three skulls becomes slightly less pronounced.

Another relevant way of testing the results of Section 4 is by comparing with results based on other sets of variables.

So far some 17 variable combinations were considered. Selection of these combinations depended mainly on the availability of scores of the skulls we wanted to compare. As a consequence, other partly complete skulls could sometimes be incorporated in the calculations.

In table 3 we give a survey of the variable combinations used so far.

A general survey of the results confirmed the extraordinary position of the three skulls. The variable set used in our original calculations, and resulting in table 1, produces a more or less "average" distance pattern. Notable exceptions are variable sets nos. 17 and 12. On using set no. 17 the distinctiveness of the three skulls was extremely large whereas, on using set no. 12 the distinctiveness of these skulls was exceptionally low. The distances of the Homo erectus skulls to the recent skull sample was in the latter calculation more or less of the same magnitude as the distances of Broken Hill and Petralona to that sample. Steinheim did not score on this set no. 12 (see table 4 below).

The Steinheim skull displays some deformation (12). Its distance scores are therefore to be regarded with some reservation. In most calculations in which it participated it was farther apart from both Petralona and Broken Hill than in table 1.

Table 2. Same as table 1 but all weighing coefficients, εg, of the variance-covariance matrix $\hat{\Sigma}$ are taken equal.
For further explanation see text.

MEASURES USED					D^2 values (all εg's equal)						
	MEASURE NO. ACCORDING TO HOWELLS, 1973				Rec.	U.P.	Ne.	So.	H.e.	Br.H.	Petr.
1	1	GOL	1	Recent (n=14)							
2	2	NOL	2	Upper Palaeolithic (n=22)	4.8						
3	5	XCB	3	Neanderthal (n=8)	37.8	33.8					
4	6	XFB	4	Solo (n=4)	43.4	40.9	9.8				
5	7	STB	5	Homo erectus (n=2)	55.5	58.4	14.5	8.5			
6	11	ASB	6	Broken Hill (n=1)	89.4	83.2	34.5	38.9	46.4		
7	36	SOS	7	Petralona (n=1)	76.1	71.3	23.8	27.5	38.1	2.7	
8	37	GLS	8	Steinheim (n=1)	81.6	87.6	36.9	41.8	32.1	13.2	16.2
9	39	FRC									
10	40	FRS									
11	41	FRF									
12	42	PAC									
13	43	PAS									
14	44	PAF									
15	48	VRR									
16	49	NAR									
17	54	FMR									

Table 3

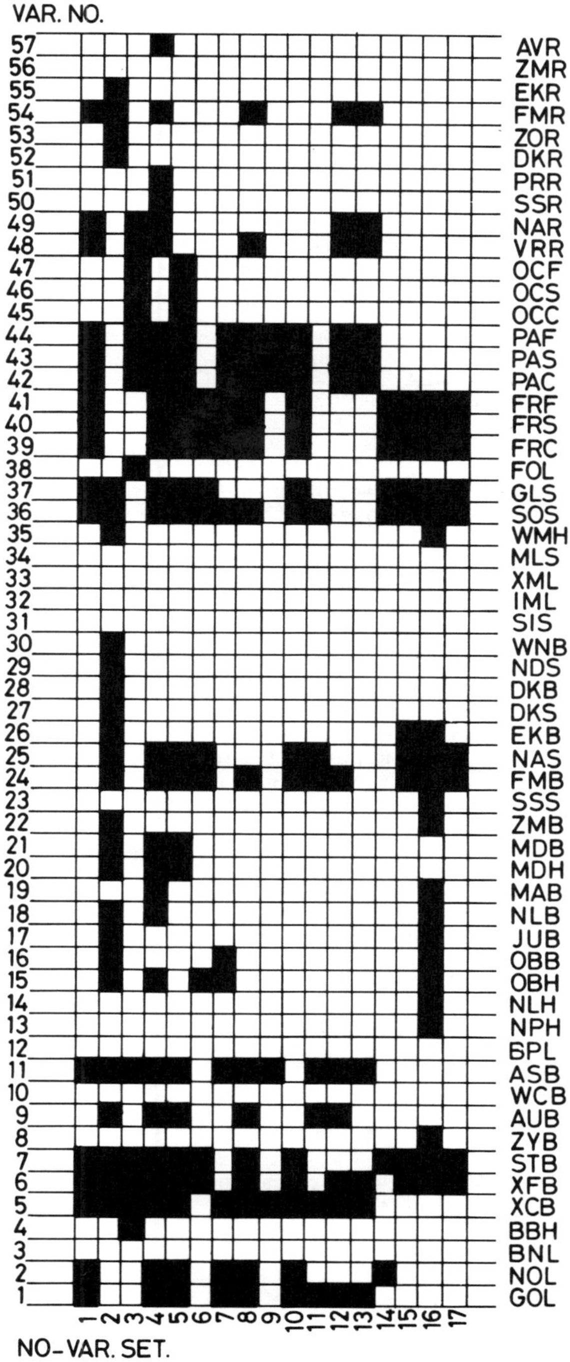
VAR. NO.
57 AVR
56 ZMR
55 EKR
54 FMR
53 ZOR
52 DKR
51 PRR
50 SSR
49 NAR
48 VRR
47 OCF
46 OCS
45 OCC
44 PAF
43 PAS
42 PAC
41 FRF
40 FRS
39 FRC
38 FOL
37 GLS
36 SOS
35 WMH
34 MLS
33 XML
32 IML
31 SIS
30 WNB
29 NDS
28 DKB
27 DKS
26 EKB
25 NAS
24 FMB
23 SSS
22 ZMB
21 MDB
20 MDH
19 MAB
18 NLB
17 JUB
16 OBB
15 OBH
14 NLH
13 NPH
12 BPL
11 ASB
10 WCB
9 AUB
8 ZYB
7 STB
6 XFB
5 XCB
4 BBH
3 BNL
2 NOL
1 GOL
1 2 3 4 5 6 7 8 9 10 11 12 13 14 15 16 17
NO-VAR. SET.

The distance of the Steinheim skull to the skulls of Petralona and Broken Hill, when tested by using the appropriate χ^2 test, was significant (note that this test replaces the usual Hotelling's F test because $\hat{\Sigma}$ is regarded as the true matrix Σ).

The overall conclusion from these calculations is that our earlier findings as to the positions of the Broken Hill, Petralona, and Steinheim skulls are supported. These skulls are very different from all other hominid skulls in our material which dates from the times of Pithecanthropus onward. This difference finds expression in various sets of measurements but more so in some sets than in others. Thus bringing the idea of a marked mosaic pattern of evolution to the mind.

6. ON THE ASSUMPTION OF EQUALITY OF VARIANCE-COVARIANCE MATRICES, AND SOME RELATED PROBLEMS

Our theory is based on the assumption that the hominid populations being compared have the same variance-covariance matrix as that computed from the recent population. The problem of testing the equality of variance covariance matrices is a complex one. The extensive literature contains elegant solutions (see e.g. Anderson (13), and Campbell (14).

Our testing problem is slightly different from the common one in that we want to test the null hypothesis that the variance-covariance matrix of some hominid population is equal to the "given" variance-covariance matrix of the present-day world population. This makes our testing problem relatively easy.

Hence suppose that Mahalanobis distances are defined with respect to the given variance-covariance matrix, and that our sample consists of n_g individuals, the i-th with squared Mahalanobis distance D_i^2 from the sample mean, and with squared distance D_{ij}^2 from the j-th individual. Finally, let p denote the dimension of the observation vector. We use as a test statistic

$$T = \sum_{i=1}^{n_g} D_i^2 = \left(n_g - 1\right) \sum_{i=1}^{n_g-1} \sum_{j=i+1}^{n_g} D_{ij}^2 = \tfrac{1}{2}(n_g-1)\bar{D}_g^2$$

where

$\bar{D}_g^2 = \sum_{i<j} D_{ij}^2/\{\tfrac{1}{2}n(n_g-1)\}$ denotes the mean Mahalanobis squared distance between individuals. Under the null hypothesis T has $\chi^2_{(n_g-1)p}$ distribution. We reject the null hypothesis if

$$T \leqq \chi^2_{(n_g-1)p;\ 1-\frac{1}{2}\alpha}.$$

or if

$$T \geqq \chi^2_{(n_g-1)p;\frac{1}{2}\alpha}.$$

Table 4. Exceptionally low distance values for the Petralona and Broken Hill skulls

MEASURES USED

	MEASURE NO. ACCORDING TO HOWELLS, 1973	
1	1	GOL
2	5	XCB
3	6	XFB
4	9	AUB
5	11	ASB
6	24	FMB
7	42	PAC
8	43	PAS
9	44	PAF
10	48	VRR
11	49	NAR
12	54	FMR

Variable set no. 12 of table 3.

D^2 values

		Rec.	U.P.	Ne.	So.	Sin.	Pith.	Br.H.
1	Recent (n=14)							
2	Upper Palaeolithic (n=14)	5.7						
3	W.E. Neanderthal (n=6)	19.2	14.4					
4	Solo (n=1)	33.0	27.3	5.6				
5	Sinanthropus (n=2)	45.4	49.9	20.3	20.2			
6	Pithecanthropus 2 (n=1)	56.0	67.1	27.4	30.3	12.8		
7	Broken Hill (n=1)	58.0	55.7	28.0	28.1	62.1	56.0	
8	Petralona (n=1)	51.9	50.7	23.3	26.7	56.4	47.3	2.3

Table 5. Within group (diagonal elements) and between group (off-diagonal elements) squared Mahalanobis distances. Individuals and variables the same as in table 2. For further explanation see text.

MEASURES USED

	MEASURE NO. ACCORDING TO HOWELLS 1973	
1	1	GOL
2	2	NOL
3	5	XCB
4	6	XFB
5	7	STB
6	11	ASB
7	36	SOS
8	37	GLS
9	39	FRC
10	40	FRS
11	41	FRF
12	42	PAC
13	43	PAS
14	44	PAF
15	48	VRR
16	49	NAR
17	54	FMR

D^2 values

		Rec.	U.P.	Ne.	So.	H.e.	B.P.S.
1	Recent (n=14)	16.62					
2	Upper Palaeolithic (n=22)	21.18	14.64				
3	Neanderthal (n=8)	56.65	53.41	14.82			
4	Solo (n=4)	58.66	53.74	27.62	25.36		
5	Homo erectus (n=2)	65.01	65.61	26.58	30.76	18.70	
6	Br.Hill + Petralona + Steinheim	105.76	94.99	45.54	58.54	59.40	10.80

We have applied this test to the groups tabled in table 5. Attention is restricted to the first eight principal components.

No significant differences were found in this table. That is, no difference in variability could be demonstrated for the groups concerned.

This is no wonder because the sample sizes are very small. Even if important differences should exist, they would not easily be detected. On the other hand, our results obviously do not contradict the assumption of a common variance-covariance matrix. This assumption might also be used for estimating the number of hominid subspecies. We shall continue this subject in Section 8.

With eight principal components, the expected value for D^2_{ij} is 2p=16, and the variance is 4p=32. The reason that the value of $\bar{D}^2_g$ of the recent sample is slightly larger than 16 is due to sampling: for technical reasons 14 individuals were selected from the total of 2,471 which have been used for the calculation of the underlying "known" covariance matrix.

The test may also be used for determining whether the recent human races are of a mono- or polyphyletic origin. As Howells puts it: "Is cranial variation of modern man great enough to support a polyphyletic origin? Understanding this variation is important" (15). Indeed, application of this test to more comprehensive samples of Late Pleistocene crania might answer this question.

We finally point to the possibility of comparing the "within groups variability" to the "between groups variability". Average squared distances, D^2_{gh}, between individuals from group g and individuals from group h may be calculated according to $D^2_{gh} = \Sigma\Sigma_{ij} D^2_{ij}/n_g.n_h$. Accordingly, a matrix can be built up with diagonal elements equal to the average squared distances, $\bar{D}^2_g$, within groups, and with D^2_{gh} off-diagonal elements equal to the average distances between individuals of two groups g and h. Table 5 is an example. It uses the same groups, individuals, variables, and number of principal components as table 1.

An advantage of this representation is that it expresses within and between group variability. One may, for example, infer from table 5 that the average individual distance within recent Homo sapiens seems less than the average distance between Upper Palaeolithic and recent Homo sapiens individuals. Though this difference might be due to sampling, it might also point to some evolution since Upper Palaeolithic times.

7. COMPARISON OF SINGLE SPECIMENS

An important advantage of using a variance covariance matrix from a (large) reference sample is that Mahalanobis distances between

Table 6

	Nor.♂	Nor.♀	Dor.♂	Dog.♀	Bus.♂	Bus.♀	Aus.♂	Aus.♀	Tas.♂	Tas.♀	Bur.♂	Bur.♀	Ari.♂	Ari.♀	Ort.	S.Te.2	S.Te.4	Oberk.♀	Oberk.♂	Chan.	L.Bas.	L.SE	MaGo.	Pre.3	Pre.4	Brn.1	Brn.3	Mla.1
Norse ♂ (n=55)																												
Norse ♀ (n=55)	2.0																											
Dogon ♂ (n=49)	2.9	1.6																										
Dogon ♀ (n=43)	6.6	2.2	1.5																									
Bushman ♂ (n=40)	6.0	3.7	6.4	8.2																								
Bushman ♀ (n=40)	9.5	4.2	7.1	6.2	1.3																							
Australia ♂ (n=49)	6.9	10.5	14.0	19.0	7.8	12.6																						
Australia ♀ (n=49)	6.3	5.5	9.7	11.3	3.4	4.9	2.5																					
Tasmania ♂ (n=42)	5.8	9.6	11.9	17.0	8.0	12.7	1.6	3.6																				
Tasmania ♀ (n=42)	6.0	6.0	9.5	11.8	4.1	6.1	3.4	1.3	2.0																			
Buriat ♂ (n=53)	3.3	3.9	6.0	8.6	7.6	10.6	12.9	10.6	10.4	9.9																		
Buriat ♀ (n=53)	7.1	3.6	5.7	5.0	8.0	7.6	19.7	12.7	16.6	12.4	2.3																	
Arikara ♂ (n=26)	5.2	6.2	9.3	11.1	9.3	11.8	6.8	5.6	5.1	5.7	4.9	8.3																
Arikara ♀ (n=26)	6.1	2.5	5.7	4.2	6.2	5.3	12.7	6.3	11.0	6.9	4.3	2.6	3.5															
Ortucchio	6.4	6.9	9.9	12.4	6.3	8.8	5.8	3.9	3.3	2.2	9.5	12.6	4.0	6.7														
San Teodoro 2	3.9	6.5	6.3	11.1	10.0	14.0	10.3	10.9	12.5	13.0	10.0	14.5	14.5	14.4	17.4													
San Teodoro 4	4.8	6.8	9.4	12.9	9.5	13.0	8.6	8.5	9.4	10.3	5.0	8.8	7.8	8.8	14.4	5.1												
Oberkassel ♀	2.2	3.5	4.9	7.7	8.9	11.8	9.6	8.3	9.3	8.3	7.4	11.0	7.0	7.8	6.7	7.7	11.8											
Oberkassel ♂	11.0	17.0	18.3	26.9	10.6	18.3	4.7	10.0	5.9	10.4	15.7	24.7	14.7	22.0	13.5	11.5	10.4	17.9										
Chancelade	10.3	13.3	14.5	18.8	19.7	24.8	21.3	22.0	21.6	22.3	13.0	19.0	15.2	17.6	17.4	17.2	23.2	5.3	28.3									
Laugerie Basse 2	3.2	8.0	8.6	14.9	9.2	14.7	4.6	7.4	4.0	6.2	9.8	16.4	10.7	14.5	9.8	4.1	6.3	7.3	6.4	18.9								
Le Roc du Sers	11.2	14.0	16.0	19.4	20.9	25.5	19.1	19.3	18.3	20.1	14.1	20.0	10.0	15.4	11.5	22.3	22.6	7.5	27.6	8.7	21.5							
Markina Gora	7.1	8.2	8.0	10.4	9.6	11.6	7.1	5.9	3.9	4.2	12.3	15.1	4.8	8.2	2.7	16.3	15.1	8.2	14.6	20.7	9.7	14.4						
Predmost 3	14.2	24.0	22.8	32.8	25.4	35.1	13.2	22.1	13.4	23.1	18.0	30.3	15.2	27.8	21.3	16.2	13.6	20.6	8.9	26.6	13.2	19.6	18.7					
Predmost 4	4.5	8.6	6.1	13.1	9.2	15.1	12.2	14.0	10.4	13.0	8.8	14.5	13.4	15.9	11.9	7.1	11.1	8.7	8.5	15.9	6.5	15.4	11.5	10.9				
Brno 1	40.1	49.5	49.4	58.2	49.8	59.7	34.7	44.6	37.7	50.0	38.7	51.9	29.5	46.0	44.1	44.5	36.0	45.1	31.3	42.4	45.1	33.0	40.3	13.5	37.7			
Brno 3	5.8	7.1	10.2	12.6	10.8	13.0	6.7	5.5	6.6	4.6	14.2	17.8	10.3	11.5	7.0	10.3	13.2	4.7	18.0	17.9	5.6	18.4	8.4	27.2	16.0	60.0		
Mladec 1	9.0	12.2	12.0	17.2	20.6	25.0	22.3	22.4	23.8	23.7	17.5	22.3	25.1	24.0	25.4	5.8	14.7	10.5	25.2	20.0	11.2	21.8	26.6	26.0	11.8	63.6	14.3	
Mladec 5	24.2	32.6	36.0	47.4	23.7	35.0	17.8	25.8	20.4	27.3	26.7	39.7	26.7	38.1	24.4	28.1	25.9	29.5	9.4	32.4	23.2	26.4	31.4	13.3	16.7	28.0	37.0	36.8
Dolni Vestonice	11.1	10.6	10.6	13.0	13.5	16.4	21.1	18.3	20.8	20.4	11.6	13.9	11.9	12.4	13.3	19.4	20.9	9.7	24.3	10.3	23.2	5.6	15.0	21.6	10.6	31.8	24.5	23.9
Engis	6.9	11.6	12.4	18.2	11.7	17.6	6.4	10.0	6.5	9.9	10.3	17.8	7.1	13.5	9.5	12.0	12.9	7.3	9.8	9.2	8.2	15.4	9.1	12.8	10.9	26.9	12.6	25.6
Cro Magnon 1	11.7	16.6	20.9	27.6	16.4	23.6	13.3	17.3	16.9	19.4	12.5	21.2	16.9	21.2	22.1	11.5	12.4	13.2	14.3	11.6	12.5	24.8	27.3	20.3	17.8	36.7	19.6	22.6
Cro Magnon 2	4.3	5.9	7.9	11.1	10.1	13.9	11.5	11.2	12.8	13.5	4.3	8.7	6.1	7.6	12.1	8.2	8.0	4.6	16.1	5.3	11.4	9.4	13.9	15.5	10.1	28.4	14.3	16.3
Cro Magnon 3	12.1	17.8	17.0	25.1	17.8	24.6	15.3	20.0	18.8	22.4	17.4	25.2	24.4	27.0	29.3	4.4	9.7	18.4	10.6	24.4	8.4	36.8	28.0	16.9	12.4	42.2	22.0	15.8
Grimaldi 6	2.4	4.8	4.8	9.2	9.4	13.4	11.2	11.3	12.0	11.9	7.3	11.5	12.4	11.9	14.5	1.6	7.2	3.9	14.2	9.7	4.4	18.1	14.7	18.9	6.9	46.8	8.3	6.6
Combe Capelle	13.3	21.8	22.8	33.1	18.5	28.6	9.9	17.7	11.5	18.4	19.2	31.6	16.3	27.7	15.1	17.2	19.5	15.1	6.8	17.2	12.7	14.6	17.7	6.6	9.2	22.2	22.1	25.1
Omo 2	33.8	43.4	46.0	54.5	50.6	59.8	33.5	42.1	39.8	49.1	41.1	54.2	36.6	49.4	47.2	30.1	27.8	39.1	34.1	48.3	35.2	31.1	46.8	16.9	35.0	29.7	44.6	29.6
Fishhoek	24.6	31.4	31.2	42.4	24.9	34.0	26.7	31.0	24.3	25.8	34.6	43.8	42.6	46.0	28.1	26.4	35.9	28.9	19.7	40.6	17.4	44.5	34.1	37.6	17.0	87.1	26.9	25.6
Monte Circeo	34.8	45.2	49.3	60.4	38.0	48.8	21.4	31.0	21.7	31.1	38.3	51.9	32.1	47.3	32.6	38.4	26.7	47.0	14.8	64.7	27.2	46.1	35.4	15.9	31.4	37.6	42.3	49.6
La Chapelle	54.2	65.5	70.0	82.3	59.6	71.9	42.6	52.9	44.3	54.8	60.5	76.1	52.8	70.4	50.2	59.4	50.2	64.0	35.4	81.0	50.3	50.0	55.6	29.2	45.8	50.2	63.4	60.1
La Quina	65.8	76.5	80.5	90.6	71.5	81.4	45.5	56.8	51.0	63.2	73.2	87.9	57.4	76.7	64.3	64.4	51.2	78.5	41.6	100.7	58.8	69.4	61.8	32.3	63.1	40.7	73.0	77.6
Gibraltar 1	39.7	46.3	50.1	58.3	38.1	46.2	27.6	33.0	25.9	33.1	39.9	49.7	31.0	44.1	30.3	49.2	33.9	51.2	22.3	69.5	37.9	42.7	33.1	22.2	34.7	37.1	50.0	61.3
Le Moustier	19.8	26.8	31.9	39.7	22.3	30.9	11.0	17.4	12.0	18.8	19.3	30.3	12.6	25.1	16.3	27.6	16.8	25.8	10.0	32.8	20.2	21.0	20.5	8.7	20.8	16.9	29.2	40.6
Spy 1	83.3	96.0	101.2	113.5	86.9	100.1	62.6	75.4	63.0	77.2	86.1	103.9	67.5	92.4	67.6	96.6	79.0	92.4	57.0	106.4	81.5	66.9	71.4	44.5	74.8	45.6	94.3	106.7
Spy 2	42.1	50.2	54.9	65.8	38.6	49.7	30.0	38.0	32.2	40.7	41.0	53.7	36.9	50.8	37.8	47.5	35.5	52.8	19.3	62.1	40.9	43.5	44.1	20.3	33.6	28.1	58.1	60.8
Saccopastore 1	39.5	46.8	47.6	55.3	43.3	50.2	28.5	34.4	26.3	34.5	43.9	53.0	33.0	46.0	34.4	45.3	32.2	52.3	25.1	76.6	34.9	48.6	30.9	21.3	36.0	41.4	47.0	57.0
Solo 1	55.4	68.8	74.1	84.3	73.2	84.2	41.5	54.7	47.5	60.6	65.2	82.6	50.3	71.1	60.1	54.4	45.3	62.5	45.0	78.6	49.9	53.4	59.0	27.9	60.2	40.0	58.0	58.8
Solo 2	38.2	43.8	50.9	59.3	34.2	41.1	22.5	27.4	24.9	29.3	39.1	47.9	35.3	43.9	37.4	37.9	22.9	52.9	18.1	75.7	29.4	62.5	41.7	30.2	41.0	54.7	42.5	54.5
Solo 3	55.0	59.8	67.9	74.3	51.4	57.8	39.7	44.6	45.9	50.9	52.6	60.7	51.3	58.6	62.5	48.1	30.0	72.9	33.4	94.4	47.2	84.3	65.5	42.2	60.1	57.6	64.8	69.4
Solo 4	45.3	54.3	56.8	65.7	52.2	60.8	34.0	42.4	37.7	47.7	50.2	62.0	41.5	55.7	49.4	43.3	31.6	58.6	29.0	79.0	40.9	53.4	47.0	20.2	42.7	33.6	56.2	53.9
Quafzeh 6	32.0	42.8	44.9	56.7	33.0	43.9	16.6	26.9	13.1	22.0	36.9	50.7	28.5	44.3	23.1	40.9	33.7	39.9	12.2	52.8	21.9	46.3	24.6	21.1	29.3	45.2	34.0	58.0
Skhul 5	21.4	29.8	30.5	39.9	23.6	31.9	10.8	18.0	6.7	13.0	27.2	37.7	20.5	32.0	15.9	28.7	24.4	28.3	9.6	42.6	12.4	40.4	15.0	19.7	21.8	48.2	21.4	45.0
Sinanthropus, E	69.6	72.5	81.2	86.5	59.8	65.9	47.7	51.2	54.1	59.2	69.7	78.1	54.6	67.1	57.9	74.4	55.7	80.4	45.2	99.9	70.2	67.4	62.9	47.3	67.8	46.1	79.1	91.8
Sinanthropus, 2	66.3	69.0	80.0	85.5	52.2	57.7	40.1	43.4	46.1	48.8	65.3	72.8	55.8	64.6	58.1	66.1	45.7	81.6	36.6	106.4	57.6	89.2	64.7	55.1	70.6	68.4	69.1	90.7
Broken Hill	104.4	123.3	121.4	140.1	108.5	124.5	75.3	95.1	72.3	90.8	115.7	136.7	101.0	127.7	96.7	105.2	93.7	124.0	59.2	152.4	81.0	129.3	90.0	62.1	91.2	91.0	108.2	127.3
Petralona	97.4	116.0	114.3	131.4	106.0	120.9	70.5	89.7	67.6	86.3	108.1	128.5	91.4	118.4	91.4	98.8	85.5	116.2	58.4	144.6	76.4	118.7	83.0	54.9	88.0	80.3	100.9	119.5
Steinheim	112.4	123.4	126.3	138.8	102.6	112.8	76.7	88.4	73.7	84.7	120.2	133.9	100.6	121.1	92.3	118.3	99.3	132.0	66.9	168.3	91.6	139.1	89.2	83.0	105.0	110.4	109.9	145.9

Note that in this table, contrary to tables 1, 2 and 5, recent sub-populations are represented by samples and not by single specimens.

Table 6 (continued)

	Mla.5	D.Ve	Eng.	Cr.M.1	Cr.M.2	Cr.M.3	Grim.	C.Cap.	Omo.2	Fish.	M.Cir.	L.Ch.	L.Qui.	Gibr.	L.Mous.	Spy1	Spy2	Sac.P.	So.1	So.2	So.3	So.4	Qat.	Skn.S	Sin.E	Sin.2	Br.H	Petra.
Dolni Vestonice	24.5																											
Engis	21.4	15.0																										
Cro Magnon 1	21.5	23.9	7.7																									
Cro Magnon 2	23.9	7.4	5.5	6.7																								
Cro Magnon 3	26.8	30.9	13.7	8.8	13.8																							
Grimaldi 6	29.9	16.3	8.5	8.6	5.3	6.7																						
Combe Capelle	4.9	15.9	8.7	14.2	13.4	18.4	16.9																					
Omo 2	31.7	40.1	42.9	41.1	33.5	36.8	37.9	27.1																				
Fishhoek	36.4	43.8	33.6	35.5	40.3	29.6	25.7	23.9	67.3																			
Monte Circeo	16.7	51.4	38.0	43.6	44.2	39.5	46.0	22.0	28.2	41.3																		
La Chapelle	22.9	58.8	63.3	69.4	63.6	66.9	68.6	32.0	25.6	55.0	10.5																	
La Quina	42.5	76.5	68.7	78.1	71.9	68.1	78.7	46.9	24.2	90.1	14.6	15.7																
Gibraltar 1	20.4	44.4	44.1	57.1	47.9	56.0	56.2	27.6	37.4	52.6	5.8	11.0	18.0															
Le Moustier	8.4	24.2	16.4	21.8	19.1	30.2	29.5	9.1	25.3	43.1	9.9	22.4	28.2	11.0														
Spy 1	39.1	77.2	80.1	97.3	87.0	103.5	104.7	48.8	49.0	96.7	23.4	12.7	20.3	14.9	29.0													
Spy 2	8.2	40.8	41.1	44.4	42.8	47.2	53.9	19.8	34.0	50.6	9.4	12.8	23.5	7.3	8.8	18.9												
Saccopastore 1	34.1	52.2	47.1	64.6	52.3	53.5	54.7	34.7	33.3	58.1	7.9	16.5	13.0	5.5	19.3	24.5	20.3											
Solo 1	45.6	71.4	59.3	63.0	59.3	60.3	64.5	41.5	10.9	86.9	22.5	22.6	11.2	34.0	29.2	34.4	38.5	26.7										
Solo 2	29.5	63.1	44.4	43.4	48.0	37.8	47.3	37.1	41.9	49.7	8.0	29.4	24.5	16.2	18.8	48.9	20.4	17.4	36.4									
Solo 3	46.7	80.8	60.9	51.8	58.8	43.2	60.7	56.4	45.2	81.6	23.0	48.9	29.8	34.6	32.6	70.0	33.2	33.8	42.6	7.6								
Solo 4	32.7	57.1	53.0	59.1	51.7	47.3	55.9	35.8	15.3	70.9	9.8	14.3	3.7	13.8	20.6	27.3	18.7	8.4	12.1	17.2	21.4							
Quafzeh 6	21.0	51.2	23.7	37.5	42.1	39.0	41.6	18.7	59.5	30.5	14.3	37.9	47.0	21.0	15.6	45.6	25.8	25.0	52.5	24.2	50.6	41.0						
Skhul 5	26.5	44.1	16.2	30.8	32.9	29.1	28.7	18.9	59.0	25.7	20.1	48.7	55.0	28.0	19.1	62.7	36.1	27.3	56.8	26.1	51.7	44.8	3.0					
Sinanthropus E	37.7	65.5	71.6	78.9	69.9	81.2	86.4	51.1	44.5	96.2	24.1	23.2	15.8	15.0	25.9	20.8	15.5	24.2	38.7	26.7	33.2	18.9	54.7	65.4				
Sinanthropus 2	45.1	86.1	67.4	66.0	72.1	64.6	77.9	59.1	64.8	82.3	22.4	44.2	30.5	26.0	31.6	53.9	27.5	32.3	53.2	7.8	10.9	29.6	42.3	48.3	18.7			
Broken Hill	72.0	135.9	97.7	118.7	125.4	97.1	118.2	76.2	93.1	88.7	31.1	50.5	41.2	42.6	63.8	52.8	55.4	33.7	64.4	45.2	67.1	46.1	34.5	45.1	72.7	60.0		
Petralona	72.8	128.6	92.4	114.1	116.9	93.7	111.9	73.7	79.4	94.4	27.7	46.2	31.7	38.7	58.1	46.9	54.6	26.5	49.7	42.3	61.2	36.6	35.5	44.5	66.9	58.4	2.3	
Steinheim	83.5	141.9	108.3	132.6	135.2	116.6	131.7	92.6	116.6	101.1	37.3	60.0	48.0	39.1	68.2	56.5	58.9	34.6	83.1	39.3	63.4	55.2	39.9	50.1	58.1	40.2	13.3	16.8

single individuals can be calculated. This does not depend on the assumption that the populations all have the same covariance matrix as the reference population.

Between-individual Mahalanobis distances were computed for all variable combinations mentioned in table 3, and for all pairs of individuals who both scored on these respective sets (individuals with two or three missing scores were included using the technique of Section 3). We have selected two examples.

The first example presents distances based on the same set of variables as that used in tables 1, 2, and 5.

The results, given in table 6, imply amongst others that

i. the Tasmanian skulls are, within "recent" Homo sapiens, the most distinct ones,
ii. the Australoid skulls are closer to the Petralona, Steinheim, and Broken Hill skulls than any other recent specimens,
iii. there are, amongst the Upper Palaeolithic specimens, skulls which are far more distinct from recent ones than any recent skull,
iv. the largest difference found in the table is that between the male Bush skull and Broken Hill, thus clearly denying an "African Neanderthal" designation for Broken Hill.

The second example presents some results obtained with variable set no. 10 of table 3.
The distances tabulated in table 7 refer to Neanderthals and presumably related populations. With the exception of the Ehringsdorf skull which is not commonly seen as a "classical" Neanderthal specimen, the European skulls form a relatively homogeneous group which seems to be less variable than recent Homo sapiens sapiens. The most distinct skull within the European Neanderthals is, curiously enough, that of Neanderthal. The Shanidar 1 skull takes a quite distinct position. This position was confirmed in other calculations.

8. ON THE PROBLEM OF CLUSTERING AND ESTIMATING THE NUMBER OF SUBSPECIES WITHIN THE HOMINID FAMILY

If one assumes that the individuals being compared derive from an unknown number of populations, all having the same variance-covariance matrix as the reference population, then one might try to estimate the number of subspecies within the hominid family or, at least, the number of subspecies within the material at hand.

As far as is known to us, methodology fully fitting this problem has not yet been developed. A simple procedure which may provide a useful estimate for the lower bound of the number of hominid subspecies is the following.

Let $\Theta = \{\mu_1,\ldots,\mu_k, a_1,\ldots,a_n\}$ denote the parameter to be estimated, a_i denoting the population (subspecies) to which the i-th specimen belongs.

Table 7. Squared Mahalobis distances between Neanderthal and presumably related specimens. Variable combination no. 10 (see table 3); 8 extracted principal components.

	Dj.Ir.	Qua.6	Qua.9	Skh.5	Amud	Tab.1	Sha.1	Ne.	M.Cir.	LaCh.	LaQu.	Gibr.	LeMo	LaFel	Spy 1	Spy 2	Sac.p.
Djebel Irhoud																	
Quafzeh 6	8.4																
Quafzeh 9	19.8	7.0															
Skhul 5	17.0	3.2	11.4														
Amud	15.6	25.8	30.2	45.0													
Tabun 1	11.0	8.6	26.8	10.4	36.6												
Shanidar 1	41.4	25.6	52.4	67.8	33.8	56.6											
Neanderthal	14.8	14.6	16.0	21.2	21.8	21.0	65.6										
Monte Circeo	9.2	14.0	18.0	28.6	5.6	21.0	31.0	9.2									
La Chapelle	14.6	26.4	27.6	46.8	6.8	38.4	19.0	28.2	7.2								
La Quina	27.6	40.2	36.2	58.6	10.0	48.0	61.4	22.4	14.4	18.2							
Gibraltar 1	15.8	21.8	26.8	37.8	11.2	24.2	38.0	26.2	9.8	13.0	15.2						
Le Moustier	14.4	9.4	8.4	21.0	14.2	21.4	29.8	11.4	4.0	13.0	24.2	12.6					
La Ferrassie 1	19.4	31.8	40.6	53.0	3.0	40.2	35.4	35.4	12.4	10.6	16.8	9.4	22.0				
Spy 1	15.0	25.0	35.0	42.2	10.2	26.0	23.3	33.8	12.0	8.8	19.6	10.8	19.6	11.0			
Spy 2	19.6	31.0	36.0	52.6	5.0	41.4	33.8	33.4	10.8	8.6	17.6	6.8	17.6	2.2	15.0		
Saccopastore	14.8	13.2	14.0	21.2	20.8	16.6	60.2	14.0	12.2	23.6	17.0	8.4	12.0	25.6	19.8	23.0	
Ehringsdorf	7.0	10.6	29.8	15.2	30.0	12.2	49.2	32.0	24.4	30.2	52.2	32.0	29.0	31.4	22.2	36.4	28.6

Assume that

a $\Sigma_1 = \ldots = \Sigma_k = I$,

b there are k_o populations (for example: Western European Neanderthals, Asiatic Homo erectus, Solo) to which a total of m specimens are assigned a priori.

Using the χ^2-test mentioned in section 3, it is tested which of the n'=n-m remaining specimens do not differ significantly at the 5% level from one of the k_o prior populations. These m'specimens are assigned to the population to which their Mahalanobis distance is smallest.

The remaining n" =n'-m' specimens are clustered by taking the following steps:

1. Test whether

$$S_1^2 = \sum_{i=1}^{n''} \sum_{j=1}^{p} (X_{ij}-\bar{X}_j)^2 > \chi^2_{(n''-1)p;\ \alpha=0.05}.$$

If this test is not significant, all remaining n" individuals are assumed to belong to one population.

2. If, however, the test is significant, then continue by calculating all $\binom{n''}{2}$ values for S_2^2, where

$$S_2^2 = \sum_{i=1}^{n_1''} \sum_{j=1}^{p} (x_i-\hat{\mu}_{a1})^2 + \sum_{i=1}^{n_2''} \sum_{j=1}^{p} (x_i-\hat{\mu}_{a2})^2.$$

3. Test whether the minimum value of $S_2^2 > \chi^2_{(n''-2)p;\alpha}$,

etc.

9. DISCUSSION

The primary purpose of this study was to test the robustness of the earlier finding that the well-known skulls of Petralona, Steinheim, and Broken Hill, while being mutually relatively close, are more distant from recent skulls than any other skulls in our material, including Homo erectus, Solo, and Neanderthal skulls. To that end, the stability of the distance pattern was examined in various ways, e.g. by trying various sets of different skull variables, and by modifying the underlying variance-covariance matrix. It was found that the extraordinary position of the three skulls was maintained under these changes. Only in one calculation the distances of the three skulls to recent ones were not obviously larger than those of Homo erectus skulls to recent ones. Thus, we may say that our first tentative results were confirmed by the present study.

These results clearly contradict existing viewpoints. This is the more striking since the other results of our calculations harmonize well with existing theory.

Our approach differs from common mathematical procedures in that we make use of much more information by incorporating a comprehensive data set of recent reference samples for estimating the variance covariance matrix. In Section 2 we showed that the latter is an essential addition. It was made clear that without it one tends to explain the error rather than biology if sample sizes are as small as in hominid investigation.

On comparison with the visualist's way of diagnosis the most essential feature of our mathematical technique is that it fully takes into account the correlations between the variables.

We think that the combination of these two features made it possible that biological structures were revealed which had remained unnoticed so far. The remarkable position of the Petralona, Steinheim, and Broken Hill skulls seems the most interesting finding obtained.

Although there is considerable argument on the exact phylogenetic positions of these three skulls, there is some kind of consensus that they have to be placed somewhere in between Homo erectus (Including Sinanthropus which Chinese scholars date between 500,000 and 400,000 years ago, and Pithecanthropus 2, which might very roughly be dated at one million years ago), the Neanderthals (ca. 100,000 - 35,000 years ago), and anatomically modern man (groups 1 and 2 of tables 1, 2, 4, and 5; possibly from 100,000 years ago onward). Apart from their morphological resemblance, phylogenetic relationships between the three skulls are supported by their datings: Broken Hill will be older than 100,000 years, but possibly not much; estimates for Steinheim range from 300,000 to 200,000 years old; according to a recent dating (16), Petralona is between 240,000 and 160,000 years old, although other datings exist. (For example, Ikeya mentioned during his communication at the 1982-EAA-congress an estimated age of this skull at between 350,000 and 250,000 years old, not to speak of datings by Poulianos who has in a number of publications (17) advocated a dating of more than 700,000 years old.)

If we ignore Poulianos' extreme dating for Petralona, the three skulls seem to take an intermediate position both according to visually determined morphological criteria, and on the timescale. Thus, in appearance they fit very well the evolutionary stage that one expects for that period. This would agree with the commonly accepted view that the evolution of the hominids, at least since the time of Pithecanthropus, took place along a single main line of development with possibly a few branches or off-shoots (18), though these are of minor significance for the overall picture. This view is obviously not supported by our calculations. We have seen that the Petralona, Steinheim, and Broken Hill skulls, though younger than the Asiatic Homo erectus finds, differ to a far greater degree from recent skulls than these Homo erectus specimens. Thus, it is very unlikely that the Broken Hill, Petralona and Steinheim skulls belong to the traditional line of development.

The most plausible explanation for the apparent contradiction between the visual and the mathematical observations is that we are dealing with two independent lineages which - and this conclusion is based on the general shape of the skulls and the relative size of the braincases - must have displayed a similar pattern of evolution, both in their rate of development and in the direction of their evolutionary change; two evolutionary lineages, therefore, which, at the time of the three skulls concerned, must have reached a comparable level of hominisation. This level, however, must have been based on two different genetic constitutions, as can be deduced from the large Mahalanobis distances between these skulls and the other groups. From a considerable body of multivariate statistical studies with craniometric data it has become evident that such distances must, to a large extent, reflect underlying genetic distances (see e.g. 19, 20, 21).

It is not yet possible to establish for how long the two supposed separate lines of development co-existed. As mentioned above, the oldest skull we investigated was the Pithecanthropus 2 skull. Since this c 1-million-year-old skull seems less different from recent skulls than the skulls of Broken Hill, Petralona, and Steinheim, it seems likely (although other models are theoretically conceivable) that the two separate lineages were already in existence at that time. This would imply that the common predecessor is even older. Further research on other skulls, and with other procedures (22), may enable us to clarify this point.

To summarize, we postulate that for a period of at least 1 million years, and up to fairly recent times, two independent human evolutionary lines have co-existed. The evolutionary development within these lines, moreover, must have been analogous, in view of the apparent correspondence in the ultimately achieved hominisation levels. One of the branches apparently became extinct some 100,000 years ago or later, or has provided only a minimal contribution to the Homo sapiens sapiens branch, for otherwise the mathematically established differences between the three above-mentioned representatives of the former branch and present-day human beings could not have been so great (23).

The phenomenon of "parallel" evolution is not unknown in evolutionary biology but, as Stanley (24) puts it, "it must be remembered that parallel directions of change are ones that we single out as curious". This seems to hold in particular for the form of parallelism under discussion, which seems to point to a more general tendency towards continuous hominisation, the driving force behind it not being directly clear. Consequently, if our hypothesis could be confirmed by further research, it might have an impact not only on present-day views on our own position in the evolutionary chain but also on views on the process of evolution as such, and thus lead to a better understanding of ourselves.

Acknowledgements

I should like to express my gratitude to Drs. W.W. Howells and W. Schaafsma for commenting on the original manuscript, to Drs. J.V. Jansen for devoting much of his spare time to the writing of the main computer programme, to Ing. P.G.M. van der Sman, Drs. Harmien Amesz-Voorhoeve, and to members of our computing team, Ir. W.G. Frederiks, Mr. J. Dijkema, Drs. S.K. Hazewindus, Ing. A. den Arend, and Mr. D. Kuizenga, for assistance with computing. Thanks are also due to Mrs. B.M. van der Meulen-Melrose, B.Sc., for the correction of the English text, to Mrs. G.T. Hoogenberg for typing the manuscript, and to Mr. T. Deddens for making the drawings.

References and notes

1 This has led to a rather negative opinion of many an anthropologist on the usefulness of mathematical multivariate techniques for hominid investigation.

As a significant example we may mention a paragraph from a letter that was sent to us by Professor Tobias some years ago: "Of course the computer has changed all that and further studies, both using D^2 and using Penrose and other approaches, are now commonplace. But the more recent studies on small groups of early hominid fossils have yielded fantastically diverse results".

2 G.N. van Vark, Some aspects of the reconstruction of human phylogeny with the aid of multivariate statistical methods. Proceedings of the Indian Statistical Institute Golden Jubilee Conference on Human Genetics and Adaptation, Calcutta, 1-5 February 1982. (1983).

3 G.N. van Vark, On the phylogenetic position of the Petralona skull. Ανθρωποζ, 9 (1983).

4 As early as 1974 an extensive survey of the literature concerned was published by Kanal (I. Kanal, Patterns in pattern recognition. IEEE Transac. Inform. Theory, IT. 20/6, 697 (1974).

5 W. Schaafsma & T. Steerneman, Classification and discrimination procedures when the number of features is unbounded. IEEE Transac. SMC 11/2, 144 (1981). These investigators showed that the diminishing performance is intrinsically caused by sampling phenomena: the vector of weights of the discriminant function is unreliable if the dimensionality is too large.

6 G.N. van Vark, A critical evaluation of the application of multivariate statistical methods to the study of human populations from their skeletal remains. Homo, 27/2, 94 (1976).

In this paper also the problem of the ordering of variables is dealt with. This problem is of no less practical importance than is variable selection. Before applying a variable selection procedure, one shall wish to order the variables according to their independent contribution to discrimination. As may be taken from the above, it is most important to incorporate as much diagnostic capacity in as few a number of variables as is practically possible, since the role of adverse sampling effects is then minimized.

7 P.A. Lachenbruch, An almost unbiased method of obtaining confidence intervals for the probabilities of misclassification in discriminant analysis. Biometrics, 23, 639 (1967).

8 Further argumentation is to be found in:
G.N. van Vark & P.G. Van der Sman, New discrimination and classification techniques in anthropological practice. Zeitschr. Morph. Anthrop., 73/1, 21 (1982).

9 According to M. Abercombie et al., A Dictionary of biology, Penguin Books (1961), by a subspecies is meant a "subdivision of a species forming a group whose members resemble each other in certain characteristics, and differ from other members of the species, though there may be no sharp dividing line".
The total present human world population may be regarded as the present representation of the subspecies Homo sapiens sapiens which may be somewhere around 100,000 years old. See also Section 8.

10 These measurements were taken from the series described in: W.W. Howells, Cranial variation in man. (Pap. Peabody Mus., 67, Harvard Univ., Cambridge, Mass., 1973), as well as from other series which were measured later by this author.

11 G.N. van Vark, The investigation of human cremated skeletal material, Part 1. OSSA, 1, 63 (1974).

12 During the "1er Congrès International de Paléontologie Humaine", held in Nice, on October 1982, Dr. Alfred Czarnetzki (University of Tübingen, West Germany) told the author that the skull was being restored.

13 T.W. Anderson, An introduction to multivariate analysis, Wiley (1958).

14 N.A. Campbell, Cranial variate analysis: some practical aspects. Thesis, Univ. of London (1979).

15 W.W. Howells, Current theories on the origin of Homo sapiens. In: D. Ferembach, ed. Les processus de l'hominisation. Colloques internationaux du C.N.R.S., 599, 73 (1980).

16 G.J. Hennig, et al., ESR-dating of the fossil hominid cranium of Petralona Cave, Greece. Nature, 292, 533 (1981).

17 The last one known to us is his brief note in Nature, 299, of 16 September 1982: A.N. Poulianos, Petralona Cave dating controversy. In this opinion, Poulianos is supported by B. Kurten.

18 Well-known candidates for a side branche are, for example, the Neanderthals. Populations of anatomically modern man may have arisen from a strictly Neanderthal population or a non-Neanderthal one. (See e.g. E. Trinkaus & W.W. Howells, The Neanderthals. Scient. American, 241/6, 118 (1979). Our figures rather seem to favour the latter viewpoint.

19 R. Mukerjee, C.R. Rao & J.C. Trevor, The ancient inhabitants of Jebel Moya Sudan. Cambridge Univ. Press, Cambridge (1955).

20 J.M. Chrichton, A multiple discriminant analysis of Egyptian and African Negro crania. Pap. Peabody Mus., 57, Harvard Univ., Cambridge, Mass. (1966).

21 G.P. Rightmire, Cranial measurements and discrete traits compared in distance studies of African Negro skulls. Hum. Biology, 44/2, 263 (1972).

22 E.g. recently (July 1983), Professor Rao told the author that he had some ideas for the further testing of our hypotheses.

23 It may be difficult to imagine two isolated human evolutionary lines which have developed in an analogous way.
Professor Oxnard (personal communication) suggested the following possible explanation:
"We usually think of human cultures and societies as allowing social intercourse, communication, trading and therefore considerable intermarriage between populations. This means that parallel and separately evolving populations are harder to envisage for humans than for non-human forms. However, though this may well be the case in the somewhat late stages of cultural and social evolution, the very earliest stages may result in an opposite phenomenon. In the earliest stages part of the adaptive value of social and cultural evolution may be in the protection of small populations from various environmental dangers; this protection could well include, of course, those from neighbouring human groups; these earliest stages therefore could sometimes result in isolation".

24 S.M. Stanley, Macroevolution, Pattern and Process. Freeman, San Francisco (1979).

MULTIVARIATE ANALYSIS AND CRANIAL DIVERSITY IN PLIO-PLEISTOCENE HOMINIDS

Alan Bilsborough

Department of Physical Anthropology, Downing Street, Cambridge, U.K.

This paper briefly reports on some multivariate aspects of a morphometric study of Plio - Pleistocene hominids. The full study includes univariate and bivariate, as well as multivariate, analysis and is directed towards the investigation of patterns of morphological diversity in hominid crania, with a particular focus on the variation displayed by Pliocene and Lower Pleistocene hominids from South and East Africa. The approach reported here utilises multivariate techniques as aids to map and quantify morphological differences between hominid groups, and to scale relative change in different cranial complexes, rather than attempting to derive clusters (phena) from the multivariate structure. Some aspects of the results are discussed below, following a summary of the hominid material incorporated in the analyses, and its chronological and phyletic framework.

HOMINID SPECIMENS

The included specimens span the period c. 3.0 m.y. ago onwards (i.e. later Pliocene-present). They represent the generally recognised species within the genera _Australopithecus_ and _Homo_, as well as specimens whose affinities are uncertain. Several samples of _Australopithecus_ are incorporated in the analyses: the gracile species _A. africanus_, regarded by many workers as ancestral to _Homo_, is represented by specimens from Sterkfontein and Makapansgat, South Africa, dated at 2.5 - 3.0 m.y. Robust australopithecines include a sample of South African _A. robustus_ from Swartkrans and Kromdraai dated at 1.5-2.0 m.y. and the contemporaneous but generally more massive _A. boisei_ from East Africa. Current palaeoanthropological concensus places these robust forms on a separate clade from that containing the genus _Homo_.

G. N. van Vark and W. W. Howells (eds.), Multivariate Statistical Methods in Physical Anthropology, 351–375.

Gracile hominids from the Lower Pleistocene of East and South Africa are generally regarded as morphologically more advanced that *A. africanus*, and are therefore usually included within *Homo*. These include a sample of *Homo habilis* (OH7, OH13, OH16, OH24) from Beds I and II Olduvai Gorge, Tanzania (?1.5-1.8 m.y.) and from South Africa, the possibly conspecific specimens SK 847 from Swartkrans (Clarke and Howell 1972) and Stw 53 from Sterkfontein member 5 (Tobias 1978), estimated on faunal evidence to be 1.5-2.0 m.y.

H. habilis is also possibly represented within the material from Koobi Fora, Lake Turkana (formerly Lake Rudolf) Kenya, which taken overall, exhibits remarkable morphological diversity. Small lightly built crania such as KNM ER 1813 appear to bear closest resemblance to the Olduvai specimens; other Koobi Fora crania (eg KNM ER 1805 and 1470) are larger and more rugged than the above material. These specimens have yet to be formally assigned to hominid species, and although included within *H. habilis* by some workers, they appear to differ significantly from the Olduvai material upon which the taxon is based, so that their affinities are better regarded as uncertain.

Other, less complete, specimens from Koobi Fora include KNM ER 407 and 732. The former specimen is a neuro cranium originally assigned to *Homo* (Leakey 1970) although more recent studies suggest australopithecine affinities (Leakey 1976; Wood 1978). ER 732 is a right hemicranium, widely regarded as a female *A. boisei* (eg. Howell 1978; Wood 1978) although some workers (eg. Wolpoff 1978) have drawn attention to similarities with gracile specimens.

The Koobi Fora hominids are dated by their stratigraphic relationships to a series of volcanic tuffs, the primary marker being the KBS tuff. Despite initial claims for a date of 2.61 m.y. (Fitch and Miller 1970) there is now a compelling body of evidence that the tuff is c. 1.8. m.y. old. Specimens from below the KBS tuff (eg. 1470, 1590) are therefore likely to be c. 1.8.-2.0 m.y., those from above the tuff (the majority) c. 1.3-1.8 m.y. depending upon their stratigraphic context, and are therefore broadly contemporary with the Olduvai Bed I / II material.

The analysis included *Homo erectus* samples differing widely in space and time. Early (c. 1.2.-1.5. m.y.) East African *erectus* is represented by specimens from Upper Bed II Olduvai (OH 9) and Koobi Fora (KNM - ER 3733, 3883), and S.E. Asian *erectus* by specimens from Trinil and Sangiran, Java. The crania from Choukoutien (Pekin) represent a later Asian *erectus* population, probably c. .5-.6 m.y. old. Several Upper Pleistocene samples (i.e. < .15 m.y.) are included in the analyses to provide a scale against which to calibrate the earlier diversity. *H. sapiens* groups include European Neanderthal and Upper Palaeolithic crania, as well as a sample of modern H. sapiens sapiens. These are separated by intervals of c. 25-40,000 years and thus give some indication of the amplitude of short term morphological fluctations within a single evolving hominid lineage.

METHODS AND TECHNIQUES

As an aid to investigating patterns of morphological diversity I have divided the skull into a number of functional regions and attempted to describe the 'total morphological pattern' of each by appropriate dimensions. The regions are (numbers of characters in brackets): upper face (20); upper jaw (11); mandible (15); cheek region and masticatory musculature (14); articular region (7); balance (6); basicranium (5); cranial vault (14). Additional details of these are given in Bilsborough (1976, 1978); the characters used in the previous studies were modified to take account of the extent of preservation and morphological features of the South and East African fossil material, so that whilst the limits of the regions remain unchanged, some of the individual characters contributing to a particular functional complex will differ from those used in the earlier studies. See also fig. 1.

Since virtually all early hominid specimens consist of unassociated crania or mandibles, this paper is confined to a consideration of aspects of cranial diversity; the results of the analyses of mandibular dimensions will be reported elsewhere.

The original data were analysed by a GENSTAT program including both Principal Components and Canonical Variate analyses; this report deals primarily with some aspects of the Q-mode CVA (Gower 1966), and further techniques derived from them. Distances between the hominid groups are expressed as D^2 values, and the canonical variates provide information on the directions of the dispersion and, through examination of the scaled loadings, the contribution of the original characters to the achieved separation. Some workers (e.g. Howells 1973) regard the use of scaled loadings for this purpose as unsatisfactory; however, I have compared the orderings of the groups on the major canonical variates based on all characters with that achieved by recalculation using only those characters with the 5 highest scaled loadings. The correspondence between the two is very striking, confirming that those variables included within the sub-set are contributing maximally to the separation achieved by the full suite of characters. See fig. 2.

The use of a reduced set of characters in this way also allows the interpolation of incomplete or distorted specimens that could not be incorporated into the full analysis - an extremely useful procedure given the fragmentary nature of the fossil record. The assumptions (and consequent dangers) of interpolation are well known: interpolated specimens have to assume the statistical

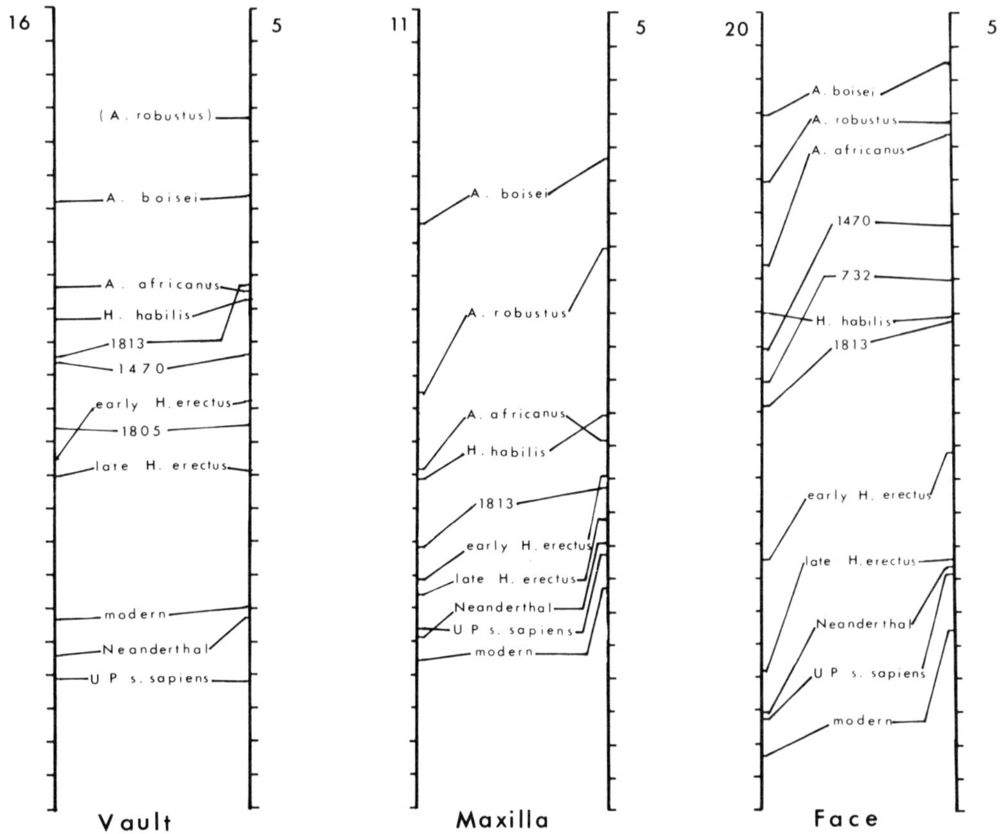

Figure 1. Canonical variate analysis. Ordinations of groups on first canonical variate based on (i) all characters (ii) characters with five highest scaled loadings for vault, maxilla and face. ( ) indicates interpolated group.

characteristics of those groups upon which the covariance matrix is based (Corruccini 1975, Campbell 1978), and the ordering so derived gives only a limited view of the specimens in multivariate space (Oxnard 1972, 1975). The complete relationship can only be elucidated by examining all canonical variates or by comparing D^2 distances, but this is, of course, impossible with incomplete material.

When data are limited, as in palaeoanthropology, it is important to derive maximal information from them, and, used judiciously, interpolation can provide valuable insights into morphological relationships. For example, S. African A. robustus could not be included in the full cranial vault analysis, since no robustus specimens possess a complete, undistorted vault. However, it is possible to obtain measurements on those 5 characters contributing maximally to the first canonical variate, and it is therefore interpolated onto that axis. Its position is close to A. boisei specimens, emphasising the basic similarity of South and East African robust australopithecines in the midvault region, and in coronal development of the frontal, as well as providing an otherwise unobtainable measure of robustus' similarity to the other hominid groups.

Most morphological contrasts will involve both size and shape differences. The interaction of these effects, and their partitioning in multivariate analysis, is a vexed issue. Several authors have emphasised that shape differences are generally of more significance than size for taxonomic studies; moreover size differences will themselves almost invariably produce shape changes due to allometric effects. Some workers (e.g. Corruccini 1975) give the impression of considering shape-based studies to be inherently superior to those which do not formally attempt to partition the differences into size and shape components. Contra this view, I consider that multivariate techniques which include variation in both size and shape are a valuable aid to analysis in hominid studies where the size differences between specimens are less than in some other applications (many primate investigations for example), and where the intention is to map morphological change rather than to facilitate taxonomic assessment. The increase in body size between early Australopithecus populations and later Homo is something real which, unless our phyletic schemes are totally erroneous, actually happened, and it is useful to have a measure of that transformation (see also Blackith and Reyment 1971 p. 29). Ideally the two approaches should be complementary, and the study of shape differences, as well as total (size and shape) variation should prove mutually illuminating.

Campbell (1978) points out that the first axis in a principal components study is very often primarily influenced by size differences and the remaining components (or at least the larger of these) represent shape differences, whereas this is rarely the

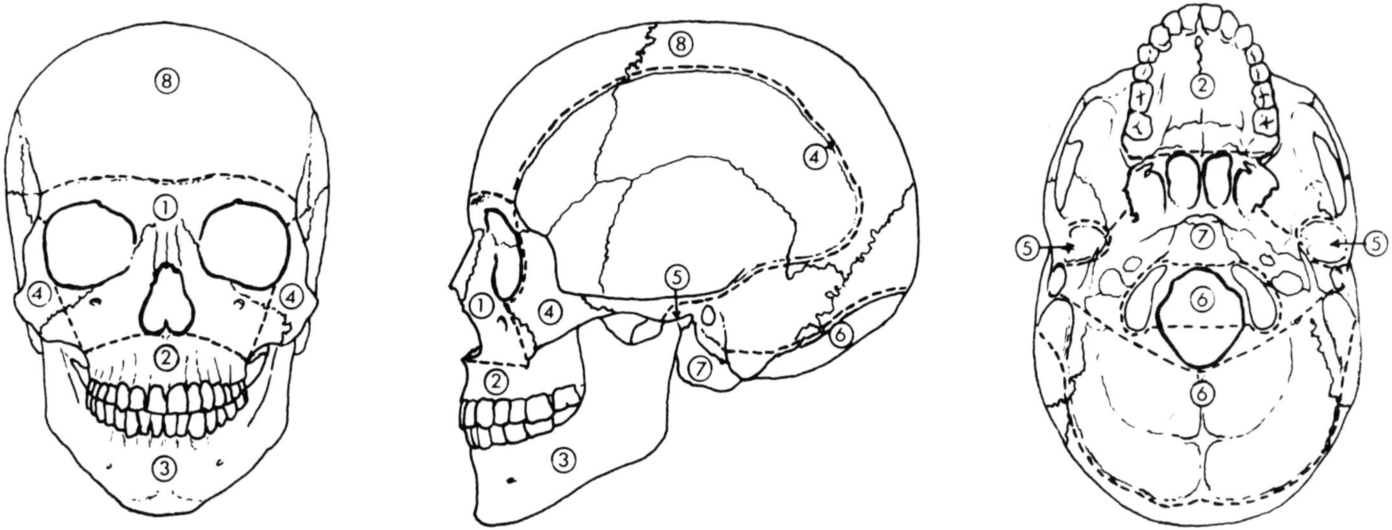

Figure 2. Character complexes (other than mandible) used in this study. 1=Face; 2=maxilla; 3=mandible; 4=cheek region and masticatory musculature; 5=articular; 6=balance and nuchal musculature; 7=basicranium; 8=cranial vault.

case in discriminant analysis. My own experience certainly supports this generalisation: in several series of investigations of hominid cranial data, each comprising analyses of eight distinct cranial regions, the loadings on the first principal component are always of the same sign, and the specimens arranged in an ordering corresponding to the general size of the structure concerned, whereas the other components include loadings of different sign and order on shape. Moreover, the first canonical variate (which usually accounts for a greater proportion of the total variance than the first p.c.) always contains loadings of different sign, and is discriminating on the basis of size _and_ shape.

The specimens may be standardised to a common size; such a procedure is, in a sense, the obverse of an allometric study - it 'freezes' shape whilst removing perturbations due to size differences. When this is done before P.C.A. the resulting first principal component as expected, now contains loadings of mixed sign and the ordering reflects shape differences. Other transformations of the original data may affect the relative contributions of size and shape differences to the analyses. Logged data produce an ordination of groups generally similar to that of the untransformed variables, and there is little difference in the relative contribution of these to the separation achieved. However, the logged data had the general effects of (1) slightly reducing the separation achieved between later hominid groups (2) increasing the diversity obtained between the early hominid specimens and (3) "stretching" the separation so that the earlier groups are further from the later ones than in the original analyses. In chronological/phyletic terms the "stretching most affects those hominids ante-dating the early Middle Pleistocene suggesting that there is greater shape heterogeneity among these specimens than among later Pleistocene hominids; unless noted otherwise, the results below, and the discussion based upon them, are derived from the analyses of the original (untransformed) data. Fuller consideration of the changing importance of size and shape differences in hominid phylogeny will be reported elsewhere.

RESULTS

Some results of the analyses are summarised as D^2 values in table 1. These reveal that gracile hominids (_A. africanus_, Olduvai _habilis_ and ER 1813) form a cluster so that for almost all cranial regions they are closer to one another than to other groups. The East African specimens are, however, closer than _A. africanus_ to later, undoubted members of the genus _Homo_, particularly in facial, maxillary and vault characteristics, and morphometric study thus supports the view that they represent morphologically more advanced hominid populations.

By contrast, robust australopithecines are widely divergent from the gracile forms and from later _Homo_, especially in those regions

Table 1. D^2 values between hominid groups

	Upper Face	Maxilla	Cheek/ Masticatory	Articular	Balance	Basi-Cranium	Cranial Vault
A. africanus-boisei	13.8	9.6	14.7	8.4	3.4	5.8	8.6
A. boisei-habilis	14.3	9.4	-	10.3	7.6	8.0	8.3
A. boisei-early erectus	18.5	17.0	-	8.5	8.0	7.5	12.0
A. boisei-late erectus	19.3	14.2	14.2	12.3	7.5	9.3	11.2
A. boisei-Neanderthal	20.0	17.2	14.6	10.2	8.0	8.2	16.2
A. boisei-modern	21.0	17.5	14.5	8.6	10.0	7.3	14.2
A. africanus-H. habilis	6.7	4.3	-	3.4	7.2	4.7	5.3
A. africanus-ER 1813	7.7	6.2	-	2.9	2.8	2.5	6.1
H. habilis-ER 1813	6.5	3.4	-	2.3	5.6	3.2	5.9
A. africanus-ER 1470	11.9	-	-	2.6	-	-	6.8
H. habilis-ER 1470	13.6	-	-	2.4	-	-	7.3
Early *erectus*-ER 1470	12.2	-	-	3.9	-	-	5.6
Late *erectus*-ER 1470	12.6	-	-	5.6	-	-	4.9
H. habilis-early *erectus*	14.2	5.6	-	4.7	7.5	3.4	7.7
ER 1813-early *erectus*	13.7	7.1	-	2.7	4.1	5.3	7.9
A. africanus-early *erectus*	16.7	8.8	-	4.2	4.3	6.8	9.4
Early-late *erectus*	9.7	7.9	-	7.4	3.8	5.0	5.6
Late *erectus*-Neanderthal	7.6	4.2	4.6	2.9	2.9	3.6	6.4
Late *erectus*-modern	6.4	4.1	6.6	4.3	4.3	5.0	6.6
Neanderthal-modern	9.5	3.6	6.3	2.7	5.3	3.0	6.0
Neanderthal-Upper Pal *sapiens*	9.6	3.3	5.7	1.7	3.7	1.9	5.7
Upper Pal *sapiens*-modern	4.8	2.6	6.1	3.1	3.1	1.7	3.7

primarily determined by masticatory activity (face, jaw, cheek, and articular complexes). Aga in this accords with those morphological studies which recognise distinctive splanchnocranial and dental characteristics in the robust australopithecines, that differentiate them from other Plio-Pleistocene Hominidae. The facial analysis, for example, discriminates principally upon the degree of prognathism and facial profile, the placement and insertion of the zygomatic process, and breadth in the toral/orbital region. Detailed proportions of the upper (supraorbital and orbital) part of the face are heavily weighted on variate II, those of the middle face (nasal region) on axis III, with the subnasal region contributing significantly to only the III and IV axes (6% and 4% of variance). The analysis isolates the robust australopithecines with their marked glabellar development, elliptical orbits, depressed nasal area and massive, anteriorly inferiorly inclined zygomatic processes and also separates other early hominids from later ones (H. erectus onwards) because of size and shape differences in middle/upper facial proportions. The gracile specimens assigned to early Homo differ from A. africanus principally in their reduced, more orthognathic middle face with more salient nasal bones and retracted zygomatic processes, and in their shorter premaxilla. They are generally closer to H. erectus except an axis II, where contrasts in the upper face, especially the more pronounced supraorbital torus and greater orbital and interorbital breadths of the latter taxon result in a greater degree of separation than on other variates.

The maxillary analysis distinguishes primarily on the length of the various tooth chords, with incisor and premolar segments having especially high weightings. East African robust australopithecines with massive cheek teeth are therefore well separated from other hominids, whilst the relatively large incisors and canines and moderately - sized cheek teeth of early Homo specimens distinguish these from A. africanus where the proportions are reversed. Axis II and III, discriminating upon the development of the anterior alveolar incisal region and anterior and posterior palatal breadths, reinforce separation between the long, relatively narrow australopithecine palates, and the shorter, posteriorly broader human ones.

The Lower Pleistocene gracile specimens occupy positions that are in many respects intermediate between A. africanus and Middle/Upper Pleistocene Homo. Check tooth dimensions and palatal breadth are characters in which the material resembles later members of the genus Homo, whilst curvature of the incisor row, combined with a relatively orthognathous premaxillary region (heavily weighted on axis III) associate the habilis material with erectus, and separate it from africanus which clusters with other australopithecine specimens ot one end of the variate. Overall, the non-robust Lower Pleistocene specimens reveal significant differences from Australopithecus, and approach human palatal proportions.

The masticatory complex includes characters describing the zygomatic region, extent of m. *temporalis* and the pterygoid plates. Only a few specimens preserve the cheek region intact, so that data for this complex are limited. Separation is based primarily upon the orientation of the zygomatic process in relation to the maxilla, the length of the zygomatic arch, depth of the infra temporal fossa and breadth of the pterygoid plate. *Australopithecus boisei* specimens are again isolated from other hominids as a result of their massive, anteriorly inferiorly sloping juga, deep, widely - flaring zygomatic arches and large pterygoids. Gracile australopithecines lie closer to members of the genus *Homo*, but are still separated from them by their longer zygomatic arches. In this region and the upper face, Neanderthals, which in other respects lie closest to morphologically modern groups, lie closer to late *erectus* specimens. This pattern accords with the generally recognised distinctive morphology of the Neanderthal face, with an inflated anteriorly positioned midfacial region, resulting in a long zygomatic arch.

The articular region reveals most variation in the lateral width of the fossa, and its depth anteriorly and medially. Three clusters emerge from the analysis: robust australopithecines with massive fossae; early gracile hominids (*A. africanus*, *H. habilis* and similar) and basal *erectus* with coronally narrower but sagitally open fossae, and Middle/Upper Pleistocene groups (late *erectus* onwards) in which the fossa is contracted both laterally and sagittally.

The balance complex distinguishes principally upon the proportions of the head lever (pre and post condylar segments) and size of the mastoid process; subsidiary discriminants are the vertical and lateral extent of the nuchal musculature. Although later groups (*erectus* and *sapiens*) show clear clustering, the spread among earlier hominids does not reveal any obvious chronological or phyletic pattern. This lack of clear separation among the earlier taxa suggests considerable overlap in the metrical characters chosen to describe this complex (see below).

Lateral development of the basicranium, especially as measured by biporionic and external biglenoid diameters differentiates the groups. Robust australopithecines, where the articular fossae extend beyond the neurocranium proper onto a lateral extension of the zygomatic process of the temporal, are again isolated; *A. africanus* is also separated from a cluster of *H. erectus/sapiens* where concurrent coronal expansion of the neurocranium and contraction of bicondylar diameter produce a distinctive pattern. The gracile East African hominids are intermediate in position between *A. africanus* and the *erectus/sapiens* cluster.

In the analysis of vault dimensions the first four axes, accounting for 92% of variance, discriminate primarily upon coronal development of the vault in the tempero-parietal regions (axis 1), together with sagittal and coronal development of: the parietals (II), frontal (III), and occipital (IV). *Australopithecus boisei* crania are maximally separated from modern man, with *A. africanus* also well away from *H. erectus* and later specimens. Divergence between gracile and robust australopithecines is considerably less, relative to total dispersion, than in many other regions, supporting the view that the major contrasts between australopithecine taxa are in the splanchnocranium, and that when the robust, forms are shorn of their 'ectocranial embellishments', (Tobias) australopithecine specimens are neurocranially similar.
A. africanus is adjacent to *H. habilis* and ER 1813, which differ from it principally in their greater degree of coronal expansion, especial in the parietal region. However, even in this region - one traditionally weighted in definitions of the two hominid genera - the divergence is less than that within the genus *Australopithecus* itself. *H. erectus* specimens form a broad cluster with Neanderthals equidistant from the later *erectus* specimens and *H. sapiens*. The earlier *erectus* crania are widely separated morphologically from the *habilis*/ER 1813 regions, even with logged data, suggesting the existence of two distinct non-robust cranial morphologies in the Lower Pleistocene. The available stratigraphic evidence from areas such as Olduvai and Koobi Fora indicates that examples of each morphology are essentially contemporaneous (at least within the range of error associated with dating specimens for this period) making it unlikely that one is ancestral to the other. The distinctiveness of the two sets also makes it improbable that the variation they display may easily be accomodated within a single population, leading inevitably to the conclusion that two taxa are represented at these sites (see below).

There is also a marked separation between early and later *erectus*, with the later crania generally closer to the *sapiens* specimens than to the earlier *erectus* material. In this case the temporal separation of the specimens allows the positing of an ancestor - descendant sequence rather than two distinct lineages. Further support for this interpretation is provided by the observation that the contrasts between the *erectus* samples - progressive expansion and kyphosis of the neurocranium, diminution of face, jaws and dentition - are features that also differentiate the *sapiens* samples, and which represent long term evolutionary trends within Pleistocene hominids. They also illustrate the major morphological changes that can exist within chronospecies, and which may be obscured by conventional Linnean nomenclature (Bilsborough 1976). Similarly, although at a lower phyletic and taxonomic level, the separation between the Upper Palaeolithic and modern groups - both referred to *H.s. sapiens* - gives some indication of the morphological "noise" generated between different populations of the same sub-species separated by relatively short time intervals (25.000 years).

On a larger phylogenetic scale, the results quantify the distinctiveness of the robust australopithecine clade. A. boisei, H. habilis and H. erectus are all approximately contemporaneous, but of these it is evident that H. habilis most closely resembles the possibly ancestral morphotype of africanus, some 1.25-1.5 m.y. earlier. A. boisei and H. erectus are roughly equidistant from africanus, but their divergence from each other emphasises that there is no correspondence in the morphological transformations that distinguish them from the earlier species. Indeed, the distances between them are greater than those between the africanus morphology and that of any of the posited descendant groups, and there is a progressive increase in the degree of divergence between robust australopithecines and later groups within the human lineage. Overall the separation is greater than that obtaining within Homo over a span of 1.5-1.75 m.y., even when the basal threshold of the genus is lowered to include the habilis material.

The data are thus consistent with the recognition of two lineages - robust australopithecine and advanced gracile hominids (habilis) respectively - derived from an earlier gracile australopithecine morphology, and a further bifurcation between habilis and erectus morphotypes.

PHYLETIC SCALING

There has, of course, been much debate about the phyletic significance of the morphological diversity represented within the later Pliocene/Lower Pleistocene hominid material, and a range of interpretations have been offered: compare, for example, Howell (1978); Walker and Leakey (1978) and Wood (1978). One advantage of multivariate techniques is that they provide a quantitative estimate of resemblance, thereby enabling one to scale subjective impressions of morphological resemblance or dissimilarity, and to compare patterns of diversity in groups from different time ranges. In this way it is possible to 'measure' the morphological implications of a given taxonomic or phyletic scheme.

For example, it is possible to compare the diversity within the problematic early hominid material with that exhibited by later Pleistocene human groups where there is a more substantial body of data and consequently a greater degree of concensus about systematic and phyletic interrelationships. As an instance of this approach I have compared the variation exhibited by early hominids with that displayed between other human groups separated in time but conventionally assigned to the same species - in this case Neanderthal and modern man.

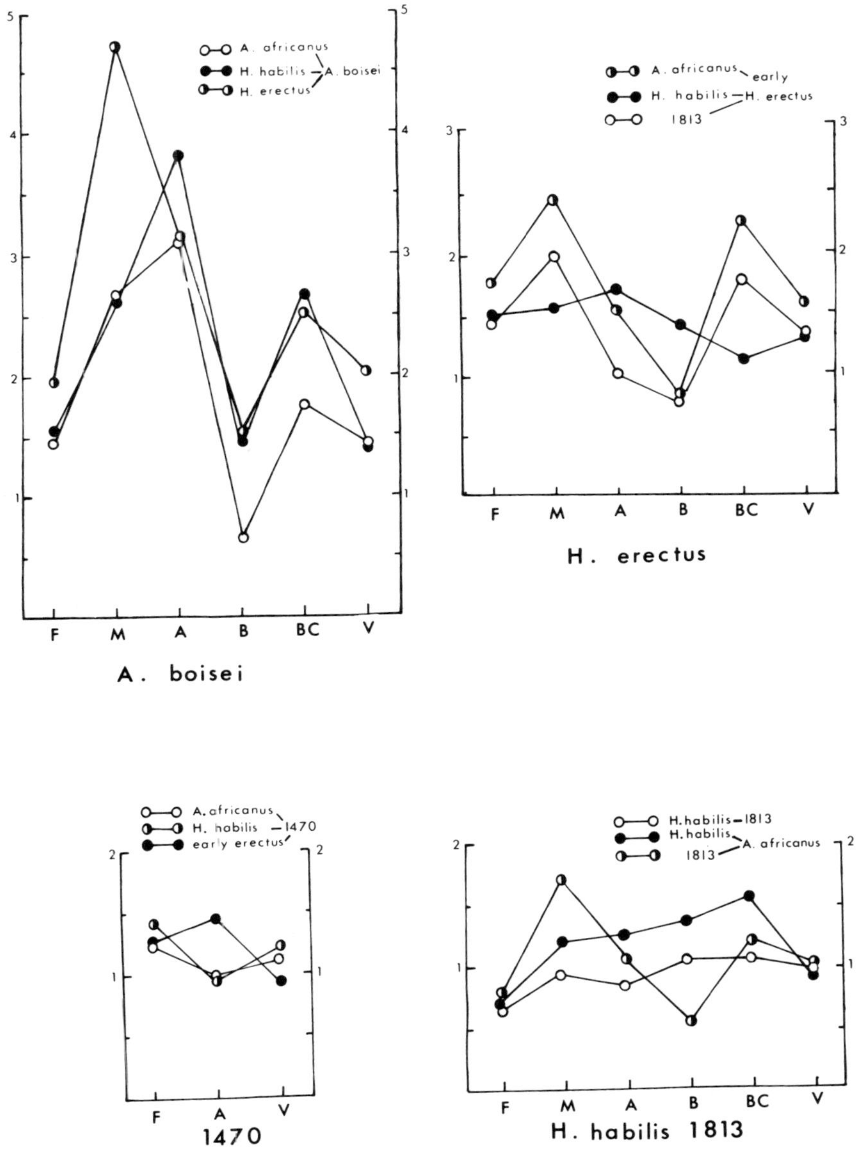

Figure 3. Distances between selected groups scaled by Neanderthal-modern distances. F=face; M=maxilla; A=articular; B=balance; BC=basicranium; V-vault. Vertical scale in appropriate Neanderthal-modern distances for each complex.

There is a substantial body of agreement that the Neanderthals, whilst possessing a characteristic cranial morphology which distinguishes them from H. sapiens sapiens, were part of the same species as the earliest representatives of anatomically modern man, and thus shared a common gene pool with them. The arguments for this conclusion are admittedly inferential, but are based upon varied non-morphological (archaeological, stratigraphic, ethnographic etc.) as well as morphological evidence, which is thus at least partly independent of the data summarised here. Moreover, the Neanderthals were relatively localised geographically and are separated from modern man by c. 50.000-70.000 years - the same order of time interval, given errors of dating, possibly separating some of the apparently 'contemporaneous' Lower Pleistocene hominid fossils. The Neanderthal - modern divergence thus provides an approximate measure of the range of morphometric variation which may be accommodated within a single clade over such a time interval as a result of local population differentiation or short-term oscillation without necessarily indicating the presence of multiple lineages.

This approach is essentially the reverse of that which uses multivariate clusters to recognise groupings de novo; instead it imposes a taxonomic framework upon the multivariate dispersion by utilising the divergence exhibited by the generally accepted later human taxa as a metric against which to scale dispersion in earlier, more contentious, hominid groups. When this is done, and the dispersion of early hominids for different cranial regions is scaled by the corresponding values between Neanderthals and modern man, some interesting patterns emerge. See fig. 3.

Homo habilis specimens and KNM - ER 1813 are very similar, and should probably be considered conspecific, and perhaps even subspecifically identical. Both groups show only slightly greater separation from the much earlier A. africanus, with dispersion for all cranial regions save the upper jaw falling within the range exhibited between Neanderthal and modern samples. This morphological stability is remarkable, given the long time interval between them, and the East African fossils are best thought of phenetically as essentially advanced gracile australopithecines, differing from the South African forms in the functionally important jaw region. Whether this divergence should be formalised taxonomically at the generic or specific level depends upon the phyletic weighting accorded to the masticatory differences, and the relevance of other evidence (postcranial, archaeological) extraneous to this analysis.

However, the habilis/ER 1813 group is very different from the early erectus sample (see above), not only exceeding the separation between Neanderthals and modern, but even that between later erectus and modern man. The magnitude of the divergence thus strongly supports the notion of two groups of hominids (apart from A. boisei) in the East African Lower Pleistocene (see above); even if the larger, more rugged KNM - ER 1805 is included within the habilis group the separation between that and early erectus still surpasses the Neanderthal - modern and late erectus - modern distances. On purely phenetic considerations therefore, this morphometric "gap" would appear to be the most appropriate place at which to draw the Australopithecus- Homo interface.

The morphology of KNM - ER 1470 which has been assigned generically to Homo, but not specifically, is distinctive. It is well separated from most other groups, but is generally closest to erectus in neurocranial features. Whilst falling below erectus minima in the coronal dimensions of the mid and rear vault, it resembles erectus crania in the parietal region and in sagittal dimensions of the occipital, and may represent an early phase in the differentiation by .3 - .5 m.y.). It appears similarly comparable to erectus in maxillary characteristics; incomplete preservation of the rear of the palate prevents 1470 being included in the full analysis of maxillary characteristics, but it is possible to interpolate its position on the first axis of reduced dimensions, where it lies closed to H. erectus specimens.

ER 1470 is, however, widely divergent from erectus in facial proportions; indeed its facial morphology appears unique, being well separated from all other early hominid groups. The upper face is a region in which Neanderthal and modern specimens also show marked contrasts; the 1470 - erectus separation comfortably exceeds that shown between any H. sapiens groups, and also that between erectus and modern man, reflecting striking dissimilarity. Further material is needed to evaluate the significance of this; whilst it is unwise to create new taxa on the basis of limited anatomical differentiation in unique specimens (see below), the magnitude in this case is such that, if confirmed by further discovery, specific distinctiveness from other Lower Pleistocene groups would be justified on phenetic criteria.

The above proposals about likely taxonomic affinities are intended as provisional hypotheses, not definitive statements. Strict correspondence between taxa is not to be expected since the mosaic nature of phyletic change means that their morphological limits will differ, and their recognition and arrangement within an evolutionary framework usually involves phyletic weighting as well as the measurement of phenetic affinity. However, within a relatively restricted group such as Hominidae some degree of uniformity of taxa is desirable, and it is illuminating, when formulating phyletic models to account for the observed diversity of the limited early hominid material, to assess their implications against the more complete later record. Multivariate techniques provide the means of achieving these aims.

MORPHOLOGICAL INTERACTION

Following the individual analyses, the above summary treats each cranial complex separately, and does not attempt to integrate them. Changes are expressed in terms of the average within-group variation for each region - an appropriate metric for evolutionary studies - and the D^2 values summarise the resultant patterning, but they are not combined into any measure of overall morphological resemblance.

Gower and Digby (this volume) summarise some methods for combining separate analyses; here I briefly discuss the application of one such technique - Generalised Procrustes Analysis (G.P.A.) - to the above data so that the interactions of the different regions may be graphed as an aid to investigation and analysis. G.P.A. takes the positions of the groups for each complex and rotates, translates and isotropically expands them about a common origin so that for each group the distances of the complexes from the group centroid are minimised relative to the distances between the groups. The procedure thus obtains the 'best fit' for each group over all complexes thereby locating the group centroids, the distances between which are measures of overall morphological difference on the basis of all included complexes. The initial and transformed position of groups may be examined for a given region, or the positions of the centroids, and of different regions relative to these, may be studied. Information from each region contributes equally to the analysis, and comparison of initial and final configurations for individual complexes will reveal the extent to which clusterings are influenced by regions other than the one under investigation - in effect a measure of morphological interaction or inertia. Similarity in the two configurations indicates congruence between that region and most others under investigation, whereas a marked transformation reveals a morphologically discrepant region, although further analysis will be necessary to determine whether this is due to exceptionally rapid or very slow morphological change. A direct measure of the morphological patterning of any group is provided by the residuals of the individual complexes from the centroid; those close to it are evolving at near modal rates, those more distant at atypical velocities.

In the present study most cranial regions regularly cluster about the centroids (Gower and Digby section 3: p. 27); this is perhaps not surprising in a structure as functionally integrated as the hominid cranium where the limits of morphological change in each region are generally circumscribed within tight limits by the interdependence of all regions. Only the balance complex appears discrepant in this respect (see above), probably because it is functionally associated with, and determined by, the primary hominid adaption of truncal erectness and bipedal locomotion (common to all the specimens included here) and as such is little influenced by changes in other cranial regions. Figures 4 and 5 show the initial and final configurations of the groups for the upper face, maxilla and cranial vault data, and their relationship to the group centroids. The initial ordinations are as on the first three canonical variates, accounting for 88-91.5% of the total variance in the three regions, and thus broadly correspond to the relative spacings revealed by the appropriate columns of table 1. In the upper face three clusters are apparent - *H. erectus/capiens* with Neanderthals as outliers, *A. africanus* and *H. habilis*, and *A. boisei*, whilst in the maxilla there is a tight cluster of later *erectus/sapiens* and grouping of *habilis* fossils, but virtual isolation of early *erectus*, and wide dispersion of *Australopithecus* specimens. Only two groupings are initially apparent on the cranial vault data - a rather loose *erectus/sapiens* one, and a tighter grouping of small brained hominids made up of the australopithecine specimens and *H. habilis*.

Procrustes transformation produces some interesting shifts to these patterns. The facial data now reveal a closer clustering of *Homo*, with Neanderthals lying closer to other later groups, and *erectus* fossils closer to *sapiens*. Separation between *Australopithecus* and *Homo* is increased, with greater dispersion of the early gracile material, and also greater separation of the *A. boisei* specimens from *africanus*. The cranial vault analysis similarly shows a tighter clustering of later *Homo*, but enhanced separation of this cluster from *H. erectus*, and of *erectus* samples from each other. *A. africanus* and *habilis* fossils remain relatively stable, but robust australopithecines are now widely separated from gracile forms, so producing three clusters - *A. boisei*, early gracile hominids, and later Pleistocene forms (final *erectus/sapiens*), with early *erectus* intermediate between the last two. A similar configuration, emphasising the distinctiveness of the later fossils and the intermediacy of early *erectus* specimens is provided by the upper jaw, but it is further modified by the similarity of the Lower Pleistocene gracile forms which in this region are well separated from *A. africanus*. *A. boisei* specimens remain isolated from all other hominids.

The interaction of the various regions is apparent from these shifts: after G.P.A. the tighter grouping of the later forms in face and cranial vault reflects their similarity in maxillary form and in other regions (masticatory, basicranium, articular) not considered in detail here, whilst the shift of *A. boisei* specimens away from *africanus* on cranial vault analysis reflects the marked contrasts in facial, masticatory, and gnathic characters between these taxa. The transformed upper jaw data provide the mirror image to these shifts: the *A. boisei* specimens are again further from *A. africanus*, whilst later Pleistocene groups are more dispersed than in the initial plot reflecting the greater morphological heterogeneity of these forms in regions other than the maxilla.

The centroid positions provide estimates of 'average' or

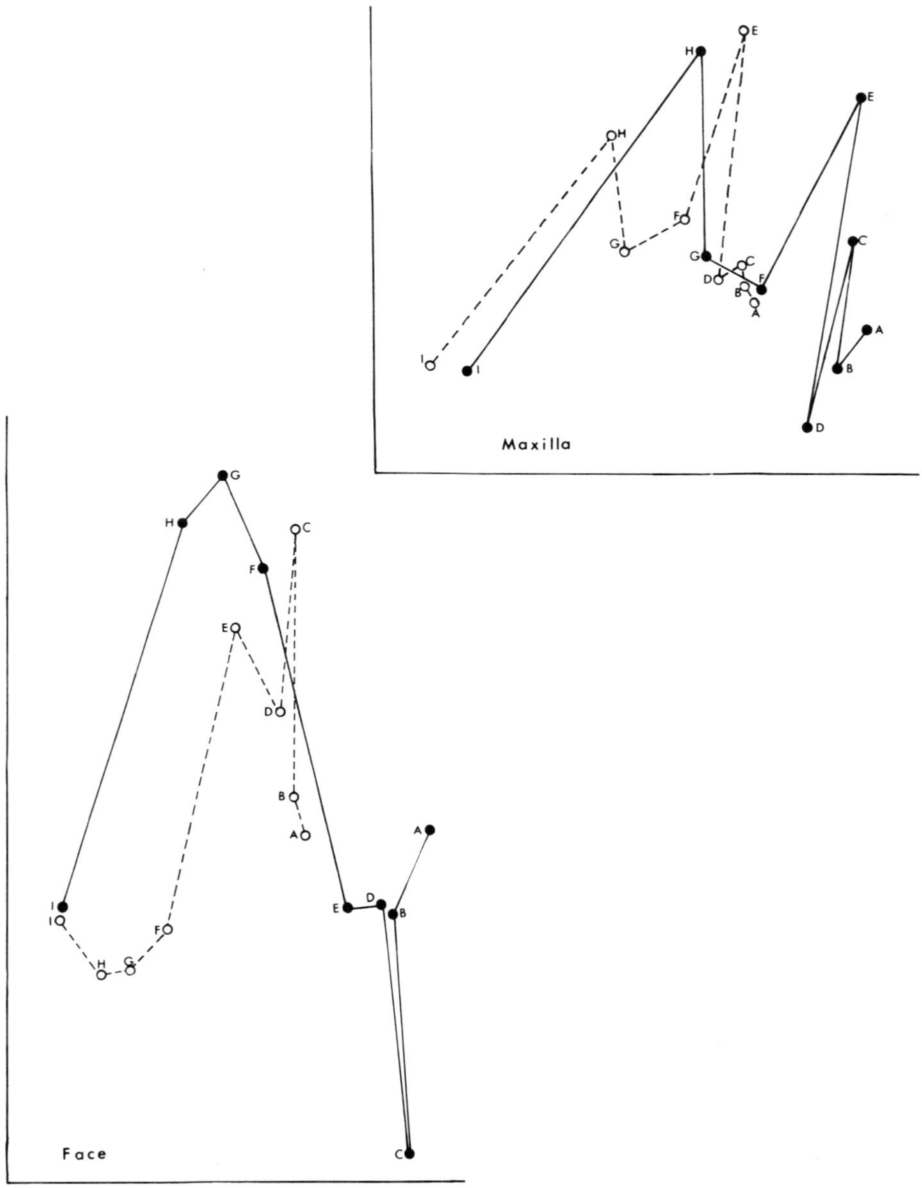

Figure 4. Generalised Procrustes analysis. Initial (open circles, dashed line) and final (black circles, continuous line) configurations of groups for face, maxilla, and vault complexes, and of centroids based on all complexes. A=modern H.s.sapiens; B=Upper Pal. H.s.sapiens; C=Neanderthals; D=late H.erectus; E=early H.erectus; F=ER 1813; G=H.habilis; H=A.africanus; I=A.boisei.

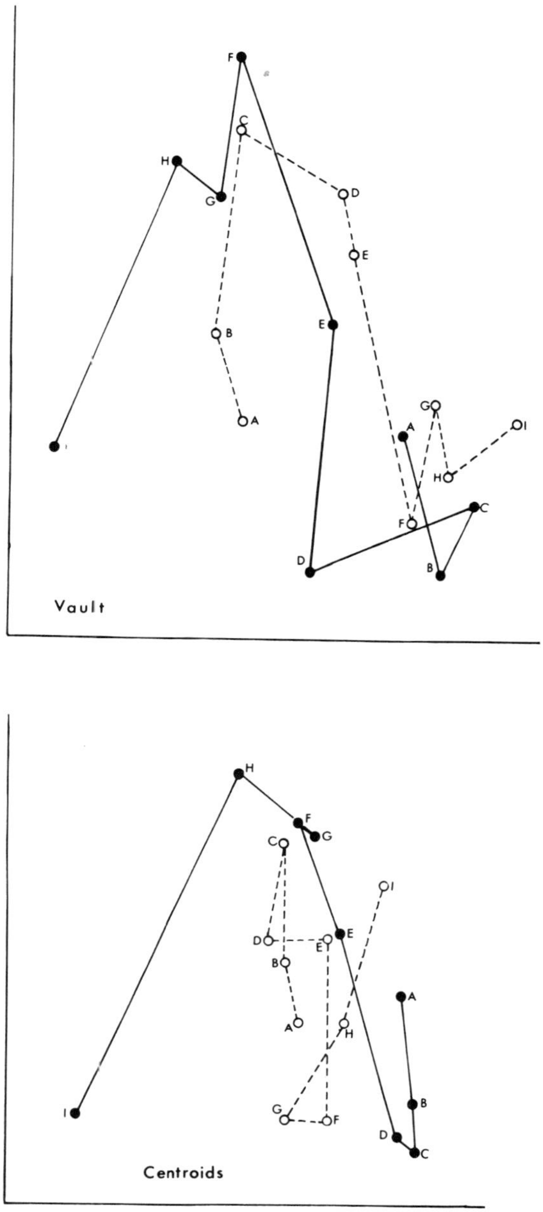

Figure 4. Continued

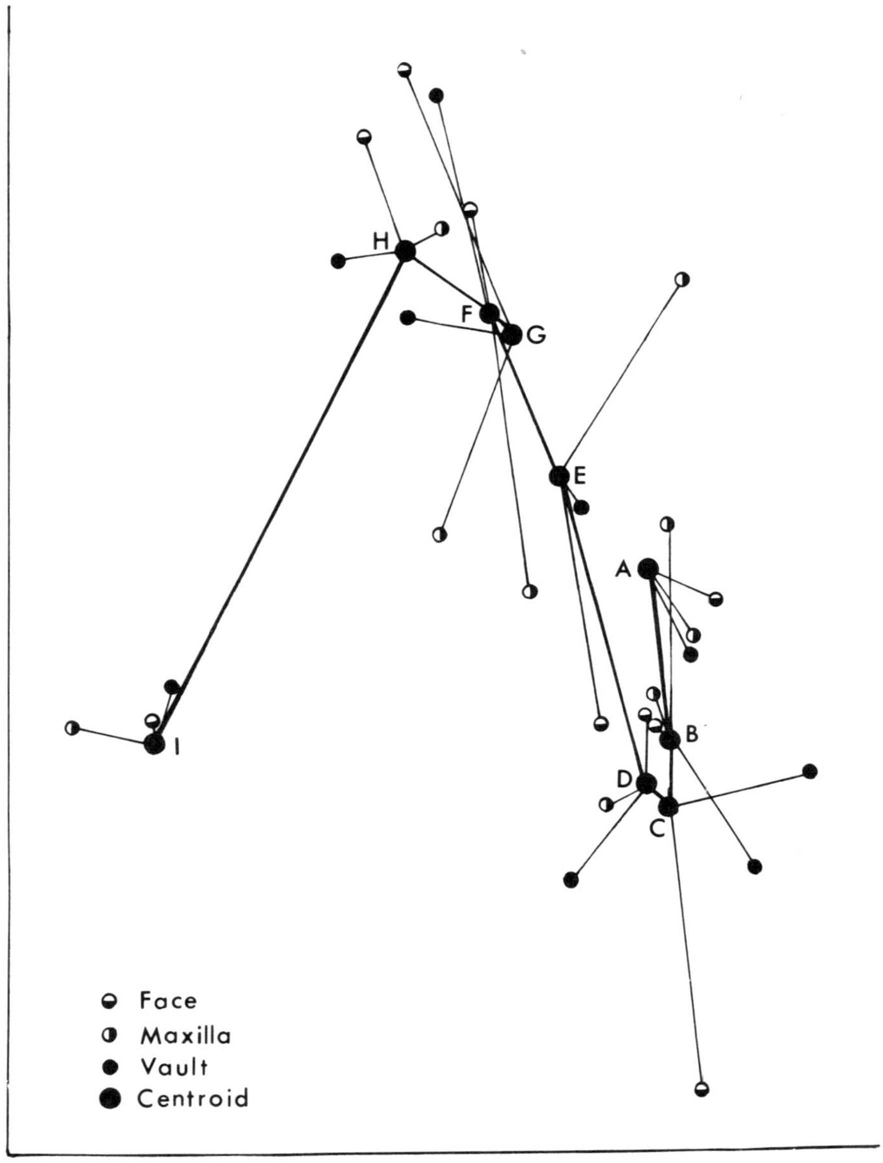

Figure 5. Generalised Procrustes analysis. Final configurations of group centroids, and relationships of face, maxilla and vault complexes to these. A=modern H.s.sapiens; B=Upper Pa. H.s.Sapiens; C=Neanderthals; D=late H.erectus; E= early H.erectus; F=ER 1813; G=H.habilis; H=A.africanus; I=A.boisei.

overall morphological change in cranial characters. The transformed positions accord broadly with the current phyletic and taxonomic concensus on Hominidae, emphasising the distinctiveness of _A. boisei_, and revealing a broad grouping of gracile hominids, but with notable separation of _A. africanus_ from the _habilis_ specimens, which differ in the direction of early _erectus_. There is a tight clustering of Middle and Upper Pleistocene remains, with the modern sample as an outlier, and considerable separation between early and late _erectus_ samples.

MOSAICISM AND PARSIMONY IN PHYLOGENY

The centroid distances may be regarded as 'minimum morphological pathways' (Oxnard 1969) required to effect the cranial changes evident from the fossil record; see Oxnard (1969, 1975) for morphological pathway approaches to the study of the shoulder and hip respectively. This approach treats morphological structures - in this case hominid crania - as freely modifiable plastic entities, an assumption that may well be unrealistic in the light of the biochemical, physiological and morphogenetic, as well biomechanical and other functional constraints to which biological systems are subject. There is no guarantee that the minimum morphological transformation obtained in each and every case; in some it may be functionally or developmentally disruptive, and some more extended pathway may therefore have obtained in reality.

However, the determination of minimum paths is valuable in constructing hypotheses to order available data, and for their predictive value; indeed such pathways are implicit in the very great majority of phyletic schemes that attempt to reconstruct ancestor - descendant relationships. For example, if the phylogeny above is correct, it is to be expected that further Plio - Pleistocene specimens will be recovered with small, lightly constructed neurocrania comparable to _A. africanus_ and _H. habilis_, but which show facial, masticatory and dental parameters intermediate between these taxa documenting the shift from a dental arcade dominated by the post-canine teeth (_Australopithecus_) to one in which the cheek teeth are only moderately sized, and with associated changes in jaw and facial anatomy (early _Homo_). Similarly if _H. erectus_ evolved from a _habilis_ -like predecessor, there should be fossil crania which display a jaw apparatus comparable to the existing _habilis_/_erectus_ specimens but with intermediate neurocranial characters.

It is difficult to envisage procedures for phyletic reconstruction which do not incorporate the principle of parsimony as a fundamental concept, but the recognition of mosaicism such as that discussed here has important corollaries for the application of parsimony in specific instances, and for the interpretation of fossil evidence, especially incomplete material.

Specimens that may be morphologically distinctive (in the sense of non-overlap with other groups) should not necessarily be assumed to represent discrete lineages or even to justify chronospecies distinction if they are represented only by complexes considered to be evolving rapidly during the relevant phase of hominid evolution. Recent suggestions that the Kromdraai robust australopithecine is specifically distinct from that represented by the much larger sample from Swartkrans (eg. Howell 1978, Grine 1981), or that a taxon other than _H. erectus_ is represented within the hominid fossils from Lower Pleistocene deposits in Java, both need to be assessed with this proviso in mind.

The recovery of material with contrasting morphologies to those predicted from posited sequent groupings suggests either that the inferred phylogeny is incorrect, or that the morphological transformation was achieved by a route other than the minimum pathway. Detailed analysis of all relevant anatomical information, together with other evidence (geographical, chronological etc.) should help in deciding between these alternatives. Congruence in the great majority of morphological features (especially if supported by concordant chronological data) suggests that the specimens do represent phyletically sequent populations, and that the discordant ones did not follow a minimum morphological pathway, rather than the proposed phylogeny is incorrect. On the other hand discordance is a substantial series of morphological features suggests that the phylogeny is in error. For example, whilst it is to be expected, on current evidence, that fossils will be found that document the evolution from a more gracile ancestor of the facial, dental and masticatory features of the robust australopithecines, the great majority of workers do not anticipate the recovery of specimens morphologically intermediate between these and the same complexes of early _Homo_. Their discovery would forcibly suggest that current phyletic schemes are seriously awry, and that the robust australopithecines are not a distinct clade.

The current debate concerning the distinctiveness and phyletic position of _Australopithecus afarensis_ provides another example of the need to take account of morphological mosaicism in early hominid studies. Some _afarensis_ dental data have been incorporated in multivariate analyses (Read, this volume); whilst it was not possible to include _afarensis_ specimens in the current cranial analyses, the results summarised above do make possible certain predictions.

Afarensis has been claimed as a stem hominid, ancestral to both _Homo_ and to later australopithecines including _africanus_ which, on this view, is not a human ancestor, but merely a phase in the differentiation of a robust morphology (Johanson and White 1979; White, Johanson and Kimbel 1981). The interpretation depends essentially upon the identification of, and primary phyletic weighting attached to, the most parsimonius pathway between _afarensis_ and early _Homo_ in a limited set of dental parameters and in facial proportions associated with them, whilst ignoring the apparently greater similarity (and hence shorter pathway) between africanus and Homo in some other regions. The evidence cited for the distinctiveness of afarensis is not unequivocal (Tobias, 1980) whilst Read's study (this volume) does not support the claim of parsimony made for the dental data.

However, if Johanson and White's interpretation is correct, afarensis specimens might be expected to show a general similarity to africanus, but with significant shifts in some complexes in the direction of Homo habilis. In particular, near identity to africanus might be expected in cranial vault, basicranium, articular and balancing complexes, but with maxillary, mandibular, masticatory and facial regions showing significant approaches to H. habilis. Moreover, in these complexes africanus should approach robustus/boisei specimens more closely than does afarensis. Since afarensis specimens show undoubted overlap with africanus in many individual characters (Tobias 1980), multivariate techniques should be especially valuable in providing a metrical expression of whether the overall combination of variables, representing total morphological pattern, is distinctive as claimed.

CONCLUSIONS

The array of multivariate statistical techniques available provide valuable tools for the investigation of morphological form and relationship. They aid analysis by reducing a complex set of variables to a more manageable sub-set, whilst also producing a quantitative estimate of resemblance which may involve the interaction of size and shape effects, or size or shape differences only, depending upon transformation of the data and/or the statistic used. Multivariate analysis may also be used for purposes of discrimination, and serve as the basis for taxonomic and phyletic schemes via the recognition of phena through clustering of similar specimens. However, given the paucity of the hominid fossil record this is likely to remain, at best, a subsidiary role in palaeoanthropology. Phylogenies are based upon many considerations, often involving besides morphological data, other kinds of evidence which cannot easily be accomodated within any multivariate analysis.

In these circumstances multivariate techniques are likely to be most useful when applied to pre-determined groups within a specified phyletic framework in order to provide a scale for that phylogeny and investigate its implications, rather than serving as the source of the phylogeny ab initio.
Such an approach provides a measure of the degree of consistency obtaining within a particular framework, and also makes possible certain statements about the likely morphological patterning of intermediate populations, which serve as predictive aids and assist

in refining the phylogeny. It is difficult to see how such information might be obtained other than through the use of techniques such as those discussed above and elsewhere within the present volume.

ACKNOWLEDGEMENTS

I thank R.E.F. Leakey and M.D. Leakey (National Museums of Kenya); P.V. Tobias and A.R. Hughes (University of the Witwatersrand); and C.K. Brain and E.S. Vrba (Transvaal Museum, Pretoria) for access to early hominid specimens in their care and study facilities; and J.C. Gower and P.G.N. Digby for statistical advice and computational assistance. Aspects of this work were supported by grants from Cambridge University, King's College Cambridge, and the Boise Fund, Oxford.

References

Bilsborough, A.: 1976, Patterns of Evolution in Middle Pleistocene hominids. Jour. Human Evol., 5, pp. 423-439.

Bilsborough, A.: 1978, Some aspects of mosaic evolution in hominids. In: Recent Advances in Primatology 3 Evolution (ed. D.J. Chivers and K.A. Joysey). pp. 335-350, London, Academic Press.

Blackith, R.E. and R.A. Reyment: 1971, Multivariate morphometrics. London & New York: Academic Press.

Campbell, N.A.: 1978, Multivariate analysis in Biological Anthropology, some further considerations. Jour. Human Evol., 7, pp. 197-203.

Clarke, R.J. and F.C. Howell: 1972, Affinities of the Swartkrans 847 Hominid cranium. Amer. Jour. Phys. Anthrop. 37, pp. 319-336.

Corruccini, R.S.: 1975, Multivariate Analysis in Biological Anthropology: Some considerations. Jour. Human Evol., 4, pp. 1-19.

Corruccini, R.S.: 1978, Morphometric Analysis: Uses and abuses. Yrbk. of Physical Anthrop., 21, pp. 134-150.

Grine, F.E.: 1981, Trophic differences between "gracile" and "robust Australopithecines: a scanning electronmicroscope analysis of occlusal events. S. Afr. Jour. Sci, 77, pp. 203-230.

Fitch, F.J. and J.A. Miller: 1970, Radioisotopic age determination of Lake Rudolf artefact site. Nature, 226, pp. 226-228.

Gower, J.C.: 1966a, Some distance properties of latent root and vector methods used in multivariate analysis. Biometrika, 53, pp. 325-338.

Gower, J.C.: 1966b, A Q-technique for the calculation of canonical variates. Biometrika 53, pp. 588-590.

Howell, F.C.: 1978, Hominidae. In: Evolution of African Mammals (ed. V.J. Maglio and H.B.S. Cooke) pp. 154-248. Harvard University Press, Cambridge, Mass.

Howells, W.W.: 1973, Cranial Variation in Man. A study by multivariate analysis of Patterns of Difference among recent human populations. Papers of Peabody Museum of Archaeology and Ethnology 67, pp. 1-259.

Johanson, D.C. and T.D. White: 1979, A systematic assessment of early African hominids. Science 203, pp. 321-330.

Leakey, R.E.F.: 1970, Fauna and Artefacts from a New Plio-Pleistocene Locality near Lake Rudolf in Kenya. Nature 226, pp. 223-224.

Leakey, R.E.F.: 1976, An overview of the Hominidae from East Rudolf, Kenya. In: Earliest Man and environments in the Lake Rudolf Basin. (ed. Y. Coppens, F.C. Howell, G. Isaac, and R.F. Leakey), pp. 476-483, Chicago: Univ. of Chicago Press.

Oxnard, C.E.: 1969, Evolution of the human shoulder:some possible pathways. Amer. Jour. Phys. Anthrop. 30, pp. 319-331.

Oxnard, C.E.: 1972, Some African fossil foot bones: A note on the interpolation of fossils into a matrix of extant species. Amer. Jour. Phys. Anthrop. 37, pp. 3-12.

Oxnard, C.E.: 1975, Uniqueness and diversity in human evolution: morphometric studies of australopithecines. Chicago and London: Univ. of Chicago Press.

Tobias, P.V.: 1978, The earliest Transvaal members of the genus Homo with another look at some problems of hominid taxonomy and systematics. Z. Morph. Anthrop., 69, pp. 225-265.

Tobias, P.V.: 1980, "Australopithecus afarensis" and Australopithecus africanus: A critique and an alternative hypothesis. Palaeont. Africana 23, pp. 1-17.

Walker, A. and R.E.F. Leakey: 1978, The hominids of East Turkana, Sci. Amer. 239, pp. 54-66.

White, T.D., D.C. Johanson, and W.M. Kimbel: 1981, Australopithecus africanus:its phyletic position reconsidered. S. Afr. Jour. Sci., 77, pp. 445-470.

Wolpoff, M.H.: 1978, Analogies and interpretation in palaeoanthropology. In: Early Hominids of Africa. (ed. C.J. Jolly), pp. 461-504, London: Duckworth.

Wood, B.A.: 1978, Classification and phylogeny of East African hominids. In: Recent Advances in Primatology: 3 Evolution. (eds. D.J. Chivers and K.A. Joysey), pp. 351-372, London: Academic Press.

FROM MULTIVARIATE STATISTICS TO NATURAL SELECTION: A REANALYSIS OF THE PLIO/PLEISTOCENE HOMINID DENTAL MATERIAL

Dwight W. Read

Department of Anthropology UCLA, Los Angeles, Calif. 90024

INTRODUCTION

Multivariate statistical techniques have had a natural application to the problem of taxa definition in hominid studies. The fundamental problem in taxonomic reconstruction of, first, sorting individual fossil specimens into homogeneous, distinctive groups and, second, of determining the taxonomic relations implied by those groupings, has a format immediately translatable into the framework of techniques such as multivariate cluster algorithms for sorting data into groups, discriminant analysis for measuring distinctiveness of groups, and factorial procedures for forming construct variables that are sensitive to morphological complexes signifying species uniqueness. Yet despite an increasingly sophisticated conceptual and analytical framework for the study of hominid fossil materials, establishing a generally agreed upon taxonomic classification for those materials has remained elusive, with alternative classifications apparently difficult to confirm or disconfirm in any generally accepted fashion as indicated by still current controversy over hominid classification.

In part this stems from three areas of weakness in the present conceptual and analytical framework for analyzing fossil hominid materials that will be examined in this chapter as illustrating the interplay between the conceptual basis for statistical analysis of data and the interpretation of those data in a taxonomic framework. These are: (1) incomplete preanalysis of data to form homogeneous groups of fossils to be used as a basis for taxonomic class definitions and taxa relations; (2) inadequate attention paid to the implications of the conceptual basis for statistical inferential techniques for group comparisons, particularly in a multivariate context; and (3) insufficient consideration given to alternative means for evaluating proposed classifications.

G. N. van Vark and W. W. Howells (eds.), Multivariate Statistical Methods in Physical Anthropology, 377–413.

Topics (1) and (2) are interrelated through use of statistical inferential techniques to form homogeneous groups of fossils to serve as the basis for ascertaining group relations in a multivariate framework. The third topic refers to means to extend confirmatory arguments for a proposed classification scheme beyond inherent limitations that are a part of arguments based on direct comparison of morphological differences between sets of fossils stemming from incompleteness of data (e.g. Pilbeam and Vaisnys, 1975). To be considered here is the use of the principle of natural selection as providing a guiding framework in a multivariate context (Read 1976; cf. Leakey, Hay, Curtis, Drake, Jackes and White, 1976: 406).

Briefly, any claimed set of taxa and taxon relations is simultaneously a proposed evolutionary pathway, whether explicitly so stated or not, and hence is an assertion about a pattern of natural selection that would have been necessary for that pattern to have occurred. The plausibility of proposed sets of taxa can be examined by translating those sets into their implied evolutionary pattern in the form of a time trajectory through the Euclidean space defined by the variables used to distinguish the taxa. The time trajectory can then be evaluated for plausibility by linking it to a pattern of selective pressures that would be needed as the driving force for creating that trajectory (cf. Rosen 1967: 90-93). This procedure will be used here for assessing alternative taxonomic schemes for the plio/pleistocene hominids using a subspace of the Euclidean space defined by mesiodistal and buccolingual measurements on teeth.

ANALYTICAL FRAMEWORK

The definition of taxa and the assignment of individual fossil specimens to taxa have generally relied for confirmatory arguments on purported morphological differences amongst fossil specimens as indicative of class distinctions in a system of taxonomic classes. The framework of using morphological distinctions to signify taxonomic class differences is predicated upon the fact of there having to be at least some morphological character differences between species, whether sympatric or allopatric. In the absence of any *a priori* definition and justification for a set of distinguishing characters, though, a method has developed of providing extensive descriptions of individual fossil specimens (e.g. White, 1977, 1980) and searching for differences in such descriptions that are thought likely to have taxonomic significance (e.g., Wolpoff, 1979). Conceptually, this procedure has gone from comparison of individual specimens to comparisons of sets of fossils based on the multivariate character of morphological features, and analytically has gone from univariate comparison of features in a sometimes simplistic qualitative identical/not identical scale to a quantitative multivariate framework that searches for significant differences in measurements imbedded in a suitably defined multivariate space.

Forming homogeneous groups of fossils is a fundamental initial step as all analysis and subsequent interpretation of taxonomic significance at the level of phylogenetic relationships is predicated upon the grouping of fossils in such a manner that a group represents a single species, whether sympatrically or allopatrically defined. Though the requirement of group homogeneity can be sidestepped by assuming one site (or one level in a deeply stratified site) = one taxon, as has been done for the hominid material from the South African sites, that assumption, in conjunction with using a statistical inferential framework for parameter estimation, produces a disconformity at the level of models for taxa. The disconformity arises from the two common assumptions of, first, taking as a probability space for statistical analysis a taxon defined by a collection of fossils found at a site, with the collection of fossils considered to be the effective equivalent of a random sample from a probability space; and second, taking the variables measured over the fossils as the equivalent of random variables defined over a probability space. Although the conceptual framework for statistical inferential techniques can include the situation where a population is defined arbitrarily (that is, defined by criteria that do not match an underlying process, or processes, thought to be explanatory for those data) by defining the set of elementary events for the probability space to be precisely the objects of the population, the probabilities associated with elementary events to be the proportions of the population represented by each distinct object and the values of a random variable to be the numerical values obtained through the measurements made on the objects, lost in this procedure is the powerful notion of a random variable(s) and its probability distribution serving as the basis for explicating a process(or experiment) through mathematical representation in the form of a model. The notion of a model as the mathematical representation of a process is inappropriate in this situation and instead a model for the population only has a heuristic role as a descriptive statement about that population (Draper and Smith 1967: 2). The model cannot be assumed to represent an underlying process explicated through the mathematical representation of the model.

For the context of statistical analysis of hominid data, models, in the form of taxa and relations amongst taxa, are to be used as a representation and explanation of the phenomenon being characterized, not merely as a convenient means to describe and classify an arbitrarily defined population, hence the disconformity between how populations are defined as taxa and the role of a model in the analysis of the data used to define taxa.

Negation of the one site = one taxon assumption implies that the solution for homogeneous grouping of fossils is dependent upon assumptions made about the form of the multivariate distribution of measurements for morphological characters when measured across the members of a single species. Clarification of these assumptions in a multivariate statistical context requires establishing the connection between the statistical framework of probability space and random variable with the empirical context of phenotypically defined measures made across a set of fossil forms as representative of hominid taxa.

Conceptual Basis for Statistical Grouping

Regardless of the particular statistical technique, univariate or multivariate, at base is a framework composed of two key concepts: probability space and random variable. By a (finite) probability space is meant a set Ω of objects called elementary events, e_i, and a set of non-negative weights, p_i, associated in a one-to-one fashion with the elementary events e_i and called probabilities when the weights p_i satisfy two conditions: (1) $0 \leq p_i \leq 1$ for each i and (2) $\sum_{e_i \in \Omega} p_i = 1$. A random variable X is a mapping from the set Ω to the real numbers, $\mathbb{R}$. A probability distribution is defined for the random variable X defined over Ω with associated probabilities p_i for elementary events e_i by setting for each x in $\mathbb{R}$, $\Pr[X = x] = \sum_{X(e_i) = x} p_i$ ("the probability that X takes on the value x is given by the sum of all probabilities associated with elementary events e_i that are mapped to the value x by the random variable X") if there is at least one elementary event e_i such that $X(e_i) = x$, and $\Pr[X = x] = 0$, otherwise. Finally, a model for one or more random variables is a statement about the form of the probability distribution for that random variable, or random variables; e.g., the random variable X is $N(\mu,\sigma)$ -- that is, has the normal distribution with parameters μ,σ -- or, e.g., the conditional distribution of the random variable Y with respect to the random variable X, where both X and Y are defined over the same probability space Ω, can be exactly defined by a regression equation of the form $y = \alpha + \beta x + \varepsilon$, where α,β are parameters and ε is a random variable defined over Ω having, say, a $N(0,\sigma)$ probability distribution independent of the random variable X or the random variable Y.

In this framework there is an ordering of concepts going from process to event to probability of an event to definition of a random variable over events as a means to characterize the process, and finally to a model as a mathematical representation of that process. In an empirical situation where the events can be described as the consequence of a naturally formed experiment or process, the framework becomes a means to characterize and mathematically represent that process or experiment. For the context of forming taxa on the basis of differences in morphological characteristics, the relevant (natural) experiment consists of possible genetic outcomes, or elementary events, for a genetic system (such as an n-loci, k-allelic system) under a specified regime of gamete recombination for the formation of zygotes (such as random recombination of gametes). With quantitative traits, a random variable of interest is a measurement of the quantitative contribution to that trait made by each allele in the allelic combination making up the possible genotypes that can result from a given mating regime (see Elandt-Johnson 1971: 13-38). The random variable so defined, along with the regime of gamete recombination, determines the expected form of the probability distribution for the genetic system underlying the trait being measured. With random recombination of gametes and a linearly additive system of allelic contribution to a quantitative trait, the above described random variable will tend to have a discrete approximation to the continuous normal distribution. This leads to the common assumption of a joint normal (or at least a unimodal, approximately symmetric) distribution for quantitative measures at the phenotypic level as the expected form for measurements taken across a single species, and used as an implicit criterion for distinguishing groups of fossils as taxa (see Pilbeam and Vaisnys 1975: 9).

Certain aspects of this assumption obviously are not strictly true; e.g., random recombination of gametes over a whole species. The degree to which violation of the assumption of, say, random recombination significantly affects the expected distribution for a measure made over quantitative traits depends on whether or not the genetic system is fixed, in the sense of unchanging allelic frequencies, for the species as defined temporally and spatially. When fossil data said to belong to a single taxon are also from a time depth that includes an episode of morphological change within that taxon, the empirical form of the frequency distribution for a trait measured over those data may vary significantly from a simple, approximately normal, unimodal distribution. An example of this phenomenon will be seen below for the *Australopithecus africanus* material from South Africa.

The framework of probability space and random variable as applied to a genetic system can be used to explicitly state the syllogism underlying the assumption of morphological differences as signifying the presence of different taxa. The syllogism is: (1) The morphological characteristics of a species are established beginning with a genetic system composed of genotypes formed through the recombination of the alleles defining those genotypes via the mating regime characterizing that species. (2) Species distinctiveness presumes different allelic, and hence genotypic, frequencies in at least some genetic systems for each species, which leads to: (3) Some morphological differences between species are due to different genotypic frequencies. This has as corollary: (4) A random variable defined over the genotypes for a genetic system which is part of species distinctiveness will have non-identical probability distributions for the members of distinctive species. Hence: (5) Random samples from the members of different species will have significantly different parameter estimates for the probability distributions of random variables measured over genetic systems that are part of species distinctiveness. Finally, the converse of (5) provides the conceptual basis for using differences in the values of variables measured over fossil remains as signifying distinctive taxa: (6) Significant differences in parameter estimates based on random samples of individuals, possibly unknown as to whether they are from the same or from different species, will, for some random variables measured over some genetic systems be indicative of species distinctiveness, if any (cf. Tobias, 1975: 17). The "some" in statement (6) establishes the need for confirmatory arguments, as discussed above, that go beyond just using metric distinctiveness of sample data.

The syllogism assumes random variables measured over a probability space of genotypes that underly morphological traits. However, the pragmatic reality of both how hominid fossil data are recovered and the still limited ability to connect phenotypically defined measures with a specific genetic system imply that the statistical framework underlying (5) and its converse (6) will likely be violated. Variables and populations as initially defined for analysis will often neither be random variables defined with respect to a probability space of genotypic outcomes, nor populations of outcomes of a specific process. Thus there arises the need for preliminary analysis to correct for confounding effects introduced by the manner in which variables and populations are defined. The preliminary analysis that may be required can be related to several modes of analysis that are defined in terms of whether or not the population and/or variable matches the ideal form discussed above.

Modes of Analysis

The conditions underlying statement (5) and its converse (6) establish two independent requirements for preliminary analysis that generate a 4-class classification of modes of analysis as defined by two states for the terms population and variable: (1) well-defined (WD) and (2) not well-defined (NWD). The distinction is whether or not each term, population or variable, matches the idealized form as given above in the notion of an experiment defining a probability space of outcomes for that experiment. A population that does not satisfy the first condition (i.e., is not a population of outcomes of an underlying process) will be said to be a NWD population and a variable that does not satisfy the second condition (i.e., is not a measure made on the immediate outcomes of an underlying process) will be said to be a NWD variable.

Given that the appropriate context for statistical inferential and inductive arguments leading from probability distributions of variables to taxon definitions will be a WD population and WD variables as indicated in statement (6), it follows that a necessary aspect of data analysis in other modes is preliminary analysis to formulate, as needed and to the degree possible, well-defined populations and/or variables prior to analysis of data for its inherent structure. The possible modes and required preliminary analysis are: Mode 1: WD population and WD variable(s) -- no preliminary analysis required; Mode 2: WD population, NWD variable(s) -- preliminary analysis for establishing WD variables using, for example, factor extractive techniques for construction of composite variables that measure underlying factors structuring the data, etc.; Mode 3: NWD population, WD variable(s) -- preliminary analysis for finding significant subdivisions of the original population likely to reflect the outcomes of a single process with the expected form of a frequency distribution for the variables under the assumption of a WD population as a criterion. A typical technique is some form of cluster analysis followed by discriminant analysis to measure the distinctiveness and homogeneity of postulated groups, and/or testing for outliers to remove idiosyncratic cases. Mode 4: NWD population, NWD variable(s) -- Here there is the double requirement of both redefining the population) and the variable(s) while using the properties of one or the other to do that redefinition. This leads to a more complex iterative approach rather than any single technique of analysis. A highly successful approach introduced in Christenson and Read (1977) is to first identify and remove outliers from the data, then use principal component analysis over the screened data to formulate a smaller, more parsimonious set of variables which also have interpretation as measuring the outcomes of a process, and finally, to examine the frequency distribution of the screened data with respect to the new variable(s) defined by the principal component analysis so as to separate the data into internally homogeneous and distinctive groups. The outliers removed from the statistical analysis are considered separately as individual cases to be compared to groupings made from the bulk of the data. In the data analysed by Christenson and Read, the raw data showed but a single, largely undifferentiated cluster when cluster analytic techniques were applied prior to restructuring the variables using principal component analysis, and the principal component analysis lead to uninterpretable variables prior to screening the data for outliers. After following the procedure outlined above, the screened data had a non-overlapping bimodal distribution for the first principal component (see Fig. 2, Christenson and Read, 1977), thereby allowing simple separation of the data into two distinctive subgroups. The outliers removed prior to the principal component analysis represented individual cases with unusual dimensions and hence were idiosyncratic.

The rationale for removing outliers from the data prior to further analysis follows from, on the one hand, the fact that outliers have a major effect on correlation coefficient estimates, and on the other hand, may represent cases that did not arise from the same natural process under investigation as did the bulk of the data; e.g., outliers may be contamination from another population, aberrant cases or individuals, etc. Outliers are not eliminated from further consideration as they are compared to the data analysed without inclusion of the outliers. It is only their extreme effects on parameter estimation that is eliminated.

Alternatively, robust estimators less sensitive to outliers can be used, as will be done here, for comparing groups using discriminant functions.

Robust Discriminant Functions

In general, robust estimators are a class of estimators formulated to be less sensitive to the biasing effects of individual cases that are not part of the assumed underlying frequency distribution for the data, and at the same time are able to preserve optimality, or at least asymptotic Bayesian optimality, properties for estimators. It has been shown (Hampel, 1968; Andrews, Bickel, Hampel, Huber, Rogers, and Tukey, 1972) that neither the sample mean nor the sample standard deviation is a robust estimator of the population mean and population standard deviation, respectively. Hence statistical tests constructed around these estimators are likely not to be robust. In particular, the usual linear discriminant function (LDF) used to test group distinctiveness is not a robust discriminant function (Ahmed and Lachenbruch, 1975; Schlain, 1978).

The robust discriminant function (RDF) that will be used here differs from the non-robust classical form by substitution of robust estimates $\bar{x}_i^*$ and s_i^* in the ith sample for the population mean and standard deviation, respectively, in place of the usual sample estimates $\bar{x}_i$ and s_i (see also Campbell, this volume). More specifically, the LDF developed by Fisher (1936) for the two-population classification problem has coefficients given by:

$$S_p^{-1}(\bar{x}_1 - \bar{x}_2) \tag{1}$$

where:

S_p = pooled sample covariance matrix

$\bar{x}_i$ = vector of sample means from population i (i=1,2),

and the cutpoint for classification is given by:

$$\tfrac{1}{2}(\bar{x}_1 + \bar{x}_2)S_p^{-1}(\bar{x}_1 - \bar{x}_2). \tag{2}$$

It is well-known that the LDF is asymptotically optimal for multivariate normal populations with equal covariance matrices.

In order to change the LDF into a robust discriminant function, Schlain (1978) has proposed that a robust estimate of location, $\bar{x}_i^*$, and a robust estimate of dispersion, s_i^*, be used in place of $\bar{x}_i$ and s_i, respectively, in the LDF. The resulting robust discriminant function has been shown by Schlain to be asymptotically Bayes optimal for multivariate normal distributions. Thus the RDF should perform almost as well as the LDF on uncontaminated data and better than the LDF on contaminated data.

The robust estimates $\bar{x}_i^*$ and s_i^* are computed iteratively using an algorithm due to Gnanadesikan (1977). Essentially the algorithm begins with a vector of trimmed estimates of the means and variances (e.g., computed by deleting the 25% most extreme values from the sample mean) as initial robust estimates and then iteratively finding new estimates based, in each iteration, on a certain proportion (the "trimming proportion") of the data points representing the smallest Mahalanobis's distance from the data points to the current robust estimate of location for that step of the iteration procedure (see Schlain 1978: 17-20 for details). The trimming proportion is equal to the estimated amount of contamination in the data.

The trimming proportions are estimated by testing for outliers and then first removing from the data set all outliers which have independent justification (i.e., are generally agreed upon as contamination from another population). The remaining outliers, the suspected outliers, are used to estimate the trimming proportions and are kept as part of the data set when using robust estimators (Schlain 1978). Correction is made for the presence of these suspected outliers in the data set by computing the RDF using the trimming proportions just estimated in the iterative procedure for obtaining robust estimates of the mean and standard deviation.

The outliers are found by using gamma plots (Gnanadesikan, 1977) based on a robust estimate of the mean and standard deviation computed using the investigator's initial estimate of the amount of contamination in the data. An initial estimate of 20% contamination was used for the South African hominid data (Schlain 1978, following a suggestion by Read 1978 (personal communication)). The outliers are found by inspection of gamma plots wherein outliers appear as deviations from a linear plot (see Schlain, 1978: 65-69 for details). The number of outliers found is determined by the characteristics of the data, not by the initial estimate.

ANALYSIS OF HOMINID DATA

The dental material to be analysed are the mesiodistal and buccolingual measurements on maxillary and mandibular canines through 3rd molars for the plio/pleistocene hominids. The material from the South African sites of Sterkfontein and Swartkrans is used to establish a baseline of comparison with the other *Australopithecus* material and the early *Homo* material from East Africa. The data from South Africa are first screened for outliers and then the groups of screened data tested for distinctiveness using the RDF as discussed above. Outliers are removed prior to analysis of the mesiodistal and buccolingual variables by principal component analysis. The principal component analysis creates, as it happens, a single variable that is close to the ideal of a WD variable by virtue of being the single dimension along which long-term change has occurred. The distribution of values for the new variable is used for further analysis of the frequency distributions for the data from Sterkfontein and Swartkrans. The results of this

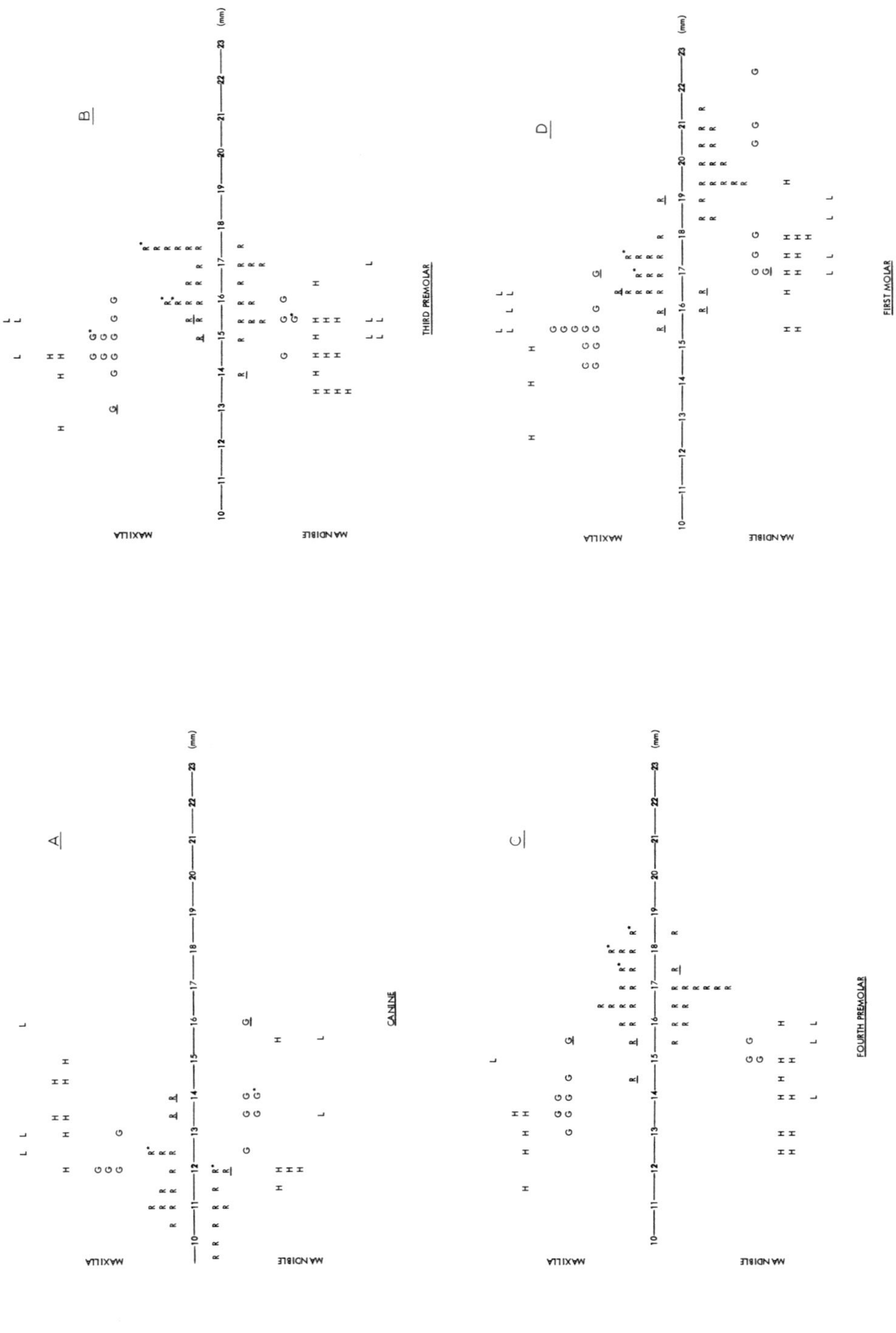
A
CANINE
B
THIRD PREMOLAR
C
FOURTH PREMOLAR
D
FIRST MOLAR
MAXILLA
MANDIBLE
(mm)

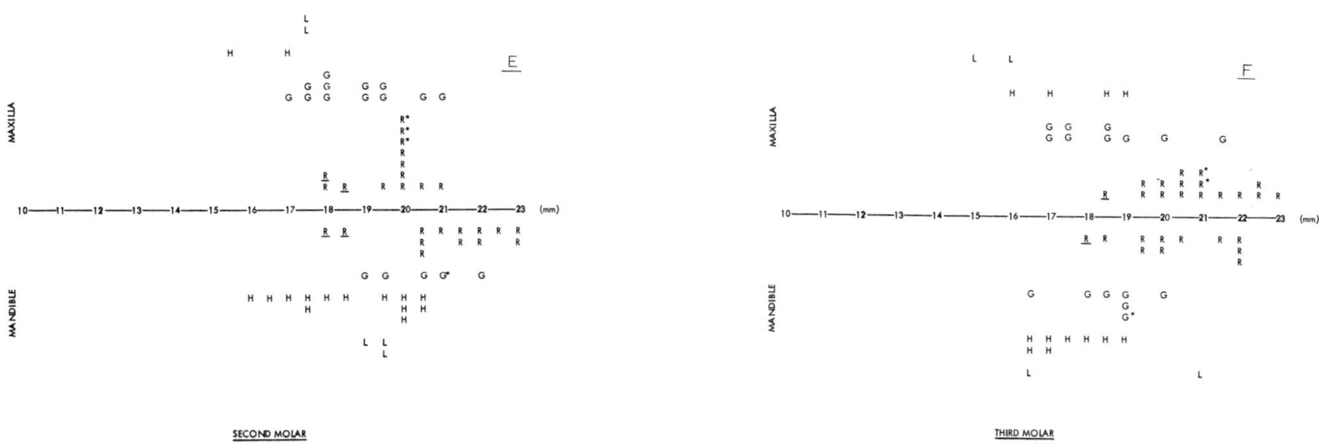

Figure 1. Histograms for each tooth stratified by site and ordered by time. The scale (center line) is the measure along the principal axis for all hominid teeth of a given kind. Confirmed outliers are underlined and suspected outliers are marked with an asterisk. (Outliers are based on the joint distribution defined by the MD and BL dimensions.) Symbols: L--Laetolil; H--Hadar; G--Sterkfontein; and R--Swartkrans. Each symbol represents one tooth, with left and right teeth averaged when both are present in a mandible or maxilla. The data have been grouped by .5 millimeter intervals.

analysis are then used as a basis for interpreting the data from the sites of Hadar and Laetolil as representing forms predating the material from South Africa, and with the material found in East Africa, and classified as either *Australopithecus boisei* or early *Homo*, as forms antedating the material from South Africa. Interpretation of these data is made using the context of implications about the pattern of natural selection that must be assumed for a postulated taxonomic model relating these groups of fossils.

Data used in the analyses are as published by site: Laetolil -- White (1979, 1980); Hadar -- (Johanson, White and Coppens, 1982); Sterkfontein and Swartkrans -- Wolpoff (1971); and East Africa: Omo -- Howell, 1969, Boaz and Howell, 1977; Olduvai -- Wolpoff, 1971; Koobi Fora -- Day and Leakey (1973), Leakey and Wood (1973), Leakey and Walker (1973). When a left and a right tooth of the same kind are present in a mandible or a maxilla, the average of the left and right tooth is taken as a single datum.

For the analysis of these teeth, the initial measurement of buccolingual (BL) and mesiodistal (MD) dimensions have been changed to size and shape measures as follows:

$$\text{size} = ((BL)^2 + (MD)^2)^{1/2} \tag{3}$$

and

$$\text{shape} = MD/BL, \tag{4}$$

as discussed in Read (1975). (Shape is given by a ratio, or in polar coordinates by the arctangent of the ratio, despite the general difficulty of using ratios to represent shape (Atchley, Gaskins, and Anderson, 1976; Corruccini, 1977) since measuring the tooth crown outline via mesiodistal and buccolingual dimensions implicitly assumes the outline can be adequately approximated by a rectangle and the shape of a rectangle is given by the ratio of its length to its width.) These two measures are closer to a measure of genetically specified events than are MD and BL since it is likely that the genotypic system underlying tooth development has phenotypic expression in size and shape, rather than tooth length and breadth. As will be shown below, though, this redefinition of the original variables is not sufficient for some of the teeth and needs modification.

When authors give both uncorrected values and values corrected for interproximal attrition for the mesiodistal dimension (e.g., White, 1977, 1980), only the uncorrected measure is used for consistency with other authors who only give the former value (e.g., Wolpoff, 1971).

Sterkfontein and Swartkrans -- Preanalysis for Homogeneity

While it is now generally agreed that the two sites of Sterkfontein and Swartkrans represent two species, *Australopithecus africanus* and *Australopithecus robustus*, respectively, the affinity of one to the other -- one lineage or two lineages -- as well as the affinity of one or both of them to later Homo forms is not so self evident. With the exception of a phylogenetic model that postulates A. africanus as ancestral to A. robustus with the lineage leading to Homo separating from that lineage subsequent to the time of these forms, all other models require that two forms existed simultaneously for at least the time period represented by the younger of the two sites, presumably Swartkrans. Both lineages would be simultaneously present in South Africa unless one further postulates that both hominid migration to and from that region, and the geographical location for the speciation that led to Homo and non-Homo lineages, ensured that exactly one lineage at a time is present in South Africa. As these constraints seem unlikely, there may be contamination in one or both of the South African sites by the form (or its descendents or ancestors as the case may be) represented at the other site, so that the "Telanthropus" material at Swartkrans, generally regarded as being from a different population (Robinson, 1953; Clarke, Howell and Brain, 1970; Pilbeam, 1972) need not be the only contaminating data. However, unlike the situation at Olduvai where two forms, Australopithecus boisei and Homo habilis, are present in the same stratigraphic bed and are distinguishable by substantially different and statistically non-overlapping size measurements on teeth; i.e., the A. boisei teeth are outliers with respect to the combined distribution of all South African Australopithecines, A. africanus forms at Omo, and Homo habiline and Australopithine forms from East Africa (Read, 1975, contra Wolpoff, 1976), the differences in tooth measurements between the forms at Sterkfontein and Swartkrans are not severe (Read, 1975) and there is overlap in the frequency distributions of the mesiodistal and buccolingual measurements. Thus contamination is not easily identified by inspection.

In Read (1975) outliers were identified using a multivariate stepwise test for outliers (Dixon, 1975), but all material classified under the genus Australopithecus or species Homo habilis was taken as a single group to be tested for outliers in accordance with a null hypothesis of but a single lineage for all of these materials. A conservative alpha value of 0.01 was used for the multivariate test. Despite the conservativeness of the data grouping and of the significance level, several outliers were identified by the procedure: P_3--Omo L7-125, KNM.ER 729, MLD 2; P_4--Omo L7-125, Omo W-23; M_1--Omo L7-125, MLD 18; M_2--Omo L7-125, KNM.ER 729; M_3--Olduvai 16, KNM.ER 729; P^3--Olduvai 24; P^4--Olduvai 16, Olduvai 24, MLD 9, STS 42; M^1--Olduvai 5, SK 46, SK 52; M^2--Omo 28-58, Olduvai 5 and M^3--Olduvai 5, Olduvai 7A and Olduvai 13, based on data published prior to 1973 (Read, 1975). As noted in Read (1975), distinctive about these outliers is the clearcut separation of all of the large toothed forms (Omo L7-125, KNM.ER 729 and Olduvai 5) as significantly ($\alpha < 0.01$) lying outside the range of values represented by the frequency distribution for the remaining material.

From the viewpoint of basing statistical comparisons on WD populations, this procedure is perhaps overly conservative. First, the original grouping of all Australopithecine and Homo habiline material is obviously a composite of several populations, where a population is defined as the members of a species over a period of time sufficiently short that no significant morphological change has occurred with respect to the variables being analysed. Second, the alpha level of 0.01 in conjunction with the "masking effect" (Murphy, 1951; Schlain, 1978), i.e., outliers not being detected when in the presence of several outliers that sufficiently inflate the sample standard deviation used in the outlier test so that the probability of a Type II error increases, makes it likely that statistically significant outliers may not be detected, as in fact is the situation here (see below).

Consequently, the procedure followed here for examining the distinctiveness of the A. robustus material as typified by the material at Swartkrans and the A. africanus material as typified by the material at Sterkfontein is to use the robust test for outliers and the RDF described above to test the degree to which the postulated populations differ as measured by the ability of the RDF to correctly reclassify the sample data.

The outliers found by gamma plots (given in Schlain 1978) are as follows: (1) Suspected outliers: C_- -- STS 3, SK 87; P_3 -- none; P_4 -- SK 826B; M_1 -- STS 7; M_2 -- none; M_3 -- none; C^- -- SK 84; P^3 -- STS 47, SK 46, SK 83; P^4 -- STS 30, SK 46, SK 79; M^1 -- STS 21, SK 47, SK 52; M^2 -- SK 47; M^3 -- SK 49 and (2) Confirmed outliers: SK 15 (P_3, M_1, M_2, M_3, M^3); SK 27 (C^-, P^3, M^1, M^2) and SK 45 (M_1, M_2). Confirmed outliers are those outliers from the gamma plots that are also generally agreed not to belong to the same groups as the remaining material as has been argued for SK 15 and 45 ("Telanthropus") and SK 27 (Clarke, 1977).

When both suspected and confirmed outliers are taken into account, a total of 27 outliers are found out of 250 data points, or about 10% of the data. Note that the number of outliers found by the gamma plots for the sites of Sterkfontein and Swartkrans is much greater than that found by the procedure used by Read (1975), underscoring the conservativeness of the latter procedure. With the confirmed outliers removed, there still remain 16 suspected outliers out of 239 data points, or about 5% of the data.

The mean shape, size and respective standard deviations for these teeth are given in Read (1975) for the data without the outliers as determined by Read's procedure of testing for outliers via a multivariate test of outliers, and in Schlain (1978) for the same data without confirmed outliers. Table 1 gives the size and shape dimensions for all outliers in comparison with the mean size and shape measurements based on all data points with only the confirmed outliers removed.

Table 1. Comparison of Outliers with Sample Means for Sterkfontein and Swartkrans

Catalogue number	Sterkfontein Size[1]	Sterkfontein Shape[1]	Tooth	Swartkrans Size[1]	Swartkrans Shape[1]
Mean	14.06	0.89	C_{-}	11.14	0.97
STS 3	16.02(1.68)	0.87(-0.64)			
SK 87				12.06(1.09)	0.85(-2.07)
Mean	14.56	0.80	P_3	15.34	0.83
SK 15				13.42(-2.42)	0.83(-.02)
Mean	15.47	0.82	P_4	17.09	0.85
SK 826B				18.47(1.62)	0.92(1.76)
Mean	19.11	1.08	M_1	20.23	1.05
STS 7	17.47(-0.89)	0.94(-1.73)			
SK 15				16.83(-4.22)	1.00(-1.59)
SK 45				16.48(-4.65)	1.03(-0.66)
Mean	20.78	1.08	M_2	21.97	1.09
SK 15				18.96(-3.00)	1.08(-0.20)
SK 45				18.12(-3.84)	1.05(-0.85)
Mean	20.68	1.12	M_3	22.15	1.16
SK 15				18.78(-2.20)	1.14(-0.33)
Mean	14.06	0.89	C^{-}	12.80	0.90
SK 27				14.85(2.27)	1.02(2.44)
SK 84				14.48(1.86)	0.84(-1.31)
Mean	15.32	0.72	P_3	17.24	0.69
STS 47	13.79(-1.87)	0.81(1.72)			
SK 27				16.43(-0.89)[2]	0.71(0.57)[2]
SK 46				15.59(-1.81)	0.63(-1.33)
SK 83				16.29(-1.04)	0.61(-1.81)
Mean	15.85	0.69	P^4	18.76	0.69
STS 30	17.42(1.94)	0.76(0.80)			
SK 46				17.26(-1.47)	0.63(-1.82)
SK 79				15.56(-3.15)	0.68(-0.35)
Mean	18.41	0.90	M^1	19.91	0.93
STS 21	20.06(2.27)	0.87(-0.85)			
SK 27				19.24(-0.77)	1.06(2.12)
SK 46				19.24(-0.77)	0.78(-2.53)
SK 47				18.25(-1.84)	1.03(1.63)
SK 52				21.98(2.36)	0.87(-1.03)
Mean	20.70	0.89	M^2	21.76	0.92
SK 27				20.44(-1.37)	1.04(3.26)
SK 47				19.94(-1.89)	0.97(1.42)
Mean	20.32	0.85	M^3	22.74	0.91
SK 49				22.04(-0.69)	0.77(-2.49)

(1) Numbers in parentheses are standardized scores.
(2) P^3 for SK 27 is not an outlier on the gamma plots, but all other teeth in the maxilla are outliers.

As can be seen from Table 1, the suspected outliers for *A. africanus* as represented at Sterkfontein do not exhibit any particular pattern: 2 are significantly larger than the average, one is smaller than the average and the other 2 appear to have an extreme size/shape configuration. For the *A. robustus* teeth, however, there is the suggestion that the suspected outliers are consistent with the range of the *A. africanus* teeth (though correction needs to be made for the time difference between the two sites, as discussed below); i.e., the size measurements for SK 46, 47, 79, 83 and 84 are consistent with *A. africanus* and SK 87 and 49 deviate from the *A. robustus* mean in the direction of *A. africanus*. Only one tooth is exceptionally large: SK 52. The last suspected outlier, SK 826B, appears to have an unusual size/shape configuration. Overall, these results for the suspected outliers seem to be equivocal as a measure for the number of lineages at Sterkfontein, but do suggest that 2 lineages are present at Swartkrans.

But regardless of the interpretation made of the outliers, their number and the magnitude of the standardized scores for the size and shape measurements on the outliers indicates that the original samples are not homogeneous as defined by joint normality of the sample frequency distribution. In the case of Swartkrans there is the possibility of contamination from a second lineage. Consequently any statistical comparison between these 2 sets of data that does not take into consideration the presence of outliers and possible contamination may lead to biased results.

The confirmed outliers are now removed from the data set and the presence of suspected outliers is taken into account through the trimming proportions used in the RDF. These proportions are based on the sample size for each tooth and the number of outliers as found in the gamma plots.

Comparison of the two sets of fossils for distinctiveness as groups was made through using the RDF to reclassify the sample data, assuming two groups. The proportion of correct reclassification for each tooth and site is given in Table 2.

As can be seen in Table 2, the proportion of correct reclassification is generally high, underscoring the distinctiveness of the two sites. Teeth with low proportion of correct reclassification are a result of overlap in the range for *A. africanus* and *A. robustus* along with more dispersed data. Or, to put it another way, these are teeth that apparently have undergone less selective pressure during the time period represented by these two sites. The mandibular tooth M_1, the outstanding example of a low rate of correct reclassification, has more widely dispersed data than the other teeth, along with mean values for the two measures that differ only slightly.

The RDF reclassification may be compared to the LDF reclassification based on the screened data as given in Read (1975). Generally speaking, the RDF performs as well as, or better than, the LDF for both *A. africanus* and *A. robustus* with the exception

Table 2. Proportion Correct Reclassification for Sterkfontein and Swartkrans Using a Robust Discriminant Function

Tooth	Sterkfontein[1])	Swartkrans[1])
C_-	1.000(6)	0.875(8)
C^-	0.750(4)	1.000(11)
P_3	0.750(4)	0.909(11)
P^3	0.800(10)	0.667(15)
P_4	0.667(3)	0.846(13)
P^4	0.817(7)	0.875(16)
M_1	0.429(7)	0.750(16)
M^1	0.909(11)	0.867(15)
M_2	0.800(5)	0.626(11)
M^2	0.667(12)	0.833(12)
M_3	0.714(7)	0.600(10)
M^3	0.714(9)	0.500(16)

(1) Numbers in parentheses are the sample sizes for each site.

of P_4 and P^4, where the LDF performed better. For some teeth, namely P^3, M^1, M^2 and M^3, the LDF was better for one site and the RDF better for the other site (see Schlain 1978 for details). Curiously, the LDF performed better in situations where more than one suspected outlier was included in the data, suggesting that there may be deviation from multivariate normality even though the MD and BL measurements are essentially univariate normal when considered separately. This raises the question of the adequacy of MD and BL as WD variables measured over a probability space.

Principal Component Analysis of Mesiodistal and Buccolingual Dimensions

In Read (1975) it is shown using principal component analysis that all hominid teeth from forms classified as *Australopithecus* to *Homo sapiens* have but a single linear dimension of change which is a linear combination of the MD and BL dimensions. Further, the principal axis for a cloud of points defined solely by modern teeth is virtually indistinguishable from the principal axis for the full set cf hominid teeth spanning some 3 million years (see Table 1 and Figs. 1a - 1f in Read 1975; see also Twiesselmann (1973). As noted by Read (1975) some of these principal axes for tooth types have an intercept that does not differ significantly from zero, indicating that for these teeth the principal axis is a pure size dimension (see Eq. (3)). For the remaining teeth, where the intercept does differ significantly from zero, the dimension of change is a combined size and shape change, yet so constrained to represent but a single, linear dimension of change. (Note that these results show it is not valid to assume the first principal component, or first factor in a factor analysis, is a pure size dimension without independent corroboration.) One tooth, M^3, differs from this general pattern by apparently having undergone two separate episodes of linear change (see Fig. 1f in Read 1975).

These results establish the following: (1) The initial measurement of hominid teeth in two dimensions can be reduced to a single dimension which is a measure of morphological change through time and hence more closely approximates the ideal of a measure defined over underlying elementary events, namely a measure of the quantitative contribution by allelic combinations that control tooth growth; (2) differences in tooth dimensions cannot be completely subsumed under allometric changes resulting from change in body size (see also Corruccini and Henderson, 1978; Wood and Stack, 1980); and (3) selective pressures on tooth size and shape change have been one-dimensional for the hominids.

The initial MD and BL measurements can be converted to a measure along the principal axis by taking the distance from the point of intersection of the principal axis with the abscissa in the graph of the MD versus the BL dimensions to the perpendicular projection of a data point in that graph onto the principal axis. Equation (4) gives the formula for this transformation:

$$z = (1/\cos(\tan^{-1}b))(-ab/(b^2+1)+by_0/(b^2+1)+x_0/(b^2+1)) \quad (4)$$

where:

x_0 = MD measurement for the data point

y_0 = BL measurement for the data point, and

$y = a+bx$ is the equation for the principal axis (with y = BL and x = MD).

The equations for the principal axes are given in Table 1 in Read (1975) and are based on four data sets: Modern teeth -- Chumash Indian village; Homo erectus -- screened for outliers; Sterkfontein -- screened for outliers; and Swartkrans -- screened for outliers.

For the following teeth the intercept for the principal axis is not significantly different from zero: C_-, C^-, P_3, P^3, P_4, P^4, and M_1. For the remaining teeth, the intercept differs significantly from zero (Read, 1975). Thus the first set of teeth have had change in size, but not shape, while the teeth in the second set have had unidimensional change that is a composite of size and shape. The slope of the principal axis, which is a measure of the rate of change from the earlier forms to modern forms, is ordered as follows from smallest to largest slope: M_3; M^1, M^2, and C^-;

M^3; C_-; P_4, P_3; P^3; P^4. The slope values separate the teeth into 3 groups: (1) Molars -- shallow slopes (0.67-0.93), (2) Canines -- moderate slopes (0.90-1.07) and (3) Premolars -- steep slopes (1.13-1.45) for the principal axes. (Note that these slopes indicate rates averaged over a time span of 3 million years. The actual rate of change for a lineage has not been constant, as discussed in Read (1975).)

In Table 3 are listed the mean values of the scores obtained through the transformation given in Eqn. (4) for the hominid fossils from the following sites or regions: Laetolil, Hadar, Sterkfontein, Swartkrans, and East Africa (separated into large toothed and relatively small toothed forms).

Table 3. Mean Value[1]) of Tooth Dimension Measured Along the Principal Axis

Site / Tooth	Laetolil[2])	Hadar[2])	Sterkfontein[2,3])	Swartkrans[2,3,4])	East Africa, Large[2])	East Africa, Small[2])
C_-	14.50(2)	12.73(4)	13.32(4)	10.53(6) 0.61	12.10(1)	12.43(2)
C^-	13.95(3)	13.69(7)	12.16(4)	11.44(9) 0.68	11.81(2)	---
P_3	15.50(5)	14.31(13)	15.32(3)	16.11(10) 0.87	19.90(2)	14.74(4)
P^3	15.15(3)	13.93(4)	14.88(8)	16.74(12) 0.81	18.58(2)	14.70(2)
P_4	15.37(3)	13.92(10)	15.29(3)	16.66(11) 0.55	21.13(5)	14.68(3)
P^4	14.90(1)	12.79(5)	13.65(5)	16.92(11) 0.82	19.54(1)	13.38(1)
M_1	18.04(4)	17.24(11)	18.75(5)	19.84(15) 0.80	23.10(3)	17.13(6)
M^1	16.03(5)	13.92(3)	15.24(10)	16.99(10) 0.60	19.63(2)	15.67(2)
M_2	19.00(3)	18.60(13)	20.26(3)	21.50(10) 0.96	25.09(5)	18.15(4)
M^2	17.44(2)	16.50(2)	18.69(12)	19.92(7) 0.98	23.90(2)	16.53(1)
M_3	17.31(2)	17.94(8)	18.49(6)	20.36(9) 1.20	23.54(5)	16.44(4)
M^3	15.60(2)	17.55(4)	18.47(9)	21.09(12) 1.07	24.29(1)	16.71(2)

(1) Units for the measure is millimeters.
(2) Numbers in parentheses are the sample sizes for each tooth and site.
(3) The mean tooth sizes for Sterkfontein and Swartkrans are computed with all suspected and confirmed outliers deleted, and all teeth with the same catalogue number as the outliers deleted. In addition, SK 34 and 52 have been deleted from the Swartkrans sample due to the extreme size of these teeth when projected onto the principal axes.
(4) The second number for Swartkrans is the standard deviation. Standard deviations are not computed for the other data as sample sizes are small (Laetolil, East Africa), more than one lineage may be present (Hadar), or data may include an episode of changing tooth size (Sterkfontein). See text for details.

It can be seen from the last two columns of Table 3 that the difference in measurement between the two East African forms is substantial and far exceeds, for example, the differences between *A. africanus* and *A. robustus*, and needs no testing for validation. The differences in mean values between Laetolil and Hadar are significant at an alpha level of 0.05 for C_{-}, P_3, P_4, P^4, and M^1 using a t-test based on the sample standard deviation computed for Swartkrans as the best estimate of a population standard deviation unaffected by extraneous factors for these data. The other teeth have non-significant differences, though the power of the t-test is low for P^3 and M^3 due to the small sample sizes for these teeth.

Shape of the Frequency Distribution -- Sterkfontein and Swartkrans

The data used to compute the means in Table 3, along with suspected outliers, are given in the form of stratified histograms in Figs. 1a-1f. The histograms clearly show that the teeth do not, in general, have simple univariate normal distributions. For several of the teeth (e.g., the premolars, which are also the teeth for which the LDF does better than the RDF) the *A. robustus* material shows marked bimodality (see also Wolpoff, 1975). The *A. africanus* material tends to either show unimodality *or* an extended distribution with several gaps in the distribution. These results demonstrate that the original MD and BL measures obscure significant patterning in the data by virtue of not being WD variables, as discussed above.

Presence of bimodality may be qualitatively tested using the transformation

$$z^* = \bar{z} - z \tag{5}$$

for each tooth and then combining all the z* values into a single sample for each site. Deviations from the mean are used to measure bimodality rather than standardized scores so that the differential effect of bimodality and unimodality on the magnitude of the sample standard deviation in comparison to the range of the data would not be included. The test is conservative in that bimodality for a variable may be obscured. But conversely, bimodality in the combined sample would not occur if the variable were unimodally distributed over all teeth. In Fig. 2 are given the histograms for the material from Sterkfontein and from Swartkrans.

A rather markedly different pattern occurs for each of the two sites, with the *A. africanus* material more or less exhibiting unimodality but with a high degree of kurtosis, while the *A. robustus* shows clear bimodality, confirming the pattern suggested in the histograms for each separate tooth. The breaks in the distribution for the *A. africanus* histograms for each tooth considered separately are also repeated in Fig. 2, suggesting that these are not the result of sampling error. The simplest

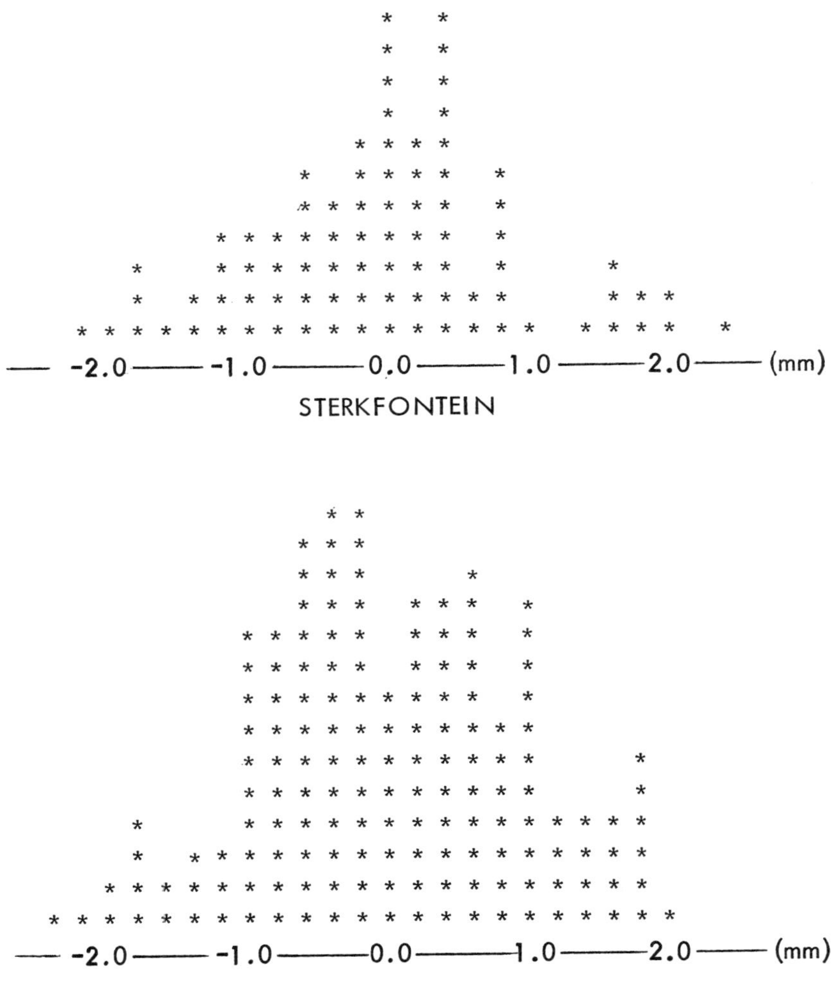

Figure 2. Histograms for Sterkfontein and Swartkrans. The histograms are a combined sample for all teeth at each site where the scale for the histogram is the deviation of the tooth measure from the mean value for that tooth at its respective site. Each asterisk represents one tooth. The data have been grouped by .2 millimeter intervals

interpretation for these results is that the bimodality in the A. robustus material is due to sexual dimorphism (cf. Wolpoff, 1975), while the A. africanus material may represent the effect of morphological change sampled unevenly through time as would occur if that material collected unevenly during an episode of changing tooth size.

While these histograms establish some of the within group properties of the frequency distributions of teeth as measured along the principal axis, direct comparison between sites cannot be made at a "fine" level due to differences in the age of the sites and the fact that each of the lineages that may be present underwent selective pressures for changing tooth size during the time period represented by these sites. One "crude" overall intersite pattern can be noted. The hominid material dating to about the time range 3-1.5 MYBP represents a period of maximal tooth size for the hominids, regardless of the postulated lineages. Ancestral forms for the hominids and descendant forms in the Homo lineage have smaller tooth size. Thus there is an overall trend through time of going from smaller teeth to larger teeth and then back to smaller teeth for the lineage leading to Homo. By around 2 MYBP there are two distinctly different sets of tooth sizes present in the fossil record as shown in Table 3, with the substantial size differences between the teeth in the 2 sets rendering sexual dimorphism an unlikely explanation for the differences, particularly in view of the argument that there is substantial sexual dimorphism within the large forms from East Africa (Rak and Howell, 1978). Hence the reasonable explanation is 2 lineages, as is now generally agreed. This leaves as problematic the pattern of hominid taxa affinity prior to 2 MYBP.

Time Trend Plots for the Plio/Pleistocene Hominids

In order to use natural selection as a basis for ordering proposed taxonomic schemes into more likely and less likely explanations, it suffices to graph the mean tooth size for each site as measured by the principal axis (with "size" for the molar teeth including a shape component as discussed above) against time, so that time trend plots for these data can be constructed in correspondance with a given taxonomic scheme. In Figures 3A and 3B are given the mean tooth sizes for each of the teeth and sites plotted against time. The dates for the sites are as reported in the literature: Laetolil -- 3.6-3.8 MYBP (Leakey, Hay, Curtis, Drake, Jackes and White, 1976; Churcher, 1978); Hadar -- 2.9-4.2 MYBP (Taieb, Johanson, Coppens and Aronson, 1976), 2.9-3.6 MYBP (Walter and Aronson, 1982), 2.8 - 3.3 MYBP (Brown, 1982); Sterkfontein and Swartkrans -- approximate ages, based on Partridge (1973); East Africa: Koobi Fora -- 1.8-1.9 MYBP (Cortis, Drake, Cerling and Hampel; McDougall, Maier, Sutherland-Hawkes and Gleadow, 1980); Olduvai Bed 1--1.75 MYBP (Leakey, Evernden and Curtis, 1961; Curtis and Hay, 1972); and Omo (L7-125 and L894-1) -- 1.9 MYBP (Brown, 1972). While the absolute placement of each site is approximate, especially

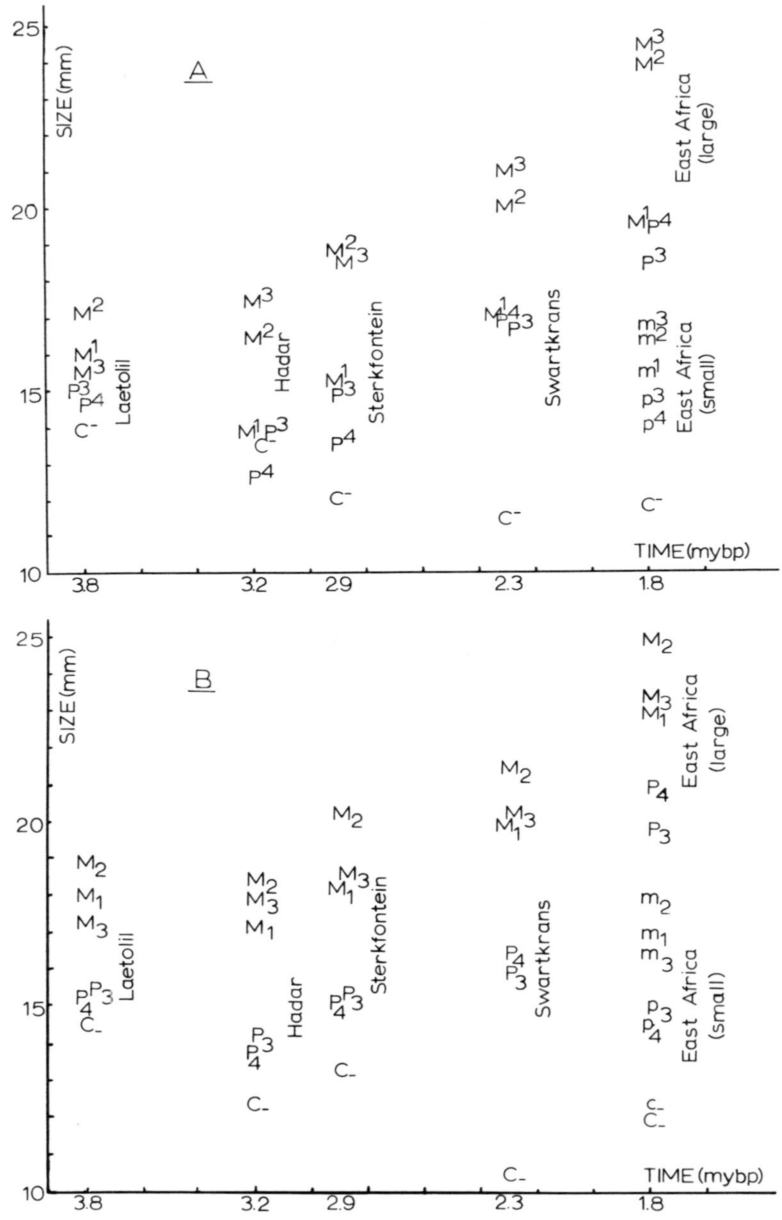

Figure 3. (1) Graphs A and B give the mean tooth size per tooth and site plotted against time for the sites of Laetolil, Hadar, Sterkfontein and Swartkrans. The teeth for East Africa are from Omo (L7-125 and L894-1 only), Olduvai and Koobi Fora.

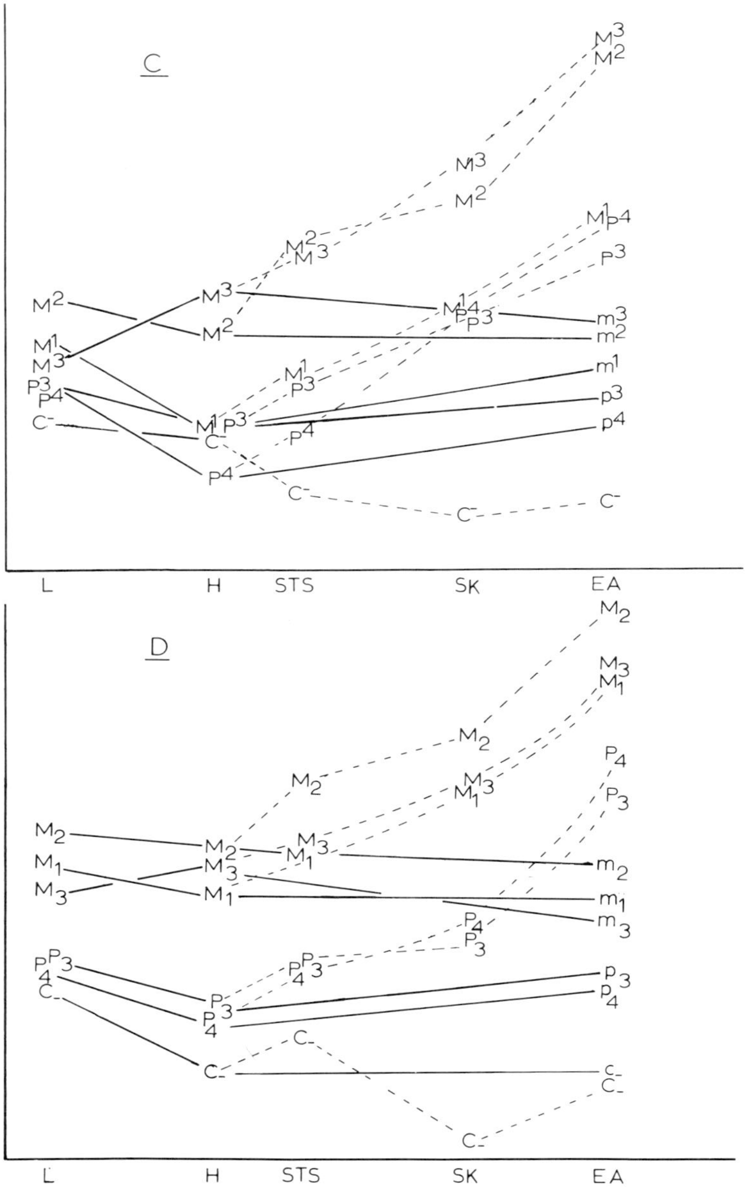

Figure 3. (2) Graphs C and D show the time trajectory plots implied by the Australopithecus aferensis model. The solid lines represent the lineage beginning with the form at Laetolil and going through the early Homo (Homo sp. indet.) form in East Africa. The broken line represents the divergence from this lineage around the time of Hadar of the lineage leading to the megadont forms in East Africa.

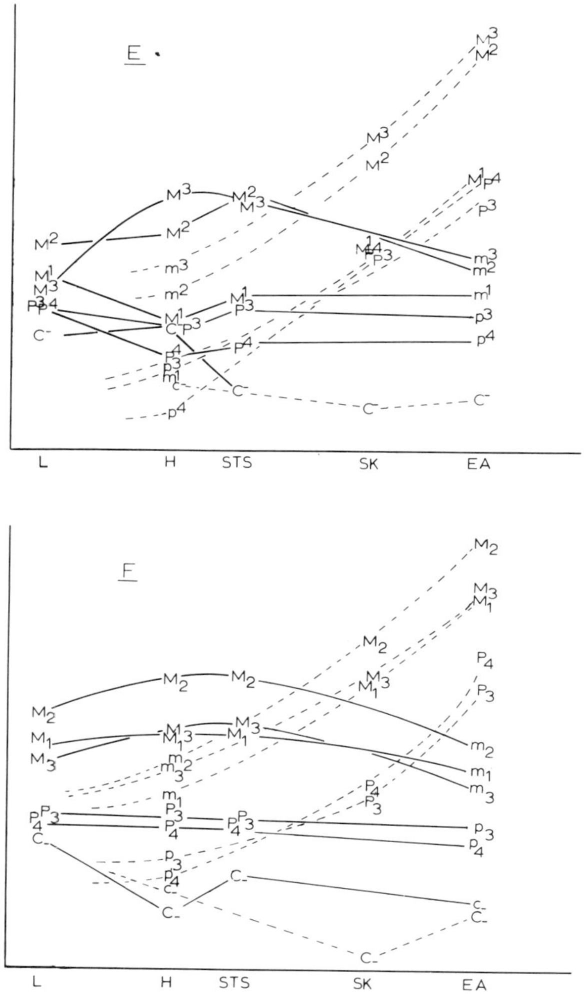

Figure 3. (3) Graphs E and F show the time trajectories implied by a two lineage hypothesis for Hadar, with the smaller form at Hadar on the lineage leading to Homo and the larger form on the lineage leading to the East African megadont forms. The solid line represents the lineage beginning with the form at Laetolil and going through the Homo (Homo sp. indet.) form in East Africa. The broken line represents the Australopithecus robustus/boisei lineage with an unknown ancestral form prior to Hadar. Symbols: Lower and upper case letters (c, C, p, P, m, and M) are used for small and large forms, respectively, at the same site or region.

for the sites of Sterkfontein and Swartkrans, the relative order is generally accepted as correct. Because of error in the absolute dating of sites, time trend plots will be examined primarily for qualitative differences that are robust with respect to reasonable range of errors in the dates of sites, than for detailed quantitative patterns.

A taxonomic model is a claim about which data points in Figs. 3A and 3B should be connected to form a postulated trajectory of tooth size change through time in a lineage. These trajectories will be assessed for likeliness using the presumption that each change in sign of the slope for a trajectory implies a new episode of selection and must be relatable to either a changing set of extrinsic or intrinsic conditions responsible for altering the adaptive value of a particular tooth size which in turn leads to the trajectory corresponding to that taxonomic scheme. To be examined here will be, first, a recently hypothesized model for taxonomic affinity of the early hominids, and then, second, an alternative model which it will be argued implies a simpler sequence of evolutionary change.

Taxonomic Model I

The first model is the *Australopithecus aferensis* model of Johanson and White (1979). This model postulates that the *Homo* lineage split away from the *Australopithecus* lineage prior to *A. africanus* as represented at Sterkfontein and subsequent to *Australopithecus aferensis* as represented at Hadar, so that the Laetolil and Hadar material are from a single, ancestral lineage -- given the taxonomic status *Australopithecus aferensis* -- while the South African Australopithecines and the subsequent East African Australopithecines are on a single lineage that became extinct. This latter lineage would be composed of the bulk of the South African material, the material from Omo, except the material from the upper beds classified as *Homo sp. indet.* (Boaz and Howell, 1977), and all of the large toothed forms from East Africa.

The trajectories corresponding to this model are given in Figure 3C and 3D. The trajectories beginning with the Laetolil material and ending with the small forms from East Africa (solid lines) would be the *Homo* lineage, while the trajectories beginning with the Laetolil material and ending with the large East African forms would be the Australopithecine lineage (solid and broken lines). The common portion of these two trajectories, going from Laetolil to Hadar, represents the common ancestral form, *Australopithecus aferensis*, for the two subsequent lineages.

For the mandibular teeth, a curve projected downward from the large East African forms through Swartkrans (as a second set of material definitely from this lineage) would clearly continue through Sterkfontein and intersect the *Homo* lineage at approximately the points represented by the material at Hadar (Figure 3D). It would seem that this taxonomic model translates into a series of trajectories that have substantial goodness-of-fit with the mean tooth size measurements for each site.

Interpretation of these trajectories for the mandibular teeth in the framework of selective patterns, however, is not simple. First, except for M_2, the Homo lineage is undergoing an essentially constant rate of reduction of tooth size with respect to time between Laetolil and Hadar with virtually identical rates for all teeth. Yet according to the A. aferensis model it is precisely while the lineage is undergoing selective pressures for reduced tooth size that divergence occurs, with the new lineage undergoing selective pressures for increasing tooth size. Second, the pattern of decreasing tooth size from Laetolil implies that a maximum tooth size must have been reached prior to the material at Laetolil since an early ancestor to the Laetolil material (e.g., a Ramapithecine-like form) must surely have had relatively small teeth. The single tooth M_3 deviates from this pattern by having increasing tooth size after Laetolil, rather than decreasing tooth size.

Putting these observations together gives the following overall pattern of change. The hominid lineage prior to Laetolil was undergoing selection first for increasing tooth size and then reached a maximum prior to Laetolil. By the time of the material at Laetolil, this lineage was now undergoing an approximately constant rate of size reduction for all teeth, except M_3. At approximately the time period represented by the Hadar material, speciation took place and the new lineage differentiated from the other lineage by, in part, being subjected to selection for increasing tooth size. In other words, this episode of speciation does not appear to be that of a form entering a new niche followed by adaptive radiation, but speciation during a period of specialization after a new niche had been invaded, as indicated by the pattern of increasing tooth size followed by decreasing tooth size.

The trajectories for the maxillary data (Figure 3C) are more problematic. Here, M^3 follows a pattern of first having increasing tooth size from Laetolil to Hadar, which continues with the new lineage, followed by selection for decreasing tooth size. With M^2 the pattern is contrary and much the same as for the mandibular teeth. For P^3 and M^1 the pattern is different with first a decrease in tooth size from Laetolil to Hadar and then an increase in size from Hadar to the small East African forms. The teeth P^4 and M^1 have the most extreme pattern with a substantial decrease in size from Laetolil to Hadar and then again a substantial increase in size from Hadar to the small East African forms.

For the non-Homo lineage the trajectories generally follow a projection backwards from the large East African forms to Swartkrans and then through Sterkfontein and finally intersecting with the Homo lineage around the time period represented by Hadar. A substantial exception to this pattern is M^2 where the pattern is decreasing size from Laetolil to Hadar, then a steep rise to Sterkfontein, followed by a slow rise to Swartkrans, and finally a steep rise to the East African large toothed forms.

Taken together, the trajectories for the maxillary teeth and the mandibular teeth seem to be suggesting contrasting directions for morphological change, with the maxillary teeth showing a complex pattern of sometimes increasing, and sometimes decreasing, tooth size. In no case do the trajectories suggest a pattern of speciation during adaptive radiation. Instead, the episode of speciation needed to form the second lineage would seem to have happened when selection is acting to reverse an earlier trend of increasing tooth size that itself is presumably the result of an earlier hominid entering a new ecological niche. For several of the teeth, the trajectories from Hadar to Sterkfontein seem to involve a questionably high rate of morphological change.

At the same time, it is evident from the trajectories that placing the Sterkfontein material on a lineage ancestral to the _Homo_ material with lineage divergence after the time period represented by Sterkfontein would lead to trajectories that are substantially more complex than those described above. The only viable alternative is that the Hadar material actually represents two lineages and that the formation of two lineages predates the Hadar material. This interpretation has been rejected by Johanson and White (1979) on the grounds that the presence of different sized individuals (large and small) can be subsumed, they argue, under sexual dimorphism. A prominent difficulty with the sexual dimorphism argument is that none of the succeeding forms exhibit the same pattern. The Sterkfontein material gives no indication of bimodality despite its close temporal relation to Hadar. Nor do the later _Homo_ material from East Africa have the bimodality needed for showing sexual dimorphism (though sample sizes are small). Hence assigning the size differences in the Hadar material to sexual dimorphism would imply the following pattern: Hadar with marked sexual dimorphism changes to the form represented at Sterkfontein without sexual dimorphism (or at least severely reduced sexual dimorphism) followed by an increase in amount of sexual dimorphism in the form represented by Swartkrans in the non-_Homo_ lineage, and from Hadar with marked sexual dimorphism to the East African small forms and then _Homo erectus_ without sexual dimorphism (Wolpoff, 1975) for the _Homo_ lineage. This pattern does not seem likely.

If sexual dimorphism is eliminated as the explanation for the different sized forms at Hadar, then the difference in size of the teeth at Hadar, amongst other characters, must be due to different lineages being present there. However, the problem immediately arises that the large forms at Hadar are strikingly similar to the material at Laetolil (Johanson and White, 1979), whereas the hypothesis of lineage divergence predating Hadar would seem to suggest that the forms on the lineage leading to the large East African forms should be diverging from the ancestral form as represented at Laetolil (if the latter is a common ancestral form).

These difficulties disappear when it is considered that the presumption of the larger form representing the non-_Homo_ lineage need not be valid. The presence of two lineages at Hadar along with the similarity of the large forms at Hadar with the forms at Laetolil can be accomodated by making the radical interpretation that the large forms at Hadar are on the Homo lineage, while the small forms are not on that lineage but rather are on the Australopithecus robustus/boisei lineage. Detailed justification for this radical interpretation would go beyond the scope of this chapter. The argument will be limited here to showing that the taxonomic model implied by this interpretation leads to a simpler and more easily interpreted set of trajectories than does the A. aferensis model.

Taxonomic Model II

The model is as follows: The Homo lineage is defined by the trajectory from Laetolil to the large forms at Hadar to Sterkfontein to the small East African forms (solid lines in Figures 3E and 3F). The A. robustus/boisei lineage is defined by the trajectory from an unknown ancestor to the small forms at Hadar to Swartkrans to the large East African forms (broken lines in Figures 3E and 3F). Observe that for this model it is postulated, contrary to the A. aferensis model, that the Sterkfontein material are part of the lineage leading to Homo.

The division of the material at Hadar into large and small forms is surprisingly simple. One tooth, M_2, exhibits marked bimodality with non-overlapping distribution between the small and large sets of teeth (see Figure 1E). Since many of the other teeth are from the same mandibles as are represented by M_2, these mandibles serve to identify the portion of the tooth size range for the other teeth that should be representing large and small forms, respectively. All teeth falling within the respective ranges just defined are then identified as large or small forms. For mandibles with several teeth (as is true for much of the Hadar dental material), each tooth is separately classified and then cross-checked for consistency of classification. It is found by this procedure that all of the teeth at Hadar, maxillary and mandibular, can be unambiguously classified as being from a large or small form. The large forms have the following catalogue numbers: Maxilla-- 200-1a, 333-1, 333-2, 333W-2, 333X-3, 333W-42, 333-86 and Mandible--145-35, 200-1b, 241-14, 277-1, 333-10, 333-74, 333W-12, 333W-32, 333W-58, 400-1a and 411-1. The small forms have the following catalogue numbers: Maxilla-- 161-40, 199-1, 400-1b and Mandible-- 128-23, 198-1, 207-13, 266-1, 277-1, 288-11, 311-1, 333W-1, 333W-57, 333W-58, and 333W-59. The mean values for each tooth for the large and small forms are given in Table 4 and the trajectories defined by the above model are given in Figures 3E and 3F.

For the mandibular molar teeth (Figure 3F) the trajectories for the Homo lineage show a pattern of increasing tooth size from Laetolil with a maximum size reached in the vicinity of Hadar and

Table 4. Mean Value of Tooth Dimension Measured Along the Principal Axis for Large and Small Forms at Hadar

Tooth	Small form[1,2])	Large form[1,2])
C_{-}	12.97(3)	12.00(1)
	2.08	--
C^{-}	12.38(2)	14.23(5)
	0.52	0.60
P_3	14.00(9)	15.65(4)
	0.60	0.59
P^3	12.71(1)	14.23(5)
	--	0.28
P_4	13.30(6)	14.98(4)
	0.65	0.62
P^4	11.35(1)	13.16(4)
	--	0.45
M_1	16.11(4)	17.90(7)
	0.77	0.85
M^1	12.58(1)	14.58(2)
	--	0.55
M_2	17.18(7)	20.10(6)
	0.83	0.33
M^2	15.43(1)	17.60(1)
	--	--
M_3	17.13(5)	18.27(3)
	0.76	0.66
M^3	16.30(2)	18.82(2)
	0.71	0.14

(1) Numbers in parentheses are the sample sizes for each mean value.

(2) Second number for each tooth is the sample standard deviation.

Sterkfontein. The premolars show little change, or at most a slight decrease in size. For the A. robustus/boisei lineage, the projection backwards from the large East African material through Swartkrans fits well with the small forms at Hadar, but does not fit with the large forms there, and continues as a lineage separate from the Homo lineage to a hypothetical point predating Laetolil.

For the mandibular teeth, then, a rather simple pattern emerges. First, the ancestral form to both lineages prior to the time period represented by Laetolil undergoes selection for increasing tooth size and during this episode of selection, speciation occurs as an example of adaptive radiation. Both lineages continue to have selection for increasing tooth size, though at different rates. The Homo lineage initially has a more rapid change to larger teeth and then reaches a maximum with the forms represented by the material at Hadar and Sterkfontein. The time period for maximum tooth size for this lineage is the time period when the teeth for the second lineage are beginning to cross the trajectories for the first lineage and are undergoing a rapid rate of increase ending with the size of teeth represented by the large East African forms. Thus the second lineage is moving in the direction of niche specialization involving large teeth (e.g., Robinson's dietary hypothesis (Robinson, 1954)). To the degree that similar sized teeth in closely related species suggest possible niche overlap, the trajectories suggest that the reversal of the pattern of increasing tooth size in the Homo lineage may result from competition with the forms represented by the second lineage. The latter are apparently moving in the direction of niche specialization centered around large molars and premolars. While the argument is largely conjectural, it establishes the fact that the trajectories for the mandibular teeth have interpretations in a well-established, general evolutionary pattern: adaptive radiation followed by niche displacement leading to a change in the direction of morphological change for one of the lineages.

The maxillary teeth (Figure 3E) exhibit essentially the same pattern, with the sole exception of M^1. For both M^2 and M^3 the Homo lineage shows increasing tooth size reaching a maximum with the Hadar and Sterkfontein material, and then decreasing tooth size to the small East African forms. Both of the premolars show a generally decreasing pattern of tooth size change. Only M^1 exhibits a different pattern: decreasing tooth size from Laetolil to Hadar and then increasing tooth size to Sterkfontein and the East African forms. However, this pattern is based on a sample of size 1 from Laetolil and a sample of size 2 from Hadar. For Laetolil, M^1 is larger than M^3, contrary to the pattern at the other sites, suggesting that the former value may be too large due to sampling error. If so, a slight decrease in the mean value for M^1 at Laetolil and a slight increase in that mean value for Hadar would shift the pattern for this molar to that for the other maxillary molars. Similarly, the somewhat uneven curve for M^2 is also possibly due to sampling error as the mean values for M^2 at Laetolil and Hadar are based on samples of size 1.

The overall qualitative pattern under Model II for mandibular and maxillary teeth is, then, as follows; (1) Homo lineage, beginning with Laetolil -- (a) The molar teeth, both mandibular and maxillary, form a unit. The pattern of change for this unit is increasing molar size reaching a maximum around 3 MYBP followed by decreasing molar size. (b) The premolars, both mandibular and maxillary, form a unit. The pattern of change for this unit is gradually decreasing premolar size. (c) Canines have a separate pattern of change which includes episodes of increasing and then decreasing tooth size. (2) A. robustus/boisei lineage -- (a) Molars and premolars, both mandibular and maxillary, form a unit. The pattern of change for this unit is increasing tooth size for the time period represented by these data. (b) Canines have a separate pattern formed of episodes of increasing and decreasing tooth size distinct from that for the Homo lineage.

For Model I the qualitative pattern of change is as follows: (1) Homo lineage, beginning with Laetolil -- (a) Mandibular molars except M_3, premolars and canines form a unit. The pattern of change for this unit is generally decreasing tooth size. (b) M_3 forms a unit. The pattern of change for this unit is first increasing and then decreasing tooth size. (c) P^3, P^4, M^1 and M^2 form a unit. The pattern of change for this unit is first decreasing and then increasing tooth size. (d) M^3 forms a unit. The pattern of change for this unit is first increasing and then decreasing molar size. (2) A. robustus/boisei lineage, beginning with Hadar -- (a) Premolars and molars form a unit. The pattern of change for this unit is similar to that of Model II, namely increasing tooth size. (b) The canines each have different patterns of change distinctive from those of the other teeth. (c) While the general trend is for increasing tooth size in the premolar-molar complex for this lineage, the pattern lacks the consistency in rates associated with the change in size of these teeth under Model II. Further, the event of speciation in Model I requires a discontinuity in the A. robustus/boisei lineage as the pattern of decreasing tooth size prior to divergence shifts to increasing tooth size after speciation. Contrariwise, for Model II speciation would seem to appear during a period of increasing molar size which then continues in the two new lineages, albeit at a different rate in each lineage.

These observations justify the claim that Model II accounts more parsimoniously for the pattern of change in the hominid lineages during the time period represented by these data than does Model I.

Model II has as implication that the split in the hominid lineage predates Laetolil. Projection backwards of both the solid and the broken lines in Figures 3E and 3F for the molars would lead to an intersection of the respective trajectories around 4.5 MYBP, making this date the estimated time of divergence assuming Model II is correct. Molar sizes for the ancestral form would also be slightly smaller than the mean molar sizes for the small form at Hadar given in Table 4. In contrast, Model I implies a larger molar size than occurs at Laetolil for the first and second molars in an ancestral form from the period preceding Laetolil, while the third molar would be smaller. Thus the two models have quite different implications for the morphology of an ancestral form.

Finally, the set of outliers identified in the preanalysis of the Sterkfontein and Swartkrans material for homogeneity can be mapped onto the trajectories implied by the two models. Both models indicate that there are two hominid lineages during the time period of either site. Both models imply that contaminating data from a second lineage should have smaller size than the bulk of the material at either site, with the sole exception of canines at Swartkrans where the contaminating data, if any, should be larger, regardless of model.

When both suspected and confirmed outliers are taken into account, 3 of the 5 outliers from Sterkfontein and 19 out of 22 outliers from Swartkrans qualitatively fit the pattern suggested by the two models if there is contamination in each of these sites from the second lineage. These data suggest that the amount of contamination at the South African sites may be more extensive than is generally argued.

CONCLUSION

Hampering quantitative study of hominid phylogeny and definition of taxonomic classes has been lack of means to imbed seemingly minor, yet statistically significant, metric differences into models which display radically different properties over part of their respective domains. The difference in measurements between small and large forms at Hadar, for instance, can potentially be subsumed under sexual dimorphism or under lineage difference. For data subsequent to, but close to the time of lineage divergence, as may be the case for the Hadar material, differences representing lineage divergence will not be pronounced and may well mimic differences expectable under sexual dimorphism. In such a situation, the metric differences do not easily discriminate between the alternative interpretations and their respective implications for hominid phylogeny and taxonomic construction. The difficulty is not resolved merely by shifting the analytical framework to a more sophisticated, multivariate context.

As has been shown here, resolution can follow from imbedding data into a context wherein the deeper implications of the alternative interpretations are clearly expressed, namely the pattern of natural selection implied by the alternative interpretations. The two alternative interpretations for the bimodal distribution in some of the data at Hadar has led to radically different implications for the pattern of hominid evolution, as discussed above. As a consequence, a variety of data are now identifiable as capable of distinguishing between the two models implied by the two interpretations of the data at Hadar.

These results suggest that the analysis of metric data will be most fruitful when the steps in the analysis are made in conformance with both a model for the process (or processes) which generated those data and the procedure of data recovery, here in the form of fortuitous discovery of fossil hominids. It is not sufficient to only require correspondence between generalized goals of analysis and equally generalized statements about an analytical technique. Rather, it is necessary to match the underlying framework for statistical techniques against the implied framework for the analysis of fossil data, as has been discussed here, in order to establish the several steps that must be taken so as to clearly express the implications of sometimes subtle, yet significant, patterning in the data being analyzed. The procedure outlined here, of going from initial screening of data (aimed at identifying and correcting for distortions of parameter estimates that may result from the vagaries of the inherent method of data acquisition) to construction or redefinition of variables (with goal ensuring that the variable values more closely approximate measures on the immediate outcomes of the underlying process structuring the data) to formulating a model sensitive to the underlying process for hominid physical evolution as a means to interpret the results of the data analysis, has led to a framework wherein substantial differences in the implication of alternative taxonomic models can be identified and then used as part of confirmatory or disconfirmatory arguments.

References

Ahmed, S.W. and P.A. Lachenbruch: 1975, Discriminant analysis when one or both of the initial samples is contaminated: Large sample results. EDV in Medizen and Biologie 1/2:35-42.

Andrews, D.F., P.J. Bickel, F.R. Hampel, P.J. Huber, W.H. Rogers, and J.W. Tukey: 1972, Robust Estimates of Location Survey and Advances. Princeton University Press, Princeton.

Atchley, W.P., C.T. Gaskins and D. Anderson: 1976, Statistical properties of ratios. I. Empirical evidence. Systematic Zoology, 25: 137-148.

Boaz, N.T. and F.C. Howell: 1977, A gracile hominid cranium from Upper Member G of the Shungura Formation, Ethiopia. American Journal of Physical Anthropology, 176: 93-108.

Brown, F.H.: 1972, Radiometric dating of sedimentary formations in the lower Omo Valey, Ethiopia. In: Calibration of Hominoid Evolution. W.W. Bishop and J.A. Miller, eds. Scottish Academic Press, Edinburgh, pp. 273-287.

Brown, F.H.:1982, Tulu Bor Tuff at Koobi Fora correlated with the Sidi Hakoma Tuff at Hadar. Nature, 300: 631-633.

Christenson, A. and D. Read: 1977, Numerical taxonomy, R-mode factor analysis and archaeological classification. American Antiquity, 42: 163-179.

Churcher, C.S.: 1978, in: Evolution of African Mammals. V.J. Maglio and H.B.S. Cooke, eds. Harvard University Press, Cambridge, pp. 509-535.

Clarke, R.J.: 1977, A juvenile cranium and some adult teeth of early Homo from Swartkrans, Transvaal. South African Journal of Science, 73.

Clarke, R.J., F.C. Howell and C.K. Brain: 1970, More evidence of an advanced hominid at Swartkrans. Nature, 225: 1219-1222.

Corruccini, R.S.: 1977, Correlation properties of morphometric ratios. Systematic Zoology, 26: 211-214.

Corruccini, R.S. and A.M. Henderson: 1978, Multivariate dental allometry in primates. American Journal of Physical Anthropology, 48: 203-208.

Curtis, G.H., R.L. Drake, T. Cerling and J. Hampel: 1975, Age of the KBS Tuff in the Koobi Fora Formation, East Rudolf, Kenya. Nature, 258: 395-398.

Curtis, G.H. and R.L. Hay: 1972, Further geological studies and potassium-argon dating at Olduvai Gorge and Ngorongoro Crater. In: Calibration of Hominoid Evolution. W.W. Bishop and J.A. Miller, eds. Scottish Academic Press, Edinbrugh, pp. 289-301.

Day, M.H. and R.E.F. Leakey: 1973, New evidence of the genus Homo from the East Rudolf, Kenya. I. American Journal of Physical Anthropology, 39: 341-354.

Dixon, W.J.:1975, BMD Biomedical Computer Programs. University of California Press, Berkeley.

Draper, N. and H. Smith: 1967, Applied Regression Analysis. John Wiley and Sons, New York.

Elandt-Johnson, P.: 1971, Probability Models and Statistical Methods in Genetics. John Wiley and Sons, New York.

Fisher, R.A.: 1936, The use of multiple measurements in taxonomic problems. Annals of Eugenics, 7: 179-188.

Gnanadesikan, R.: 1977, Methods for Statistical Data Analysis of Multivariate Observations. John Wiley and Sons, New York.

Hampel, F.R.: 1968, Contributions to the Theory of Robust Estimation. Unpublished Ph.D. Dissertation. University of California, Berkeley.

Howell, F.C.: 1969, Remains of Hominidae from Pliocene/Pleistocene formations of the lower Omo basin, Ethiopia. Nature, 223: 1234-1239.

Johanson, D.C. and T. White: 1979, A systematic assessment of early African hominids. Science, 203: 321-330.

Johanson, D.C., T. White and Y. Coppens: 1982, Dental remains from the Hadar formation, Ethiopia: 1974-1977 collections. American Journal of Physical Anthropology, 57: 544-603.

Leakey, M.D., R.L. Hay, G.H. Curtis, R.E. Drake, M.K. Jackes and T.D. White: 1976, Fossil hominids from the Laetolil beds. Nature, 262: 460-466.

Leakey, L.S.B., J.F. Evernden and G.H. Curtis: 1961, Age of Bed I, Olduvai Gorge, Tanganyika. Nature, 191: 478-479.

Leakey, R.E.G. and A.C. Walker: 1973, New Australopithecines from East Rudolf, Kenya (III). American Journal of Physical Anthropology, 39: 205-221.

Leakey, R.E.G. and B.A. Wood: 1973, New evidence of the genus Homo from East Rudolf, Kenya. II. American Journal of Physical Anthropology, 35: 175-186.

McDougall, I., R. Maier, P. Sutherland-Hawkes, and A.J.W. Gleadow: 1980, K-Argon estimate for the KBS Tuff, East Turkana, Kenya. Nature, 284: 230-234.

Murphy, R.B.: 1951, On Tests for Outlying Observations. Unpublished Ph.D. Dissertation. Princeton University, Princeton.

Partridge, T.C.: 1973, Geomorphological dating of cave openings at Makapansgat, Sterkfontein, Swartkrans and Taung. Nature, 246: 75-79.

Pilbeam, D.: 1972, The Ascent of Man. Macmillan Publishing Co., New York.

Pilbeam, D. and S.J. Gould: 1974, Size and scaling in human evolution. Science, 186: 892-901.

Pilbeam, D. and J.R. Vaisnys: 1975, Hypothesis testing in paleoanthropology, In: Paleoanthropology. Morphology and Paleoecology. R.H. Tuttle, ed. Mouton Publishers, The Hague, pp. 3-13.

Rak, Y. and F.C. Howell: 1978, Cranium of a juvenile Autralopithecus boisei from the lower Omo basin, Ethiopia. American Journal of Physical Anthropology, 48:345-365.

Read, D.: 1975, Hominid teeth and their relationship to hominid phylogeny. American Journal of Physical Anthropology, 42: 105-125.

Read, D.: 1976, Quantitative analysis of fossil data and the principle of natural selection. Yearbook of Physical Anthropology, 18: 163-171.

Robinson, J.T.: 1953, Telanthropus and its phylogenetic significance. American Journal of Physical Anthropology, 11: 445-501.

Robinson, J.T.: 1954, Prehominid dentition and hominid evolution. Evolution, 8: 324-334.

Rosen, R.: 1967, Optimality Principles in Biology. Butterworth Press, New York.

Schlain, B.R.: 1978, An Asymptotically Optimal Robust Discriminant Rule. Unpublished Ph.D. Dissertation. University of California, Los Angeles.

Taieb, M., D.C. Johanson, Y. Coppens, J.L. Aronson: 1976, Geological and palaeontological background of Hadar hominid site, Afar, Ethiopia. Nature, 260: 289-297.

Tobias, P.: 1975, Discussion. In: Paleoanthropology. Morphology and Paleoecology. R.H. Tuttle, ed. Mouton Publishers, The Hague, pp. 16-17.

Twiesselmann, F.: 1973, Evolution des dimensions et de la forme de la mandibule, du palais et des dents de l'homme. Annales de Paléontologie, 59: 173-277.

Walter, R.C. and J.L. Aronson: 1982, Revisions of K/Ar ages for the Hadar hominid site, Ethiopia. Nature, 296: 122-127.

White, T.D.: 1977, New fossil hominids from Laetolil, Tanzania. American Journal of Physical Anthropology, 46: 197-230.

White, T.D.: 1980, Additional fossil hominids from Laetolil, Tanzania. 1976-1979 Specimens. American Journal of Physical Anthropology, 53: 487-504.

Wolpoff, M.:1971, Metric Trends in Hominid Dental Evolution. Case Western Reserve University, Cleveland.

Wolpoff, M.: 1975, Sexual dimorphism in the Australopithecines. In: Paleoanthropology. Morphology and Paleoecology. R.H. Tuttle, ed. Mouton Publishers, The Hague, pp. 245-284.

Wolpoff, M.: 1976, Multivariate discrimination, tooth measurements and early hominid taxonomy. Journal of Human Evolution, 5: 339-344.

Wolpoff, M.: 1979, Anterior dental cutting in the Laetolil hominids and the evolution of the bicuspid P_3. American Journal of Physical Anthropology, 51: 233-234.

Wood, B.A. and C.G. Stack: 1980, Does allometry explain the differences between "gracile" and "robust" Australopithecines? American Journal of Physical Anthropology, 52: 55-62.

Subject index